Sustainable Housing

Edited by Amjad Almusaed
and Asaad Almssad

Published in London, United Kingdom

IntechOpen

Supporting open minds since 2005

Sustainable Housing
http://dx.doi.org/10.5772/intechopen.95680
Edited by Amjad Almusaed and Asaad Almssad

Contributors
Sharov Maksim Igorevich, Aleksandra Sikorska-Lewandowska, Rodolfo Jiménez Cavieres, Javier Carrasco Eade, Camilo Valdebenito Monsalve, Guillermo Yorel Noriega Aquise, Olusola Oladapo Makinde, Ibiwunmi Saliu, Evangelisca Akiomon, Maged Attia, Average Chigwenya, Prisca Simbanegavi, Andreas Koch, Kazutoshi Fujihira, Lorenzo Franco Escamirosa Montalvo, Carlos Uriel del Carpio Penagos, María de Lourdes Ocampo García, Ángel René Estrada Arévalo, Arturo López González, Roberto Arroyo Matus, Taufiq Choudhry, Syed S. Hassan, Sharosh Shabi, Hendrik Marten Koolma, Catharina Frederika van Dreven, Uyi Ezeanah, Amjad Zaki Almusaed, Asaad Almssad

First published in London, United Kingdom, 2022 by IntechOpen
IntechOpen is the global imprint of INTECHOPEN LIMITED, registered in England and Wales, registration number: 11086078, 5 Princes Gate Court, London, SW7 2QJ, United Kingdom
Printed in Croatia

British Library Cataloguing-in-Publication Data
A catalogue record for this book is available from the British Library

Additional hard and PDF copies can be obtained from orders@intechopen.com

Sustainable Housing
Edited by Amjad Almusaed and Asaad Almssad
p. cm.
Print ISBN 978-1-83969-647-3
Online ISBN 978-1-83969-648-0
eBook (PDF) ISBN 978-1-83969-649-7

For EU product safety concerns:
IN TECH d.o.o., Prolaz Marije Krucifikse Kozulić 3, 51000 Rijeka, Croatia, info@intechopen.com or visit our website at intechopen.com.

Meet the editors

Prof. Amjad Almusaed has a Ph.D. in Architecture (Environmental Design) from Ion Mincu University, Bucharest, Romania. He completed postdoctoral research in 2004 on sustainable and bioclimatic houses at the School of Architecture, Aarhus, Denmark. His research expertise is sustainability in architecture and urban planning and design. He has carried out a great deal of research and technical survey work and has performed several studies in these areas. He has edited many international books and is an active member of many worldwide architectural associations. He has published more than 170 international academic works (papers, research, books, and book chapters) in different languages.

Associate Prof. Asaad Almssad has more than thirty years of experience in industry, academia, and research at Umeå University, Sweden; Karlstad University, Sweden; and various European and non-European institutions. His research focuses on building structures, materials, sustainable building, and energy efficiency in building systems. He has authored and co-authored more than fifty research papers and many books. Currently, he is employed as a docent at Karlstad University.

Contents

Preface

The right to housing is a vital human right. It represents a fundamental component of a peaceful, dignified, and safe life. Sustainable housing provides residents with a comfortable and healthy living environment and can effectively solve issues of housing quality in the modern era. This book analyzes the concept of sustainable housing, focusing on the realization of competent sustainable housing based on practical cases.

The OECD Horizontal Project on Housing 2021 study documents trends in cost and affordability growth, international experience, and guidance needed for housing policy development. It focuses on three broad dimensions: inclusiveness, efficiency, and sustainability. Housing is becoming a more pressing economic, social, and environmental issue in OECD countries. Rising house prices and rents have eroded affordability and led to social exclusion. The COVID-19 crisis, which has resulted in significant job and income losses in the most vulnerable groups, exacerbates the difficulties in ensuring access to quality and affordable housing. Housing is clearly among the most important factors of quality of life and human welfare. Several documents promoted by various international bodies (UN, UNECE, UNHCR, UN-HABITAT, Council of Europe, etc.) emphasize that the right to housing is a fundamental tool in promotion and transposition in the practice of access to housing (as a complex phenomenon), but it is not limited (only) to simple possession of a home (adequate or not). It includes many other rights, such as the right not to be discriminated against (homeless people suffer severe discrimination and marginalization) and, consequently, the right to equal treatment and the right to self-determination and self-realization. Those excluded from adequate housing have, in fact, the same basic housing needs as any other individual, not only in terms of shelter but also in terms of access to basic utilities and services such as water, hygienic conditions, sanitation, and safety to ensure at least a minimum of security and dignity as foundations and premises of normal social integration and participation. Opting for a narrow view of perceiving these needs only in terms of physical needs would lead to an underestimation of the symbolic importance of housing in the daily life of individuals, the community, and society and would undermine the fundamental notion of human dignity on which the whole philosophy of human rights is based.

This book focuses on housing as the most important human living space in modern society. It starts from legislation and ends with unit living conformation. Therefore, sustainable housing must ensure that the habitation area is functional, comfortable, and environmental to meet both today's needs and future visions. Housing must be sustainable with natural systems. Therefore, it is characterized by environmentally friendly and certified materials. The book provides a comprehensive analysis of the modern housing design trend of creating sustainable cities, which poses complex problems for architects and urban planning.

We would like to thank all the contributing authors for the quality, range, diversity, and richness of the chapters. We offer a special thanks to Author Service Manager Ms. Iva Ribic at IntechOpen for her assistance and efficiency in managing the publication of this book.

Amjad Almusaed Ph.D.
Professor,
Department of Construction Engineering and Lighting Science,
Jönköping University,
Jönköping, Sweden

Asaad Almssad Ph.D.
Associate Professor,
Department of Engineering and Chemical Sciences Construction Technology,
Karlstad University,
Karlstad, Sweden

Section 1

Introduction to the Thematic Area

Chapter 1

Introductory Chapter: Sustainable Housing – Introduction to the Thematic Area

Amjad Almusaed and Asaad Almssad

1. Introduction

One of the most critical phenomena in the real estate sector to reduce the environmental impact and climate change is sustainable houses. "All nature strives for self-preservation," said the philosopher Cicero. And residents of megalopolises, too, increasingly began to think about the future of cities and how to improve the environmental background around their place of residence. One of the most innovative urban developments in the twenty-first century is the design of buildings and entire neighborhoods in sustainable architecture. Ebenezer Howard, whose 1902 book was entitled *Garden City of Tomorrow* and whose political and social agenda has recently made a comeback [1]. The idea of sustainability involves enhancing the quality of life, thus allowing people to live in a healthy environment with improved social, economic, and environmental conditions [2]. In addition, this type of building facilitates the most respected lifestyle with the territory, reducing the ecological footprint.

Sustainability is an essential subject in the housing research area, and it is a challenging theme for city inhabitants, builders, and architects around the world. And it is a complex theme for city inhabitants, builders, and architects around the world. In general, a sustainable building is environmentally friendly high-tech architecture. It strives to minimize the negative impact on the environment through the efficient and thoughtful use of materials, energy, space, and the ecosystem. Sustainable housing design focuses on energy conservation, environmental protection, and many other supporting factors. Sustainable housing aims to provide affordable, sustainable housing for all city inhabitants. Therefore, sustainable housing provides and discusses sustainable housing in the context of affordability. Housing costs include initial construction costs and building operation and maintenance costs. Therefore, low-cost housing based on the entire life cycle is the real low-cost housing; high-quality housing means that the housing needs to have good building quality and focus on providing users with high-quality living conditions. In the final analysis, the house is for the occupants [3]. A healthy and livable indoor and outdoor environment is more conducive to the physical and mental health of the occupants.

A sustainable house is not just a fashion trend or a trend from the West. In today's world, rising energy prices and the finite nature of these resources raise the acute problem of the operating costs of housing. The development of energy-efficient buildings is now being actively pursued in Europe, where government programs have been developed to bring all facilities to a low level of energy consumption. The sustainable house has ecological materials, uses recyclable systems, and is supplied by clean energy. Sustainability is a broad and complex concept, which has become one of the significant issues in housing buildings. In the process of developing ideas in the field of sustainable housing design, dozens of specialists, such as architects, urban planners, ecologists, sociologists, and others, calculate how houses will interact with the surrounding architectural environment, for example, whether a building will cast a shadow on nearby buildings. In the present day, sustainable homes are invariably linked to making buildings as comfortable and convenient as possible worldwide. The building sector is increasing by investing 30–40% of total global essential resources [4]. Buildings and users consume energy and resources that create waste on a large scale, and we are tied to the current construction methods about resource and energy consumption, waste emissions, and environmental damage. The primary objective of sustainable design and construction is to minimize buildings' negative ecological, social, and economic impacts. Fast development in many countries across the globe has made significant incontrollable construction waste, thus creating considerable adverse effects on the environment such as increased soil, water, and air pollution, which contribute to climate change, health hazards, and ecological imbalance [5]. The concept refers to buildings planned with a sustainable development concept, including building materials. Buildings, the size of urban areas, etc., to the functional, economic, social, cultural, and ecological factors related to these. When our housing is poorly designed, it leaves a legacy with adverse social, economic, and environmental side effects for the next generation. The complex and challenging plan of sustainability requires a fundamental change in our understanding of nature, the purpose of the buildings, the architects and the builder's role, and the users of the facilities. Users' habits, behavior, and lifestyle in everyday practice are topics for discussion due to the environmental challenges facing the world. It becomes at once our own culture that is the focal point of the debate. If our own culture can change because of environmental problems, it will sooner or later influence the architectural design to reflect inhabitants' cultural and social values. One of the most critical phenomena in the housing sector to reduce the impact on the environment and climate change is the sustainable housing concept: houses that use ecological materials, use a recycling system and are supplied with clean energy.

Additionally, these housing types promote the environment and the most respected lifestyles, reducing their ecological footprint. The house is a living space, home, room, or apartment in which one or more people permanently live [6]. A house usually contains areas for cooking, hygiene, and sleep. In addition, often spaces for socializing, family life, and solemn occasions. The house not only fills the requirements, but it is also a pleasure. From the child's stacking of bricks to learning how to master the heavens and space to self-expression, construction is something that concerns us all deeply [7].

A residential building is usually defined as a house where at least half of the area is intended as a living space. Housing can be defined as an "industry of protection and comfort." In the first place, it protects humans from various aggressions against which they feel the need to defend themselves. In a history of societies that have made human settlements (unequally) safer, the essential requirement of shelter is outweighed by comfort and well-being. During the 1980s, the assumption that development and the environment were incompatible was increasingly questioned [8]. Views that a future growth based on sustainable utilization of natural resources was an option gradually emerged. Therefore, it was decided in 1983 at the UN that a World Commission on Environment and Development should be set up. Since the Rio Declaration, which is also the origin of the Environmental Summit, the term sustainable has referred to economic activities that consider the global environment [9]. Sustainable housing is a long-lived housing that is easy to live in and will be passed down to the next human generation.

2. Sustainable housing concept and designs

Housing is a shelter that provides primary living conditions such as safe housing, drinking water, and healthy food for humans [10]. Even in developed countries, low-income families often have no housing for economic reasons or face health and safety problems caused by poor housing quality [11]. Sustainable architecture is mainly embodied in the overall planning of ecological, economic, and social and cultural sustainability. Sustainable housing is guided by the concept of sustainable development, carries out reasonable planning and design, effectively utilizes resources in the process of construction and use, minimizes the impact on the environment, and provides residents with health and comfort. A safe space, a living carrier that enables people to be satisfied in an elegant environment [12].

The new housing design concerns the thoughtful activity of the actor who creates the artifact providing man with the place of his protection and comfort. This acceptation of the term conception, thus understood as cognitive activity, can admit another, broader one, making it possible to also understand this cognitive activity within a framework of thought specific to a given period. We will approach the design of housing by examining how it went from the self-design of its inhabitant to a complex elaboration, both in terms of the resources it calls on, the knowledge and techniques it uses, and the skills cooperation it uses. Sustainable housing has an important role to play in ensuring an adequate quality of human life [13]. The positive impact of housing can be enhanced by applying conservation principles, economic efficiency, social inclusion, and public participation, and adequacy in terms of culture. Sustainable housing development in the EU region faces many challenges, driven primarily by globalization, demographic change, climate change, and the economic crisis. The challenges are that sustainable public housing is typically different from ordinary construction. Sustainable construction can be different in heating, venting, technologies used, etc. [14]. This implies that these houses are too different regarding the necessary knowledge and handling of the operational conditions. Without using petroleum energy, which causes global warming, we will create homes that use natural clean energy such as solar heat and wind power. In addition, because it is a residence that considers the cycle of tree growth and regeneration, it also leads to the effective use of recycled materials such as dismantled old folk houses [15]. In addition to making the house last longer, sustainable housing is also characterized by consideration for building a house that can reduce waste when dismantled and reuse building materials. Sustainable housing where people can live comfortably forever. In future home building, the idea of sustainable housing will be strongly required. In sustainable housing, it is considered to create a house that is friendly to people and the earth everywhere, such as the structure, floor plan, equipment, and building materials used. By being conscious of building a house where you can live for a long time, you will not only have the advantage of not damaging the environment, but you will also be able to build a house that is kind to the residents [16]. This time let us think about sustainable housing that is kind to both people and the earth. To talk about sustainable houses, it is required to build using sustainable materials. It should also be able to generate its own energy and use its own resources. One of the main features of sustainable energy is the use of solar generators. Solar panels allow you to use solar energy. Sustainable houses are those that, being respectful of the environment, take advantage of all available resources to reduce energy consumption and, therefore, help save on household bills, something that is always appreciated [17]. These types of houses are integrated into nature and are in such a way that they make the most of their environment: light, water. Another aspect of sustainable housing is the reuse of rainwater. You can also build a generator that can store energy from the wind. But it's not very pretty. There are several problems with sustainable rates. Perhaps the most important is the price of the material. And this price is much higher than that of traditional materials. Even so, the cost is not always high. Some architects and designers have recycled materials, so they do not cost too much. It should also be considered that sustainable housing begins to reduce supply costs. Energy efficiency is essential when building this type of house [18]. It is also important to include the use of renewable resources. Construction and materials used must be completely ecological. Water must be managed intelligently. It is important to achieve comfort in both ventilation and air conditioning in a sustainable house. Recent environmental issues have attracted worldwide attention. This has stimulated a response in many countries, which has led to a more in-depth review of energy conservation strategies for traditional fossil fuels. One way to reduce the energy consumption of buildings is to design buildings that are more economical in their use of energy for heating, cooling, ventilating, and lighting [19]. Comfortable lighting conditions can be achieved, and the highest level of sound insulation can be achieved without the use of many consumables. That is why a sustainable house must account for energy efficiency and savings in supply and construction costs over its entire service life. After obtaining the first building materials, a house can have many years of service life and maintenance until it is reused. In addition to all of this, it is essential to enjoy a sustainable home based on its architecture. At the heart of maintaining sustainable homes are the three R's of sustainability: reduce, recycle, and reuse. Resource utility and technical efficiency must be maximized. However, it's not all that beautiful. Sustainable rates have several problems. Perhaps the most important is the price of the materials. And this is because the price is much higher than that of traditional materials. The costs should not always be higher. There are some architects and designers who have managed to recycle materials so that the expenses are not too high. In addition, it must also be considered that sustainable homes start saving supply costs the principle compared with a traditional home. In the long run, this is a big saving economically and environmentally [20]. To create a competitive advantage using environment-friendly construction practices, the whole life cycle of buildings should, therefore, be the context under which these practices are carried out.

3. Sustainable housing under EU platform

The principles and strategies for standard urban policies in the EU countries, including guidelines for affordable and sustainable housing and social housing, are set out in the Leipzig Charter on Sustainable European Cities [21]. As part of the EU Housing Agenda Partnership on Housing, EU countries and cities work with the European Commission and other stakeholders to promote good quality, affordable housing. Among the measures proposed are those in public housing support, funding opportunities, and general housing policies and knowledge. To talk about sustainable houses, it is required to build using sustainable materials. It should also generate its energy and use its resources. One of the main features of sustainable energy is the use of solar generators. Solar panels allow you to use solar energy. Sustainable houses are those that, being respectful of the environment, take advantage of all available resources to reduce energy consumption and, therefore, help save on household bills, something that is always appreciated. These types of houses are integrated into nature and are in such a way that they make the most of their environment: light, water. Another aspect of sustainable housing is the reuse of rainwater. You can also build a generator that can store energy from the wind. But it's not very pretty.

4. Conclusion

This chapter, directly or indirectly, deals with the problem of housing and sustainable housing requirements. Several priorities have been identified among the wide range of specific tasks, conditions, and means of implementing Sustainable Development. These include promoting the sustainable development of human settlements (solving the housing problem). There are several problems with sustainable rates. Perhaps the most important is the price of the material. And this price is much higher than that of traditional materials. Even so, the cost is not always high. Some architects and designers have recycled materials, so they do not cost too much. It should also be considered that sustainable housing begins to reduce supply costs. Principles compared with traditional assumptions. This is a great long-term economically and environmentally friendly savings. For sustainable housing to be attractive to the public, the cost factor is very important. The consumer knows that he will recover the investment in a certain number of years, so he can bear higher construction costs. It should reflect the amount of electricity and water bills that will be saved over time.

Despite all the points discussed, the economic aspect is paramount. Here an excellent architectural and design management comes into play so that the cost may not be very high. In addition, it is interesting that the design of the house is quite attractive. Creating such an optimal building concept with minimal energy consumption and environmental impact represents a complex optimization task of elaborate planning, where partial aspects are pushed to the background in favor of the overall performance of the building or used in the whole context correctly: such integrated planning, resp. The design offers a real chance to reduce material costs and mainly operating costs because, e.g., the planning of the technical equipment of buildings does not begin with the planning of the building but with the planning of its external space. Therefore, it becomes essential to select technologies that can be used to create sustainable housing by an objective process. This applies not only to individual houses but also to settlements of any scale. Living in such houses and settlements will create the preconditions for the formation of ecological consciousness and give an additional chance for a sustainable healthy life of our civilization.

Author details

Amjad Almusaed[1]* and Asaad Almssad[2]

1 Department of Construction Engineering and Lighting Science, Jonkoping University, Sweden

2 Faculty of Health, Science and Technology, Karlstad University, Sweden

*Address all correspondence to: a.amjad@archcrea-institute.org

References

[1] Lehmann S. Green Urbanism: Formulating a Series of Holistic Principles. SAPIENS. 2010;3(2). Online since 12 October 2010, connection on 04 December 2021. Available from: http://journals.openedition.org/sapiens/1057

[2] Akadiri PO, Chinyio EA, Olomolaiye PO. Design of a sustainable building: A conceptual framework for implementing sustainability in the building sector. Buildings. 2012;**2**:126-152. DOI: 10.3390/buildings2020126

[3] Sodagar B, Fieldson R, Gilroy-Scott B. Design for sustainable architecture and environments. The International Journal of Environmental, Cultural, Economic, and Social Sustainability: Annual Review. 2008;**4**(4):73-84. DOI: 10.18848/1832-2077/CGP/v04i04/54505

[4] Kaushik ACG. Renewable energy technologies for sustainable development of energy efficient building. Alexandria Engineering Journal. 2018;**57**(2):655-669. DOI: 10.1016/j.aej.2017.02.027

[5] Amaral REC, Brito J, Buckman M, Drake E, Ilatova E, Rice P, et al. Waste management and operational energy for sustainable buildings: A review. Sustainability. 2020;**12**:5337. DOI: 10.3390/su12135337

[6] Almusaed A, Almssad A. Introductory chapter: Housing policy matters. In: Housing. UK: IntechOpen; 2018. DOI: 10.5772/intechopen.81622. Available from: https://www.intechopen.com/chapters/64126

[7] Gyurkovich J. Living space in a city—selected problems of shaping modern housing complexes in cracow—a multiple case studies: Part 1—the case study of urban villas. IOP Conference Series: Materials Science and Engineering. 2019;**471**:092015

[8] Asefi-Najafabady S, Villegas-Ortiz L, Morgan J. The failure of Integrated Assessment Models as a response to 'climate emergency' and ecological breakdown: The emperor has no clothes. Globalizations. 2021;**18**(7):1178-1188

[9] Clémençon R. From Rio 1992 to Rio 2012 and beyond: Revisiting the role of trade rules and financial transfers for sustainable development. Journal of Environment & Development. 2012;**21**(1):5-14. DOI: 10.1177/1070496512436890

[10] Krieger J, Higgins DL. Housing and health: Time again for public health action. American Journal of Public Health. 2002;**92**(5):758-768

[11] Almssad A, Almusaed A. Environmental reply to vernacular habitat conformation from a vast area of Scandinavia. Renewable and Sustainable Energy Reviews. 2015;**48**:825-834. DOI: 10.1016/j.rser.2015.04.013

[12] Moghayedi A, Awuzie B, Omotayo T, Le Jeune K, Massyn M, Ekpo CO, et al. A critical success factor framework for implementing sustainable innovative and affordable housing: A systematic review and bibliometric analysis. Buildings. 2021;**11**(8):317. DOI: 10.3390/buildings11080317

[13] Almusaed A, Almssad A. Building materials in eco-energy houses from Iraq and Iran. Case Studies in Construction Materials. 2015;**2**:42-54. DOI: 10.1016/j.cscm.2015.02.001

[14] Dutil Y, Rousse D, Quesada G. Sustainable buildings: An ever evolving target. Sustainability. 2011;**3**(2):443-464. DOI: 10.3390/su3020443

[15] Almusaed A. Biophilic and Bioclimatic Architecture, Analytical Therapy for the Next, Generation of

Passive Sustainable Architecture. England, London: Springer-Verlag Limited; 2011. p. 341. DOI: 10.1007/978-1-84996-534-7

[16] White K, Habib R, Hardisty DJ. How to shift consumer behaviors to be more sustainable: A literature review and guiding framework. Journal of Marketing. 2019;**83**(3):22-49. DOI: 10.1177/0022242919825649

[17] Almusaed A. Intelligent sustainable strategies upon passive bioclimatic houses: From Basra (Iraq) to Skanderbeg (Denmark). Aarhus School of Architecture. 2004;**10**:27

[18] Mohammad SA, Krarti M. Energy efficiency of residential buildings in the kingdom of Saudi Arabia: Review of status and future roadmap. Journal of Building Engineering. 2021;**36**:102143

[19] Omer AM. Built environment: Relating the benefits of renewable energy technologies. International Journal of Automotive and Mechanical Engineering (IJAME). 2012;**5**:561-575. DOI: 10.15282/ijame.5.2012.3.0044

[20] Almusaed A, Yitmen I, Almssad A, Homod RZ. Environmental profile on building material passports for hot climates. Sustainability (Switzerland). 2020;**12**(9):3720. DOI: 10.3390/su12093720

[21] Eltges M. Leipzig charter on sustainable european cites—a work in progress. European Spatial Research and Policy. 2010;**16**(2):63-78. DOI: 10.2478/v10105-009-0013-5

Section 2

Sustainable Development and Housing Strategies

Chapter 2

Towards Sustainable Housing: Waste Management in Informal Settlements in Masvingo City, Zimbabwe

Average Chigwenya and Prisca Simbanegavi

Abstract

Urban informality is fast becoming a permanent feature in cities not only in the global South but also in the global North, hence the need to include it in the delivery systems of the city for sustainable urbanity. However informal settlements are left out in the service delivery systems of cities and this has created spatial deprivation in the city. The exclusion of urban informality is not only a denial of their right to the city but also a denial of environmental justice to these people, which also constitute unsustainable housing. Including informal settlements in the urban fabric will result in sustainable housing because the housing delivery is very important in attaining sustainable development goals. All cities therefore need to provide its services to all the city inhabitants for inclusive and sustainable city. Informal settlements demands safe and liveable spaces for their well being and the sustainability of the city. Effective waste management in informal settlements does not only reflect the inclusivity of the city but is also an important pillar for sustainable city. The research used a mixed methods approach to data collection, where both qualitative and quantitative methods were used to collect data. The research find out that informal settlements in Masvingo city are excluded from the waste collection services and this is imposing an impending danger to their lives because of life threatening environments that surround informal settlements.

Keywords: informal settlements, waste management, inclusive cities, sustainable cities

1. Introduction

Informality has been growing in cities; not only of the Global South, but even in the Global North [1]. As a result, informality has become a permanent feature of urban development. Urban informality has actually developed to be the form of urbanisation in many cities [2]. This, therefore, calls for the integration of the informal sector into the main stream development of cities. This inclusion will lead to inclusive and sustainable cities. Slum settlements are lacking a lot of services and this is against the sustainable development goals, which seek to reduce the number of people living in slums by 2030. Informal settlements have grown in magnitude in cities of the global South but services such as sanitation and waste management are severely lacking thereby exposing them to life threatening environments. People in informal settlements are living in epicentres of diseases outbreaks and this weighs heavily against city sustainability [3]. Informal settlements are playing a critical role in the provision of housing to the low income urban dwellers and this provides urban resilience and sustainability. So city authorities should take it upon themselves to provide services to these otherwise neglected spatialities for sustainable housing. Excluding informal settlements in the delivery of services creates spatial deprivations, which does not auger well with sustainable housing. Sustainable housing aims to improve the quality of housing through slum upgrading and provision of services that will lead to safe and liveable spaces in informal settlements, [4]. Waste management is one of the important services that should be offered by the city authorities and it should be enjoyed by every resident as a way of giving them their right to the city. This chapter examines waste management in the informal settlements and its impact on the hygiene of spaces occupied by the informal settlements and the city at large. It analyses waste management practices in the informal sector and evaluates its implications on people's right to the city and sustainability of housing in the informal settlements. Waste management in the informal sector has different ramification on the city's general hygienic conditions and also reflects the inclusivity of the city in terms of how the services are spread to the diversity of urban inhabitants. This research is important in the sense that it contribute to the growing literature on the sustainable housing in cities. It examines sustainable housing from the informal settlements.

2. Urban informality and neo-liberal policies in Masvingo

The growth of informal activities in cities of the global South is associated with a lot of problems, most of which are to do with the management and delivery of services. In most cases the informal settlements are found in areas where city services do not reach, as a result most informal settlements are living with minimum or no services. This neglect is usually a result of neo-liberal urbanisation policies pursued by cities especially in the global South. There is a general obsession for clean cities that appeal to international standards and as a result urban informality is regarded as urban filthy, which needs to be eradicated, [5, 6]. Most cities are therefore trying by all means to portray cities without urban informality because of these neo-liberal urbanisation processes. These neo-liberal urbanisation policies are therefore responsible for marginalisation of informal settlements in the delivery of services in cities of the global South. Contemporary cities have developed to be highly diverse and this call for rethinking in the way cities does their business. The urbanisation of poverty in many cities calls for new ways of managing cities because cities can no longer afford to do business as usual, [2, 7, 8]. Cities of today cannot afford to develop without urban informality, hence there is need to integrate urban informality in the delivery of services and the development of the city for sustainable cities. Urban societies in most cities of the global South are however managed by rigid regulatory frameworks that take very long to reform in order to accommodate new land uses that are coming up due to urbanisation of poverty. Urban informality is a reality in contemporary cities, which therefore requires a paradigm shift in the way cities are managed in order to create sustainable and liveable cities. Creation of safe and liveable spaces in informal settlement is a critical component for sustainable housing and the general city sustainability. There is need to provide quality housing to people in the informal settlements, where their living conditions are improved socially, economically and environmentally. Neo-liberal urbanisation policies therefore do not suit cities of the global South, which have high prevalence of poverty. Most of these neo-liberal urbanisation policies are adopted from cities which do not have the experience of urban informality hence they do not provide for urban informality, [9, 10]. For example Zimbabwe has borrowed the engineering type of planning from the British type of planning and this planning approach does not tolerate any unplanned activities such as urban informality, [9]. However urban informal settlements are growing despite efforts worldwide to reduce by half people living in slums. By 2018 people living in slums had increased to 24% and these settlements are lacking critical services such as water, sanitation waste management that will allow them access to safe and livable spaces, [3]. The lack of such important services does not auger well with sustainable housing as it force people in informal settlements live in hostile environments.

2.1 Impacts of neo-liberal policies on waste collection in informal settlements

In the city of Masvingo informal settlements have been without was waste management services and this has created unsafe living conditions. They have been excluded from receiving such services and their areas are characteristically unhygienic and filthy. Waste has been thrown away indiscriminately in informal settlements thereby creating spaces that are prone to disease out breaks. Waste is dumped everywhere including road sides, open spaces and river banks. The **Table 1** below shows frequency of waste collection in informal settlements in various slum areas in Masvingo city.

The growth of informal settlements in cities of the global South requires a paradigm shift in the way cities are managed. There is need for inclusion of informal settlement in the city wide service delivery systems for inclusive cities and sustainable housing. The growth of informal settlements has however not been complimented with reforms in the regulatory framework, which had remained rigid and unaccommodative to urban informality in the delivery of services. Informal settlements have remained outside city's service delivery systems and this exclusion is not good for inclusive and sustainable cities. Informal settlements have therefore remained illegal settlements, which do not warrant access to services. As a result service provision in most areas occupied by informal settlements is non-existent. These people have therefore never been afforded their right to the city especially their right to receive city services for safe and liveable spaces. Informal settlements have in most cases been found on margins of the city where city services rarely reach, [11]. Informal settlements are commonly found in ecologically deteriorating spaces, [12].

One of the issues that the cities of the global South have been trying to grapple with is the rapid urbanisation and the failure by the formal sector to respond to the problems that are associated with this rapid urbanisation especially issues to do with creation of employment and provision of shelter. Cities of the global South have been failing to address issues associated with urbanisation of poverty especially in

Area	Frequency
Mucheke	No collection
Garikai	No collection
Mudhadhadha	At last once a month

Source: Survey 2021.

Table 1.
Waste collection frequency in informal areas.

the housing sector where people in informal settlements have been experiencing exclusion and living in bad environments, which threaten their lives, [13]. This exclusion is mainly due to their planning systems that have failed to respond to the changing urbanity that calls for inclusion and adaptive planning that suit existing environments in cities. Urban informality is a reality in cities of today, which call for their inclusion in the city's service delivery systems, [5, 8]. The urban poor and their ways of living have always been left out of the city provisioning systems and this is a denial of their right to the city. According to Lefebvre [14] every city inhabitant has the right to receive city services that will allow them access to safe and live-able spaces.

2.2 Poor waste collection and livability in informal settlements

The city of Masvingo has been found to be struggling to embrace informal settlements in their service provisioning systems. Waste collection in informal settlements in Masvingo as highlighted in the table above is nonexistent; hence the people in the informal settlements are living in life threatening environments. Most of the informal settlements n the city of Masvingo are close to formally planned areas, these areas are not receiving services that are enjoyed by the former land occupiers. For example Settlements behind Mucheke bus terminus is less than 20 meters from the formally planned R section of Mucheke but this areas has never received waste collection ever since. The Garikai informal settlement is between two formal settlements of Runyararo West and Victoria Ranch but there is no waste collection in the area. Trucks that collect waste in these formal areas actually pass very close to this informal settlement to the dumping site but they do not collect waste in the area. The informal settlements are therefore excluded in the waste collection services in Masvingo City. This is not sustainable because sustainable housing calls for safety, inclusion and resilience in cities, where there are safe and affordable housing, [15]. Smets and van Linder [15] further argued that sustainable housing should see upgrading of slum dwellings and creation of housing that withstands natural disasters. The case in Masvingo however does not show anything towards the direction of sustainable housing. There are informal settlements that are very close to Mucheke River and these people suffer a lot during the winter season as they are exposed to very cold conditions. Their houses are made of collected waste materials such as plastics and cardboard boxes, which cannot withstand cold conditions. In case of any flood they are the first victims as they are less than 10 meters from the river bank. These spaces are also their work spaces, where they make their livelihoods. As a result there is a lot of waste that they are producing especially the burning of rubber for collection of wire for resale. This waste is just thrown away indiscriminately and most of it is washed into Mucheke River. This creates further environmental problems as it contaminates the water bodies that supply water to the city. The water in Mucheke River is heavily polluted and the water is now blue-black in colour showing heavy pollutants. The washing away of these wastes into Mucheke River is also affecting the aquatic life in the river because the heavily polluted water is now unable to support aquatic life. Plate 1 below shows a heap of waste that is dumped in the Mucheke River banks. The heap is composed of an assortment of waste which include dippers and other degradable waste. When rain comes all this is washed into Mucheke River thereby polluting the water. There is need for proper waste management so that the environment is protected (**Figure 1**).

One of the respondents in the informal settlement said that there river can no longer support any life. He actually said ...

'*... we used to fish in this river but nowadays we cannot because we no longer can catch any fish in this river...*'.

Figure 1.
Waste dumped in Mucheke River bank.

The river can no longer absorb the pollutants that are being washed into it and this is no longer sustainable. If there was proper collection of waste in the informal settlements this would have minimised the amount of waste washed into the river. The pollution of the environment therefore constitutes unsustainable housing in Masvingo City. Asked why the city is not providing waste collection in informal settlement the city authorities said that there is no provision in the city by-laws and to provide services in the informal settlements. They said:

'..*Informal settlements are outside the city's jurisdiction therefore they do not provide our services...providing services will mean legalising the illegal...*'.

This exclusion is therefore a result of city regulatory framework that has remained rigid and failed to accommodate new land used that are coming up as a result of urbanisation of poverty such as informal settlements. Their service delivery system is mainly for the traditional land uses that are provided for by the Regional Town and Country Planning Act. Informal settlements, which are new land uses that are coming up due to urbanisation of poverty, have therefore remained outside the city's provisioning system of the city. This does not auger well with sustainable housing which calls for inclusion and creation of ecologically friendly environments, [13, 15]. Sustainable housing calls for inclusivity in cities where all city inhabitants are included in the city's governance system and provisioning systems. Exclusion of informal settlement in the waste management services creates life threatening environments for people in the informal settlements and this is not sustainable. Environmentally unfriendly conditions are not only dangerous to the people in the informal sector but can also affect the whole city as it can expose the whole city to disease out breaks. If one sector of the city is not healthy, it impacts negatively on the sustainability of the whole city. In order for the city to be sustainable it should be able to cater for the marginalised and vulnerable groups of the city [16, 17]. Failure to provide services to the marginalised groups also does not auger well with environmental justice, which calls for every citizen, including even the marginalised groups of the society to enjoy the same environmental goods and bads of the city, [17, 18]. If some sectors of the city experience too much environmental bads it is not sustainable. Every city inhabitants should enjoy the same degree of protection from environmental and health hazards, [19].

The need for safe and liveable spaces for the informal settlements is often neglected in many cities as many people in slum settlements are lacking a lot of services, which expose them to disease out breaks [3]. Cities are marginalsing people in the informal settlements in the provision of services resulting in the poor housing in the informal settlements, [20]. Cities prefer to deliver services to the formal sector rather than the informal sector and this does not auger well with dictates of rights to the city and environmental justice, which calls every city inhabitant to have the same right to receive city services as basic needs [19]. The city of Masvingo therefore has been failing to offer waste collection services in informal settlements as shown by **Table 1** above and this is environmental injustice perpetrated on informal settlements.

In contrast formal areas such as City centre, Mucheke high density areas, Industry and Rujeko high density areas are receiving waste collection services on average 5 times a week, while other areas such as the Mucheke bus terminus is receiving waste collection nearly every day including weekends. However such services are not extended to informal settlements and hence waste is indiscriminately dumped all over their areas, which create environmentally unfriendly conditions. Such conditions threaten their quality of life as bad environments reduce innovation and life expectancy [13]. The **Table 2** below shows waste collection schedule in formal areas. The table shows that there is strong bias towards formal settlements than the informal sector in their delivery of waste collection services in the city of Masvingo and this does not auger well with the dictates of environmental justices, which calls for fair distribution of environmental bads and goods in the city.

There is discrimination against informal settlements in terms of waste collection and this does not auger well with environmental justice and sustainable housing. This exclusion is also a denial of the people in the informal settlements' right to the city. According to the theory of right to the city, every city inhabitant has the right to receive services from the city, [14, 21–24]. Therefore the practice obtaining in the city of Masvingo where people in the informal settlements are sidelined in waste management is not sustainable. Adamec et al. [12] argued that sustainable housing should fight social isolation and promote inclusivity in the city, where all city inhabitants are treated the same and afforded the same quality of housing. The exclusion in waste management services creates poor environments in spaces occupied by informal settlements, which usually results in elevated risk of contracting diseases, [3]. The deprivation of waste management services in informal settlements also creates hostile environments characterised by high prevalence of disease causing agents. This then calls for inclusion of informal settlements in the delivery of waste management services for sustainable city.

The Masvingo city has remained against informal activities despite the wide prevalence of informal activities in the city. Urban informality in the city of Masvingo constitutes more that 80% of the [25, 26], but the city is doing nothing to integrate informality in their mainstream economy. The city up to now does not have a policy on urban informality, a thing that is hindering the growth and development of urban informality in the city. Issues of urban informality are still being dealt on a piece-meal basis without any comprehensive plan. The city does not even have by-laws that deal with urban informality and worse it is still being managed by

Area	Frequency
City centre	5times a week
Mucheke high density residential area	5 times a week
Industrial area	4 times a week
Rujeko high density residential area	5 times a week
Mucheke bus terminus	6 times a week

Source: Survey 2021.

Table 2.
Frequency of waste collection informal settlements.

outdated colonial statutes that were promulgated when cities had no experience of urban informality. These statutes do not recognise urban informality as a legal land use hence they remain illegal settlements and activities. However nowadays urban informality is a permanent feature of contemporary cities, which calls for integration into the city wide system for sustainable development. This therefore requires a more flexible management system that is adaptive. Cities of today need to adopt adaptive planning that takes into account the realities of their cities. UN-Habitat [7] argued that cities need not to do business as usual but they have to react to the different realities that are obtainable in their cities. The urbanisation of poverty requires a paradigm shift in terms of how cities manage their situations, [2, 5, 8]. The rigid regulatory systems in most cities in the in sub Saharan Africa has forced people in informal settlements to live with minimum services and are mostly confined in areas without important services, [27]. Kamete [27] further argued that people in the informal settlements are always found in spartialised enclaves where the rights of the inhabitants are completely stripped off to resemble people in a detention camp. Cities do not have budget for provision of services in the informal settlements and as such the rights of these people to receive services from the city authorities is highly dramatised [6]. This has resulted in most informal settlements living in conditions that are sub-human and this is contrary to the provisions of environmental justice which calls for fair distribution of environmental goods and bads.

2.3 Initiatives taken by informal settlements to manage waste management in Masvingo City

The people in the informal settlements in the city of Masvingo are trying to do whatever they can to create better living conditions in areas they are stay. They are employing various initiatives to clean their areas and these initiatives which include burning of waste, recycling material or just heaping waste outside their working spaces. **Figures 2–4** show some of the ways that the people in the informal settlements were using to manage their waste.

Figure 2.
Burning of waste at Garikai informal settlement.

Figure 3.
Open dumping of waste in open spaces in and around the city.

Figure 4.
Recycling of metal waste in Mucheke.

The pictures above show the rudimental waste management practices that are undertaken by the people in the informal settlements in Masvingo city. These initiatives are attempts to keep their environments clean and reduce the threats that are associated with unclean environments.. These methods are however rudimental hence are likely to cause further environmental damages in the city. For example method such as burning is likely to create green house gases that damage the ozone layer thereby exacerbating problems of climate change. The heaped scrap metals in Mucheke are too close to the residential (less than 8 m) and can be home to disease causing agents such as mosquitoes and rats. One of the respondents in the informal settlements said that the heaped metal are home to mosquitoes, rats and even dangerous animals such as snakes hence they live in constant threats of danger. He said:

"...mosquitoes give us sleepless nights during rainy season and rats are a big problem again due to these heaped metals.."

The people in the informal settlements in Masvingo city are therefore living in environments that continuously threaten their lives due to the waste that surround their areas. They therefore live in environments that keep them vulnerable to disease outbreaks [28]. Wilkinson further argued that the environments in informal settlements are severely curtailed by poor waste management, which is often done in streets, which in turn escalates their risks of disease contamination. Poor waste management practices are not sustainable as it pollutes the environment and expose people to diseases. Failure to provide waste management services to the informal settlements does not only expose people in the informal sector at risk of disease outbreaks but the whole city thereby affecting the sustainability of the whole city. For example during the outbreak of Cholera in 2005, the R-Section of Mucheke was severely affected and this put the whole city at risk of contamination.

Masvingo city needs to recognise informal settlements as an alternative and viable housing option and extend their services to their areas for the sustainability and inclusivity of the city. They need to incorporate these informal areas into the city's planning systems and provide services to them so that they improve the living conditions in these slum areas. This will reduce social injustices and also improve the inclusivity of the city, which will result in sustainable housing. Masvingo city like any other city in the global South need to adopt pro-poor development initiatives that are tailored to improve the conditions of the marginalised groups of the society, However Masvingo City still believes in neo-liberal development intervention, where elitist planning approaches are used and the net effect is the marginalisation of urban poor. Most city services therefore do not reach informal settlements and this is a denial of these people's right to the city [7, 18, 27]. Denying the people in the informal settlements good quality living conditions is also a denial of environmental justice and right to the city. Fisher et al. [29], argued that environmental justice aims to deliver among other things access to city services for better, safe and liveable spaces. Right to the city is a right that is offered unselectively but each every city inhabitant has to enjoy it and special attention should be given to marginalised groups of city who should be proritised in giving city inhabitants their rights to the city [21, 23, 24, 30, 31]. When the marginalised people of the society are included in the development initiatives of the city, this will result in sustainable development [16]. Environmental injustice often experienced by the urban poor are a result of unfair distribution of environmental bads and goods where the urban poor are forced to live in areas characterised by vulnerable ecological environments due to poor services but the most affluent groups of the city are the only ones that enjoys the environmental goods [18, 32]. The environmental threat that has been caused by poor waste management in the informal settlements should be a wakeup call to the city authorities of Masvingo city to try and plan for waste management service delivery in the informal settlements because it has a strong bearing on the sustainability of the whole city. Environmental management is very critical component in the sustainability of the city hence Masvingo city should device environmental management systems that cater for every city inhabitant including people in the informal settlements. **Table 3** sumerises some of the initiatives that are employed by the people in the informal sector to create safe and livable work space.

Recycling is the most popular way of waste management in informal settlements in the city of Masvingo. Most informal settlements are associated with livelihoods of the poor hence their spaces are characterised by a lot of waste from their livelihoods. Most of this waste is recycled for sale or reuse. Most of the scrap metal is

Initiative	Percentage
Burning	13
Just through waste way	36
Recycle our waste	51
Total	100

Source: Field Survey 2020.

Table 3.
Initiatives taken to manage solid waste in informal settlements in Masvingo City.

recycled for sale to scrap metal dealers in the city. Plastics are also recycled for sale to plastic dealers and in this way they are making their places cleaner and environmentally friendly surroundings. Recycling efforts are good local initiatives in the reducing waste in areas settled by urban informality because it forms bottom-up initiatives that are more sustainable than top-down [20].

2.4 Including urban informality for sustainable housing

The initiatives taken by the people in the informal settlements to help in the management of waste is therefore a very noble initiative that can be adopted by Masvingo city for sustainable waste management. The city can support these initiatives by providing protective clothes and availing simple machines that are affordable to the urban poor, [33]. Asibay *et al.* [33] further argued that there is need to create synages between the people in informal settlements and city authorities for effective participatory waste management in informal settlements. Participatory waste, management approaches provided an opportunity for integration and inclusion of informal settlements into the city-wide development interventions [34]. The inclusion will therefore lead to sustainable housing, because sustainable housing promotes inclusivity, [12, 15, 35]. These participatory initiatives could also be supported through injecting financial resources that will see informal settlements being included in city budgets for collection of waste. Such initiatives will result in inclusion of informal settlements in waste collection services of the city and this will go a long way in creating safe and livable spaces in the informal settlements. Participatory waste management practices are also a sustainable way of managing the environment in the city where there is promotion of locals in the management of waste. The people in the informal sector can be critical players in waste management and their participation in waste management is an essential component of sustainable development, [34, 36]. Hahn [36] further argued that there are great opportunities in involving the urban poor in development initiatives but these opportunities are often neglected and this has resulted in unsustainable development. Such participatory approaches in waste management are glaringly missing in the city of Masvingo because they city is not supporting local initiatives in waste management. This is therefore not sustainable because sustainable development gives that all the people the chance to contribute their ideas in development [22, 23]. The people in the informal settlements in Masvingo city have initiated a lot of initiatives to address problems of waste management in their areas but these initiatives are not complimented by the city of Masvingo, hence they have remained rudimental and ineffective. For example the city can come to collect all the waste that has been heaped at various points and transport it to dumping site, or even build communal waste collection centers in informal settlement for collection of waste generated in informal settlements. These centers will collect waste in the informal settlements that will later transported to designated dumping sites by the city. Such partnerships will help to create safe environments not only for the people in the informal settlements but also for the city as a whole. In addition such efforts will go a long way in giving the people in the informal settlements their rights to environmental justice and access to city services [29].

2.5 Waste collection in the formal and informal settlements: a comparison

As already alluded above, there is no waste collection in informal settlements but formal settlements are receiving these services nearly every working day of the week. Such discriminatory practices are not sustainable and does not auger well with environmental justice, which calls for equitable environmental treatment of all city inhabitants. There should be equitable share of environmental ills and risks in the city in order to achieve environmental justice [17]. However the urban poor are always found in environmentally hazardous places and this place a special burden on them because the poor environments add a cost on the way they do their business [13]. Environmental justice calls for reduction of environmental ills and elimination of all environmental threats that harm people [37]. Schoenfish and Johnson [37] further argued that environmental justice is for all citizens and environmental discrimination exposes the discriminated people to high risk of harm. Reduction of such risks will go a long way in improving the lives of urban poor who always find it difficult to access important city services or are working where important services are inadequate [38]. This does not auger well with the dictates of sustainable development, which calls for poverty alleviation efforts among the poor through improving their living conditions. Environmental justice is a right for all to be protected from environmental degradation and advocates for every citizen to live, work and play in healthy environments [18, 37]. This is again critical for sustainable development as it allows for social inclusion and social justice [12].

There is a big disparity in service delivery between formal and informal sector in the city of Masvingo. There is preference of the formal than the informal sector and this is a direct violation of people in informal settlements' right to the city and environmental justice. Rights to the city calls for all the city inhabitants to be accorded the same treatment in terms of service provision [14, 24]. Coggin and Pietersen [39] argued that such discriminatory practices divide the city between the propertied and privileged on one hand and the property less and the underprivileged on the other and this is not good for a sustainable city. Cities of the 21st century should celebrate urban diversity where all city inhabitants are given equal treatment [30, 31]. Coggin and Pietersen [39] further argued that rights to the city promotes inclusivity in the city, it endeavours to dismantle all structures that produces exclusionary practices in the city and this will create sustainable cities. The city of Masvingo therefore needs to acknowledge and recognise people in the in formal settlements as citizens of the city and provide the important services that will lead to safe and liveable spaces. Denying them waste collection services is a denial of their rights to the city and is not sustainable because it excludes other section of the city in waste collection services. The wide spread urbanisation of poverty in contemporary cities call for paradigm shift in the way cities do their business. It calls for inclusion of urban poor in the development interventions of the city [21–23]. The city is not only for the propertied people of the city but it is also for the poor [5]. Contemporary cities are now political collectivity and a place where public interests are defined and realised, which means that the interests of all city inhabitants should be observed [22, 23, 30, 31, 40]. Failure to embrace urban diversity is usually a result of insistence on old and rigid regulatory frameworks that are failing to realise the realities in their cities [2, 8]. Roy [8] further argued that the diversity in cities of today calls for a new planning theory that realises the prevalence of urban poverty and plan for it. Cities of today cannot afford to develop without urban informality because this phenomenon has developed to be a permanent feature of urbanity [5]. Urban policies need to reform so that they embrace the new forms of urbanism, which are coming up as a result of urbanisation of poverty [2, 7, 8, 41]. The rigid urban planning policies with their deep entrenchment in neo-liberal urbanism have failed to cater for new ways of life that are coming up in cities as a result of urbanisation of poverty [10, 30, 39]. These traditional planning systems have long been overtaken by events of 21st century urbanism, which call for flexible planning and inclusion of all city inhabitants in the development of the city. These new land uses such as the informal activities need to be included in the provisioning systems of the city so as to allow them to enjoy urban life [14].

3. Conclusions

Informal settlement in Masvingo city has been left out in the waste management system of the city and this does not auger well with sustainable housing. They are living without important services such as waste collection and this is not sustainable. Waste is indiscriminately dumped in open spaces, roadsides and river banks and this is polluting the environment. Pollution of the environment does not constitute sustainable housing. The water bodies are no longer able to sustain aquatic life as it is heavily polluted. The people are living in unhealthy environments characterised by rudimental waste management practices. Waste collection services are non-existing in informal settlement because the city of Masvingo has not included informal settlements in their waste collection schedule. This has prompted the people in informal settlements to employ rudimental ways of managing their waste. These rudimental ways include, burning of waste, dumping, and recycling. These rudimental ways of managing waste however cause further environmental problems for example burning of wastes exacerbates climate change problems. The local initiatives that are employed by people in informal settlements therefore, are worsening the environmental dilemma of the city. Those that just throw their waste indiscriminately pollute the environment and those that heap waste are creating habitants for disease causing agents such as mosquitoes and rats. This again escalates the threat to safety and liveability of their spaces and is not sustainable.

Author details

Average Chigwenya[1]* and Prisca Simbanegavi[2]

1 National University of Science and Technology, Department of Property Studies and Urban Studies, Zimbabwe

2 University of Witwatersrand, School of Construction Economics and Management, South Africa

*Address all correspondence to: chigwenyaaver@gmail.com

References

[1] Alijev H., 2015, Post-soviet informality: Towards theory building, International Journal of Sociology and Social Policy, vol. 35, no. 3/4, pp. 182-198.

[2] Roy, A., 2005, Urban informality, Towards an epistemology of planning, Journal of American Planning Association, vol. 71, no. 2, pp. 147-158.

[3] UN, 2020, The Sustainable Development Goals Report, UN, Nairobi.

[4] Yang J and Yang Z 2015, Critical factors affecting implementation of sustainable housing in Australia, Journal of Housing and Built Environment, vol. 30 pp. 275-292, DOI: 10.1007/S10901-014-9406-5.

[5] Huchzermeyer, M., 2011, *Cities with Slums: From Informal Settlement Eradication to a Right to the City in Africa*, Cape Town, UCT Press.

[6] Kamete AY, 2019, Neither friend nor enemy: Planning ambivalence and invalidation of urban informality in Zimbabwe, Urban Forum, 57, (5), 927-943.

[7] UN-Habitat, 2010, *Planning sustainable cities, un-HABITAL practices and perspectives*, Nairobi, Kenya, UN.

[8] Roy, A., 2009, Why India cannot plan its cities: Informality, Insurgence, and the Idiom of urbanization, Planning Theory, vol. 8, No.1, pp.76-87.

[9] Kamete A.Y. 2007, Cold hearted, negligent and spineless? Planning, planners and the rejection of filth in urban Zimbabwe, International Planning Studies, vol. 12, no. 2, pp.153-171.

[10] Potts, D., 2008, The urban informal sector in sub-Saharan Africa: From bad to good (and back again), Development Southern Africa, vol. 25, no. 2, pp. 151-162.

[11] Monacada E. 2013 The politics of urban violence: Challenges for development in global south, Studies in Comparative International Development, 48, pp. 217-239.

[12] Adamec T, Janouskova S, Hak T, 2021, How to Measure Sustainable Housing, a Proposal for an Indicator Based Assessment Tool, Sustainability, 13, 93), 1152

[13] Safronova, N, Nezhnikova E, and Kolhidov A, 2017, Sustainable housing development in conditions of changing living environment, Matev web conference 106, DOI. 10.1051/matecconf/201710608024, Moscow, Russia.

[14] Lefebvre, H.1968, *Le Droita la Ville (the Right to the City)*, Paris, Antropos.

[15] Smets D and Van Linder 2016, Sustainable housing and urban poor, International Journal of Urban Sustainable Development, 5,(1), 1-9.

[16] Chambers, R., and Connway, G.R., 1991, *Sustainable rural livelihoods: Practical concept for 21st century*, IDS discussion paper 296.

[17] Schlosberg, D, 2004, Reconciling environmental justice, global movements and political theories, Environment Politics, vol. 13, no. 1, pp. 517-540.

[18] Agyeman, J. and Evans, B, 2004, Just sustainability: The emerging discourse of environmental justice in Britain, The Geographical Journal, vol. 170, no. 2, pp. 155-164.

[19] Laurent, E., 2011, Issues in environmental justice within European

Union, Ecological Economics, vol. 70. Pp. 1840-1853.

[20] Ross N, Bowen PA, and Lincoln D, 2010, Sustainable housing for low income communities: Lessons from South Africa, Construction Management and Economics, 28, (5), 443-449.

[21] Harvey, D., 2003, Debates and development. The right to the city, *International Journal of Usrban and Regional Research*, vol. 27, no. 4, pp. 9-44.

[22] Harvey, D., 2008, The right to the city, New Left Review, vol. 53, pp.23-40.

[23] Harvey, D., 2012, *Rebel Cities, from the Right to the City to Urban Revolution*, London, Verso Books.

[24] Lefebvre, H., 1996: *Writings on Cities*. Cambridge, Blackwell.

[25] Chigwenya A, 2019, Low income housing problems and low income housing solutions: Opportunities and challenges, Journal of Housing and Built Environments, DOI:10.1007/5/10901/019-0967-w

[26] Government of Zimbabwe, 2002, *Small and Medium Enterprise Act* (Chapter 24: 12), Harare, Government Printers.

[27] Kamete, A.Y, 2017, Governing enclaves of informality: Unscrambling the logic of the camp in urban Zimbabwe, GeoForum, vol. 81, pp.76-86.

[28] Wilkinson, A, 2020, Local responses to health emergencies: Key considerations for addressing Covid-19 pandemic in informal urban settlements, Environment and Urbanisation, 30, (2), 503-522.

[29] Fisher, R., Katiya, Y., Reid, C., and Shragge, E., 2013, We are radical: The right to the city alliance and the future community organization, *journal of sociology and social welfare*, vol. XL, no. 1, pp. 156-182.

[30] Fainstein, S.S., 2005, Planning theory and the city, Journal Planning Education and Research, vol. 25, pp. 121-130.

[31] Fainstein, S.S., 2006, *Planning and the just city*, a paper presented at a conference on searching for the just city, GSAPP, Colombia university 29 April 2006.

[32] Masika, R., and Jeokes, S., 1997, *Environmentally Sustainable Development and Poverty: A Gender Analysis*, Sussex, Institute of Development Studies.

[33] Asibay MO, Amponsa LO and Yeboa V 2019, Solid waste management in informal urban neighbourhoods: Occupational safety health practices among tricycle operators in Kumasi, Ghana, international journal of environmental Health Research, 29, (6), 7-02-717.

[34] Asibay MO, Lykke MA, and King RS, 2020, Understanding the factors for increased informal electronic waste management in Kumasi, Ghana, International Journal of Environmental Health Research, DOI: 10.1080/09603123.1755016.

[35] Ssmugo C, Wafula ST, Lubega GB, Ndejjo R, Osuret J, Halage AA and Masoke D, 2020, Status of household solid waste management and associated factors in slum community in Kampala Uganda, Journal of Environment and Public Health, DOI:10.1155/2020/6807630.

[36] Hahn, R., 2009, The ethical rational of business for the poor: Integrating the concept- bottom of the pyramid sustainable development and corporate citizenship, Journal of Business Ethics, vol. 84, pp. 313-324.

[37] Schoenfish- J.K. and Johnson, G.S., 2010, Environmental Justice and Health: An Analysis of persons of color injured at the work place, *Race, Gender & Class*, vol. 17, no. 1/2 , pp. 270-304.

[38] Rocha, S., 1997, Sustainable development and poverty reduction goal, IPEA, Rio de Janeiro, www.ipea.gov.br.

[39] Coggin, T. and Pieterse, M., 2012, Right and the city: An exploration of interaction between socio-economic rights and the city, Urban Forum, vol. 23, pp. 257-278.

[40] Mustafa, D. and Leitte, G., 2002, Right to the city: Homage or new societal ethics? Capitalism, Nature and Socialism, vol. 13, no. 2, pp. 58-74.

[41] Dube, D. and Chirisa, I., 2012,Informal city: Assessing the scope, variants and directions in Harare, Zimbabwe, Global Advanced Research Journal of Geography and Regional Planning, vol.1, no.1, pp. 16-25.

Chapter 3

Sustainable Housing Design: System Control Strategy

Kazutoshi Fujihira

Abstract

Current housing design faces various environmental, social, and economic issues, such as climate change, aging population, and workplace needs. Dealing with such issues and pursuing sustainability is a target-oriented challenge. The science of system control can be utilized for all target-oriented tasks. Therefore, applying system control, we have been developing methods for sustainable design. Based on our finished research and practice results, this chapter shows how to design sustainable homes. Section 2 briefly illustrates the methods with two figures: (1) the control system for promoting sustainable housing design, (2) the process of producing and revising sustainable housing design guidelines. Section 3 demonstrates a concrete process of creating sustainable design guidelines. First, it identifies global and general problems related to current homes and specifies requirements for sustainable housing design. Next, it converts these requirements into a tabular form of "housing elements, variables, and their desired values." The completed table has turned out compact "sustainable housing design guidelines" for general use. The methods have four significant features: (1) total visualization for promoting sustainable design, (2) user-friendliness, (3) comprehensiveness, (4) flexibility toward optimization.

Keywords: system control, sustainable design guidelines, climate change, aging population, workplace needs

1. Introduction

Present housing design faces various environmental issues, including climate change. In 2014, the Fifth Assessment Report of the Intergovernmental Panel on Climate Change (IPCC) said, "without additional efforts to reduce greenhouse gas emissions beyond those in place today, global warming in 2100 ranges from 3.7°C to 4.8°C above the average for 1850-1900" [1]. The Paris Agreement of 2015 has aimed to limit global warming to well below 2°C, preferably to 1.5°C, compared to pre-industrial levels [2]. In order to curb global warming, the building sector must reduce CO_2 emissions drastically [3]. Meanwhile, no matter whether the Paris Agreement is fulfilled or not, the global mean temperature will inevitably rise from the present level. Accordingly, today's housing also needs to prepare for more severe extreme weather events caused by climate change [1].

On the other hand, changing social and economic situations also urge housing design to be altered. For example, aging population, which is remarkably progressing in this century [4], increases the proportion of people with disabilities in the population and households [5]. Thus, houses should adopt fundamental accessible and universal design features [5, 6]. Meanwhile, along with the development of information technology and network, the number of people who work at home or from home has been increasing [7, 8]. Currently, homes need to be recognized as significant places for working and economic activities [9].

Dealing with environmental, social, and economic issues and pursuing sustainability is a target-oriented challenge. The science of system control can be used for all target-oriented tasks [10]. Besides, that science has brought fruitful results in many fields, including engineering [10]. Therefore, applying system control, we have been progressing in research on sustainable design.

Our finished study results include the "control system for promoting sustainable housing design" and the "process of producing and revising the sustainable design guidelines." Utilizing these methods, we have already developed and revised sustainable housing design guidelines, mainly for use in Japan. Based on these accomplished results, this chapter shows how to design sustainable homes. First, the next section briefly illustrates the methods. After that, the main section demonstrates a concrete process of creating sustainable housing design guidelines.

2. Methods

2.1 Control system for promoting sustainable housing design

The control system for promoting sustainable housing design is shown in **Figure 1**. The upper and lower sections separated by the dotted line are the "theoretical world" and the "practical world."

In **Figure 1**, "controlled objects" are homes, which include both new and existing homes. "Disturbances" mean adverse effects on controlled objects originating from environmental, social, or economic issues. Instances of the disturbances are impacts of pollution and extreme weather events caused by climate change. The route from "disturbances" to "sustainability" is "adaptation." Recently, the

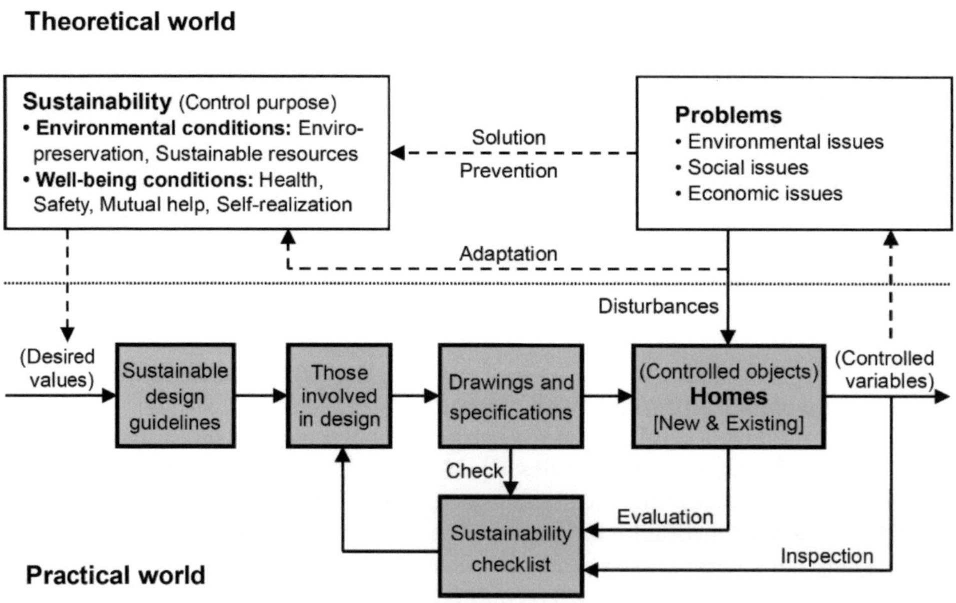

Figure 1.
Control system for promoting sustainable housing design.

necessity of adaptation to disturbances has become widely recognized, along with the progress of climate change [1]. The IPCC has stated that achieving sustainability also requires adapting to climate-related impacts, in addition to mitigating climate change [11].

The control purpose is the accomplishment of "sustainability." The upper-left box in **Figure 1** shows that achieving sustainability requires fulfilling both environmental and well-being conditions. Environmental conditions are "environmental preservation" and "sustainable use of natural resources." Meanwhile, well-being conditions are "health," "safety," "mutual help," and "self-realization," which are significant factors for people's well-being [12].

"Controlled variables" are the variables that are related to controlled objects and should be controlled for mostly solving or preventing the issues or adapting to disturbances. On the other side, "desired values" are derived from the control purpose, namely sustainability. The control objective of this control system is to adjust the controlled variables to their desired values [13, 14].

In the practical world, the subjects of control are "those involved in design," including homeowners, designers, architects, and homebuilders. In order to adjust the controlled variables to their desired values, those involved in design utilize the "sustainable design guidelines" and "sustainability checklist." Both of the design guidelines and checklist have almost the same expressions, that is, elements, variables, and desired values. However, the checklist is formed to smoothly compare measured or estimated values of the variables with the desired values and search for controlled variables [13, 14].

When objects are new homes, information about the desired values reaches "those involved in design" through the "sustainable design guidelines." Those involved prepare "drawings and specifications" so that the variables of the home's elements can satisfy their desired values to full potential. At significant phases in the design work, those involved in design check the drawings and specifications by seeing the "sustainability checklist" [13, 14].

When objects are existing homes, the design work starts with "inspection" on the home as an object. Using the sustainability checklist, "those involved in design" measure or estimate each element's variables of that home. Next, they compare the variables' measured/estimated values with the desired values. As a result, the variables whose measured or estimated values fall below the desired values should be "controlled variables." After the inspection, those involved in design usually prepare "drawings and specifications" for improvement so that the controlled variables meet their desired values to the maximum [13, 14].

2.2 Production and revision process of sustainable design guidelines

Figure 2 demonstrates the process of producing and revising the sustainable housing design guidelines and sustainability checklist. The upper section of the diagram is the "theoretical world," and the lower section is the "practical world."

The four blocks on the left side are the items that system designers refer to when producing or revising the sustainable housing design guidelines. The central part shows the flow of planning and using the sustainable design guidelines and checklist. First, the system designers produce or revise the design guidelines through the three-stage process. Subsequently, system users employ the design guidelines and checklist. Finally, the residents use the completed homes that have been designed with the design guidelines and checklist. Meanwhile, the two items on the lower right show the feedback loops from the system users and home residents.

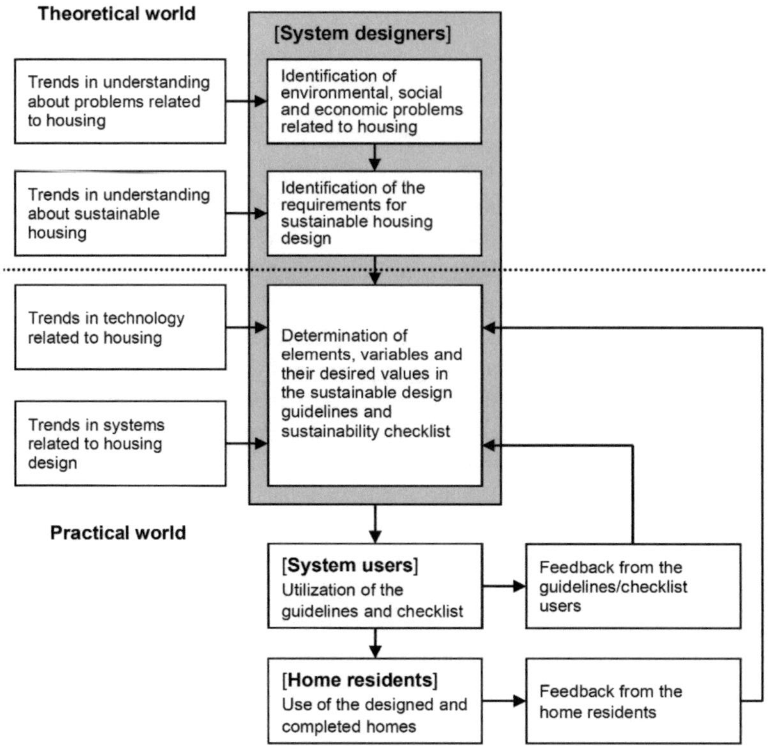

Figure 2.
Process of producing and revising the sustainable housing design guidelines and sustainability checklist [9].

2.2.1 Production process of the sustainable design guidelines

The production process of the sustainable housing design guidelines is made up of three stages: (1) identification of environmental, social, and economic problems related to housing, (2) identification of the requirements for sustainable housing design, (3) determination of elements, variables, and their desired values in the sustainable design guidelines [9].

In the first stage, system designers identify environmental, social, and economic issues related to housing, while checking trends in understanding about such issues. The basis for the identification is that the issues affect the well-being and environmental conditions shown in **Figure 1**, such as health, safety, and environmental preservation. System designers can specify local/particular issues in their country or region, in addition to global/general issues [9].

Next, based on the specified housing-related issues, the system designers identify the requirements for sustainable housing design. When identifying such requirements, it is useful to observe trends in understanding about sustainable housing [9].

In the third stage, the requirements for sustainable housing design are converted into a table of "element-variable-desired value," which can be found in the design guidelines. First, the system designers select "elements" so as to cover significant and necessary parts of homes. When selecting elements, it is helpful to base two factors on: "material" and "space." "Material" considers homes as the complexity of material elements, including framework, exterior, interior, windows and doors, and plumbing. "Space" regards homes as the complexity of spatial elements, such as rooms and areas [13, 14]. After identifying elements, system designers specify the elements' "variables" that can show the directions of the requirements for sustainable design. Subsequently, they determine the variables' "desired values" to meet the requirements.

The most significant advantage of converting the requirements into the "element-variable-desired value" table is practical convenience to system users. "Elements" in the table correspond to actual parts of homes. "Variables" and their "desired values" show the design points and their targets, respectively. Therefore, the system users can easily understand what should be designed and design steps [9].

2.2.2 Revision process of the sustainable design guidelines

The sustainable design guidelines are necessary to be revised, for adjustment to changing situations and improvement toward higher validity and user-friendliness. We have created the revision process by integrating three factors: (1) changes in the theoretical world, (2) changes in the practical world, (3) user feedback [9]. After making preparations from these three spheres, system designers amend the "element-variable-desired value" expressions.

As demonstrated in the upper left of **Figure 2**, changes over time in the theoretical world are reflected into the design guidelines [9]. First, observing the latest trends in understanding about housing-related problems, system designers amend the problem list. After that, they also modify the list of the requirements for sustainable housing design. When modifying the requirement list, it is also necessary to search for changes in understanding about sustainable housing.

Changes over time in the practical world also need to be taken in the design guidelines. In order to reflect such changes, the system designers observe current trends in housing-related technology and systems related to housing design [9]. In addition, systems related to housing design include compulsory systems, such as building codes, and voluntary systems relating to sustainable housing (assessment and rating systems, standards, guidelines, etc.) [15].

Meanwhile, "user feedback" is significant for the improvement of the design guidelines and checklist. In this case, there are two feedback loops: from system users and home residents. The feedback from the system users is information about reactions to the guidelines and checklist, such as comments on their adequacy and user-friendliness. On the other hand, the feedback from the home residents includes responses to the completed homes, such as remarks about the homes' amenities and sustainability performance [9].

3. Sustainable housing design guidelines for general use

Using the methods illustrated in the previous section, we have already produced and revised sustainable housing design guidelines, chiefly for use in Japan. Based on these studies and practical results, this section anew demonstrates how to produce sustainable housing design guidelines for general use. In line with the process of producing guidelines, this section consists of two subsections: (1) housing-related problems and requirements for sustainable design, (2) conversion into housing elements, variables, and desired values.

3.1 Housing-related problems and requirements for sustainable design

Housing-related problems and requirements for sustainable housing design have been demonstrated in **Table 1**. We have first selected main global and general issues, considering the conditions for sustainability. The first two items, "global

Housing-related problems (Main global/general issues)	Requirements for sustainable housing design	Conditions for sustainability
• Global warming and climate change	• Energy saving • Use of renewable energy • Increase of green spaces	• Enviro-preservation
• Environmental issues related to building materials (environmental destruction, biodiversity loss, environmental pollution, climate change, resource depletion, waste)	• Material saving • Extension of housing lifespan • Use of low-environmental-load materials	• Enviro-preservation • Sustainable resources
• Harmful effects on housing caused by climate change (more-severe heat waves, heavy precipitation events, cyclonic storms, floods, drought, wildfires, etc.)	• Adaptation measures (measures against increasing temperatures, more impact-resistant exterior, securing emergency water and energy, etc.)	• Health • Safety
• Urban heat island • Urban flooding due to rainwater runoff	• Increase of green spaces • Reduction of impervious surface areas	• Enviro-preservation • Health • Safety
• Blackout risks due to increasing solar and wind power generation	• Storage of electricity	• Health • Safety
• Water shortage risks	• Water saving • Use of rainwater	• Health • Safety
• Issues resulting from aging population (Increase of medical and nursing care expenses, etc.)	• Accessible and universal design	• Health • Safety
• Poor indoor thermal performance	• Healthy indoor thermal performance	• Health
• Indoor air pollution caused by toxic substances (volatile organic compounds, etc.)	• Use of non-toxic substances for the occupants	• Health
• Insufficient considerations for working at/from home and learning	• Spaces and equipment for working and learning	• Self-realization
• Issues resulting from insufficient communication	• Planning suitable for communication	• Mutual help • Self-realization
• Neglected landscape	• Consideration for the landscape	• Health
• Crimes (burglaries, etc.)	• Security measures	• Safety

Notes: (1) Housing-related problems listed in this table are only main global and general issues. If intending to create design guidelines for a region or country, system designers need to add local and particular issues of that region or country to the problem list [9]. (2) This housing design study deals with issues that can occur after housing sites are determined. Therefore, matters concerning housing land development or site selection are excluded from this table. Instead, such issues are handled in the study on urban design [16].

Table 1.
Housing-related problems and requirements for sustainable design.

warming and climate change" and "environmental issues related to building materials," are significant problems linked to environmental conditions. The other issues are chiefly related to well-being conditions, such as health and safety. Choosing five items from the table, this section briefly explains the outlines of the problems and the requirements for sustainable design.

3.1.1 Global warming and climate change

The IPCC's Fifth Assessment Report has concluded that cumulative emissions of CO_2 largely determine global mean surface warming [1]. The amount of CO_2 in the atmosphere has been obviously increasing due to human emissions [1, 17]. In order to cut CO_2 emissions, the housing sector must reduce energy originating from fossil fuels. Therefore, two mitigation measures, namely saving energy and using renewable energy, are significant as the requirements for sustainable design [1]. Moreover, the necessity of enhancing the sinks of CO_2 requires increasing green spaces [1].

3.1.2 Environmental issues related to building materials

Building materials over their life cycles, from resource acquisition, throughout construction, demolition, and disposal or recycling, are related to various environmental problems. Such issues include environmental destruction, biodiversity loss, environmental pollution, climate change, resource depletion, and waste.

The strategy of reducing environmental impacts related to housing materials can be divided into three: (1) material saving, (2) extension of housing lifespan, (3) use of low-environmental-load materials [18]. "Material saving" aims to reduce the amount of material in a housing structure. The "extension of housing lifespan" requires tactics for long-life homes, such as selecting durable materials, adopting deterioration prevention measures, and considerations for adapting to future layout changes. Meanwhile, the "use of low-environmental-load materials" includes various options, such as using locally produced materials, utilizing recyclable or recycled materials, and sustainable use of renewable materials [18].

3.1.3 Harmful effects on housing caused by climate change

Along with increasing atmospheric CO_2 concentrations, climate-related impacts on the built environment are also becoming more severe [1]. Such impacts include heat waves, heavy precipitation events, cyclonic storms, floods, drought, and wildfires [1]. Accordingly, the necessity to adapt housing design to climate change is also increasing. Adaptation measures in the building sector include measures against rising temperatures (insulation, external shading, cross ventilation, etc.), more impact-resistant exterior, and securing emergency water and energy [19, 20].

In addition, measures to reduce climate-related risks differ across regions [1]. Thus, those engaged in sustainable design in each region should adequately predict future risks and plan effective measures.

3.1.4 Issues resulting from aging population

Aging population is becoming a crucial challenge to economic and social sustainability. An increasing number of the elderly inflates public expenditures on pensions, and medical and nursing care [21]. Meanwhile, since disability rates increase with age, the aging population raises the percentage of people with disabilities in the population and households [5].

Current demographic changes require housing to adopt accessible and universal design [5, 6, 22]. Incorporating universal design features into homes improves safety, helping to prevent accidents caused by falls and slips [6, 22, 23]. In addition, universal design helps everyone with assistance needs, such as people with disabilities, the elderly, small children, pregnant women, and people with a temporary injury or illness [24, 25]. Furthermore, including universal design principles in advance leads to a drastic reduction of future renovation costs. According to the Victorian Government of Australia, the cost of incorporating fundamental universal design features into a new home is more than 20 times cheaper than retrofitting such features into an existing home [26, 27].

3.1.5 Insufficient considerations for working at/from home and learning

Recently homes have been becoming more significant as working places [9]. The development of information technology and networks has been promoting home-based businesses [7]. Besides, an increasing number of firms are adopting working from home [7, 8]. Meanwhile, life longevity is increasing the necessity of lifelong learning [28]. As a result, housing is also gaining importance as a place of learning for adults as well as children.

Previously, houses were not recognized as significant places of working and lifelong learning. Accordingly, if attempting to start working or learning at home, people often encounter difficulties due to a lack of space and facilities. In fact, unexpected demands for working from home forced by the COVID-19 pandemic revealed such difficulties. For example, in April of 2020, a housing-related firm in Japan, Recruit Sumai Company Ltd., conducted a questionnaire to office workers living in the Greater Tokyo Metropolitan area and gained 1390 valid responses. As a result, many workers from home answered that there were various insufficiencies in their houses, such as a lack of space or room for working and equipment shortages [29].

3.2 Conversion into housing elements, variables, and desired values

We have converted the requirements for sustainable design into the structure of housing elements, variables, and desired values. First, based on the above-mentioned two factors, namely "material" and "space," we have selected a total of 23 housing elements. After specifying the elements, we have determined variables and their desired values, as shown in **Table 2**.

3.2.1 Material element design

In **Table 2**, material elements are from "entire building" to "outdoor facilities." Choosing four from these 14 items, this section illustrates material element design for sustainable housing.

3.2.1.1 Entire building

The shape of the entire building closely relates to the "energy saving" and "material saving" shown in **Table 1**. Preferring compact forms to sprawling ones reduces the building envelope surface area and decreases thermal transfer through the surface [18, 30]. Moreover, pursuing compact shapes also leads to lower embodied energy and environmental impacts related to materials for constructing the envelope itself [30]. Therefore, we have identified the entire building's key variable and its desired value as "shape: compactness" and "compact," respectively.

The most common quantitative indicator of compactness is the ratio of the envelope surface area (S) to the enclosed volume (V) [30]. Accordingly, we have adopted the "surface-to-volume ratio (S/V)" as the index of the variable and determined its desired value as a "smaller S/V ratio." For example, the S/V ratio of the house shown in **Figure 3** is 0.888, which is a considerably small figure among S/V ratios of shapes with the same enclosed volume as this house. When planning this home, the owner and designers pursued a more compact form in the restriction of the shown narrow land.

Element	Variable	Desired value
Entire building	Shape: Compactness (Index: Surface-to-volume ratio [S/V])	Compact (Not sprawling) (Index: Smaller S/V ratio)
	External appearance	Consideration for the landscape
Framework	Durability	Long service life (Superior deterioration resistance against decay or corrosion)
	Materials	Low-environmental-load materials (Priority of wood over steel and concrete)
Exterior (roof, wall, etc.)	Resistance against impacts (fires, storms, etc.)	Sufficient resistance against anticipated impacts
	Durability	Long service life (Longer expected lifespan, easier replacement of the features)
	Materials	Low-environmental-load materials
Thermal insulation	Thermal insulation performance	Sufficient thermal insulation performance for occupants' health
Windows and doors	Thermal insulation performance	Sufficient thermal insulation performance for occupants' health
	Sunlight adjustment capability	Sufficient capability of taking the sunlight in winter and reducing it in summer (Primary related factors: Types of window glass, solar shading materials such as blinds, and eaves)
	Protection of glass against impacts	Shutters
	Security measures	Sufficient intrusion prevention measures
Interior (floor, wall, ceiling, etc.)	Toxicity	Non-toxic for occupants' health
	Materials	Low-environmental-load materials
Lighting fixtures	Type of light	LED (Lighting fixtures in living spaces are fitted with lighting controls)
Energy-using equipment (heating, air-conditioning, hot-water supply, ventilation, etc.)	Energy efficiency	High energy efficiency
Plumbing (water pipes, drainage pipes, gas pipes, etc.)	Durability	Long service life (Longer expected lifespan, easier replacement of the piping)
Equipment for harnessing renewable energy	Harnessed renewable energy	100% or more of the total energy usage
Storage battery	Type	Household battery or Electric vehicle battery
	Linkage	Interconnection with the home electrical system
Equipment for rainwater use	Rainwater-using equipment	Rainwater tank
Water-using equipment (toilet bowls, faucets, shower heads, etc.)	Water saving performance	High water saving performance
Outdoor facilities (fence, etc.)	Form	Secure unhindered sightlines
	Appearance	Consideration for the landscape
	Materials	Low-environmental-load materials

Element	Variable	Desired value
Areas relating to water use and hot-water supply	Areas in the home	Placing them closer
Living/dining room and kitchen area	Place in the home	Between the entrance and private room area
	Type of kitchen	Open or semi-open
Areas for working and learning	Places in the home	In or near the living/dining room and kitchen area
	Equipment	Table/desk and shelf (fixed or movable) and Internet connection
Bedroom for the disabled or elderly	Floor space	10 m^2 or more, with one wall a minimum length of 3 m
Accessible route	Areas connected with the accessible route	Bedroom for the disabled or elderly, toilet and bath area, living/dining room, kitchen, area for working and learning, entrance, street, (parking)
	Passages' surface on the route	Flat or gently sloping (Gently sloping: 1/12 gradient max.)
	Doorways' thresholds on the route	No level differences
	Passages' width on the route	90 cm or more
	Doorways' width on the route	75 cm or more
Toilet and bathroom	Wheelchair maneuverability space	Sufficient clear space from the rim of the toilet bowl and the bathtub
	Handrails help toilet users sit and stand	Installed
	Handrails help bath users go in and out of the bathtub	Installed
Stairs	Gradient of steepness	40 degrees or less
	Handrails	Installed
Position and area of windows	Positions of windows in each living space (Natural ventilation)	Two or more places on walls in each living space (Cross ventilation)
	Ratio of total window area to floor area in each living space (Daylighting)	20% or more
Green space	Ratio of the green space in the outside area	High ratio of the green space

Notes: Material elements are from "entire building" to "outdoor facilities" and spatial elements are from "areas relating to water use and hot-water supply" to "green space".

Table 2.
Sustainable housing design guidelines for general use.

3.2.1.2 Equipment for harnessing renewable energy

Responding to "use of renewable energy," a requirement for sustainable housing design, we have identified "equipment for harnessing renewable energy" as a material element. After that, we have determined the key variable and its

Figure 3.
An example of material element design for sustainable housing [31].

desired value to be "harnessed renewable energy" and "100% or more of the total energy usage," respectively. This desired value means aiming at net-zero-energy or surplus-energy housing. Achieving the desired value usually needs both energy saving and a considerable equipment capacity to harness renewable energy.

The most common equipment for harnessing renewable energy on housing sites is solar power generation systems. For example, the house demonstrated in **Figure 3** is equipped with 49 solar panels on the single-pitch roof. The combination of this larger-scale photovoltaic generation system and various energy-saving schemes has enabled this home to reach an amazing 500% of self-sufficiency in energy [31].

3.2.1.3 Windows and doors

Openings, especially windows, are related to many requirements for sustainable housing design, including healthy indoor thermal performance. Improving indoor thermal performance requires windows to secure sufficient "thermal insulation performance" and "sunlight adjustment capability," both of which contribute to energy saving by decreasing the demand for heating and air-conditioning [32]. Besides, such tactics also become adaptation measures because they can reduce temperature-related impacts, including overheating during heat waves [19, 33]. Therefore, we have identified the desired value of thermal insulation performance as "sufficient thermal insulation performance for occupants' health." Meanwhile, the desired value of sunlight adjustment capability has been determined to be "sufficient capability of taking the sunlight in winter and reducing it in summer." Primary factors influencing the sunlight penetration ratio are types of window glass, solar shading materials such as blinds, and eaves. Concerning the windows on the southern façade of the home in **Figure 3**, the window glass, lace curtains, the balcony, the roof with the pendent eave, and the deciduous tree cooperate to control the sunlight.

Moreover, considering an adaptation measure, that is, more impact-resistant exterior, we have specified "protection of glass against impacts" as another variable of windows. Subsequently, we have determined its desired value to be "shutters." Protecting the most vulnerable part of housing external surfaces, namely window-panes, helps prevent damage caused by climate-related extreme events. As shown in the lower right of **Figure 3**, shutters cover the windows and protect the glass against impacts, such as fire, hurricanes, and flying objects. In addition, covering windows with shutters also helps upgrade intrusion prevention measures.

3.2.1.4 Lighting fixtures

Regarding "lighting fixtures," we have specified the key variable and its desired value as "type of light" and "LED," respectively. LED lamps are significantly more energy-efficient than others, including fluorescent lamps and incandescent light-bulbs; therefore, using LEDs can satisfy energy saving, one of the requirements in **Table 1**. Besides, LEDs are much more durable than other light sources. Furthermore, LEDs do not contain toxic materials, such as mercury [34]. Accordingly, using LED also contributes to reducing environmental impacts related to building materials.

Meanwhile, we have added notes to "LED," saying "lighting fixtures in living spaces are fitted with lighting controls" (**Figure 3**, lower left) [9]. Dimmers and other controls can reduce brightness and help consume only the amount of electricity needed. Thus, LED with lighting controls is highly energy-efficient. Moreover, brightness and color adjustment functions are beneficial for occupants' health and well-being. For example, avoiding bright lights and blue light before bedtime contributes to preventing sleep-quality-related diseases [35, 36].

3.2.2 Spatial element design

The items shown in the latter part of **Table 2**, namely from "areas relating to water use and hot-water supply" to "green space," are spatial elements. Selecting five from these nine items, this section illustrates spatial element design for sustainable housing.

3.2.2.1 Areas relating to water use and hot-water supply

"Areas relating to water use and hot-water supply" include a kitchen and hygiene-related area, such as a bathroom and laundry, as well as a place for a water heater (**Figure 4**). These areas should be placed closer, in order to reduce plumbing-related materials and energy. This spatial planning leads to reduction of the total length of water and hot-water plumbing and waste pipes. This consideration also contributes to cutting down the heat loss from hot-water supply pipes. In addition, this arrangement leads to comfort in occupants' daily life. As the distance from the water heater to the faucet decreases, the time until hot water comes out shortens [31].

3.2.2.2 Areas for working and learning

As described in Section 3.1, housing has been gaining importance as a place of working and learning. Considering such recent situations, we have selected "areas for working and learning" as a spatial element. After that, we have determined the key variable and its desired value as "places in the home" and "in or near the living/dining room and kitchen area," respectively. In addition, the above-mentioned survey conducted under the COVID-19 pandemic has also supported the appropriateness of this space planning. The answers to the questionnaire have included two types of needs: (1) want to work in a larger living room, (2) want to work in a space or room for exclusive use [29]. Accordingly, in **Figure 4**, we have partially overlapped the "areas for working and learning" with the "living/dining room and kitchen area."

3.2.2.3 Bedroom for the disabled or elderly

A "bedroom for the disabled or elderly" means a private area used or expected to be used by the residents limited in their movements, such as wheelchair users and the elderly. As shown in **Figure 4**, this room needs to be connected with the accessible route. Therefore, the room is usually placed on the entrance-level floor, unless the home is equipped with an elevator.

The "bedroom for the disabled or elderly" is capable of responding to various needs. If there are no occupants with disabilities at first, it can be utilized as an ordinary bedroom. Meanwhile, this room can also be used for working or learning. Thus, in **Figure 4**, the area of this room has been outlined by the broken green line,

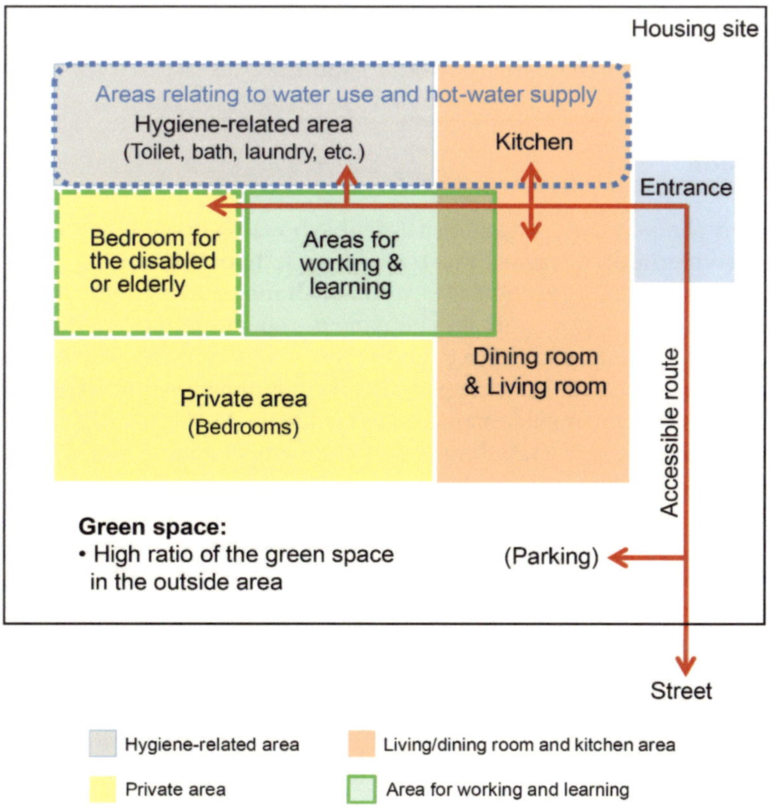

Figure 4.
Concept of spatial planning for sustainable housing.

as well as colored with yellow. While homes are used, occupants' number, age, and health conditions change inevitably. Therefore, housing should have the flexibility to manage such changing situations. Adaptability to possible future layout changes contributes to extending housing lifespan and reducing the environmental burden.

3.2.2.4 Accessible route

An "accessible route" means a course on which disabled people and the elderly can move around easily. As demonstrated in **Figure 4**, the accessible route should be connected to a bedroom for the disabled or elderly, a toilet and bath area, an area for working and learning, a living/dining room, a kitchen, an entrance, and the street. Moreover, if there is parking in the housing site, it should also be linked with the accessible route. The surface of the passages on the route must be flat or gently sloping. Similarly, the doorways on the route should have no level differences. Moreover, the passages and doorways on the accessible route must be wide enough for a wheelchair to pass.

3.2.2.5 Green space

A "green space" is an area with plants, such as trees, shrubs, herbs, and grasses. Increasing green spaces contributes to environmental preservation, health, and safety in various ways. First, it mitigates global warming and climate change, enhancing the sinks of CO_2. Moreover, increasing green spaces also mitigate both urban heat island and urban flooding risks. As implied in **Figure 4**, extending green spaces in the outside area of the housing site leads to reducing impervious surface spaces, typically areas covered with concrete and asphalt. These two factors reduce urban heat island effects through increasing evapotranspiration and decreasing heat absorption. Meanwhile, the increase of pervious surface spaces can decrease urban flooding risks by reducing rainwater runoff.

4. Conclusion

This study showed how to design sustainable housing by employing the system-control-based methods. Section 2 illustrated the methods with the two diagrams: (1) the control system for promoting sustainable housing design, (2) the process of producing and revising sustainable housing design guidelines. Using these methods, Section 3 demonstrated a concrete process of creating sustainable housing design guidelines. After identifying global and general problems related to homes, it specified requirements for sustainable housing design. Subsequently, these requirements were converted into a tabular form of "housing elements, variables, and their desired values." The completed table has turned out compact "sustainable housing design guidelines" for general use.

The proposed methods include four significant features: (1) totally visualized for promoting sustainable housing design, (2) user-friendly, (3) comprehensive, (4) flexible toward optimization. The first feature originates from the two diagrams shown in Section 2, namely **Figures 1** and **2**. These schematizations help understand the whole picture for promoting sustainable housing design.

The second characteristic, "user-friendly," results from the tabular form of "element-variable-desired value." Elements in the table are equivalent to actual parts of homes. Accordingly, system users can easily compare the design guidelines with actual homes or drawings. Meanwhile, the third characteristic, "comprehensive," originates in the understanding about sustainability and housing elements. As shown in **Figure 1**, we inclusively examined the sustainability conditions from two aspects: environment and well-being. Moreover, when selecting housing elements, we analyzed homes based on two fundamental factors: material and space. Consequently, the completed design guidelines could deal with various environmental, social, and economic issues, such as climate change, aging population, and workplace needs.

The fourth feature, "flexible toward optimization," results from the process of revising the design guidelines. As shown in **Figure 2**, the revision process has been created by combining the two aspects: (1) updates of related knowledge and information, (2) feedback from users. Through the updates of related knowledge and information, the system designers can adapt the design guidelines to deal with the changing necessities of the situation. On the other hand, the feedback loops help make the design guidelines more accurate and user-friendly. Therefore, repeatedly using that revision process enables the system designers to optimize the design guidelines to pursue sustainability.

Our future work includes further research and practice on sustainable housing design. First, following the revision process demonstrated in **Figure 2**, we are planning to revise the sustainable housing design guidelines for general use as well as those for use in Japan. Moreover, we must secure the coordination between the guidelines for housing design and urban design. Through such future work, we are aiming to refine the system-control-based methodology for sustainable design.

Author details

Kazutoshi Fujihira
Institute of Environmentology, Inagi, Tokyo, Japan

*Address all correspondence to: fujihira@kankyogaku.com

References

[1] IPCC. Climate Change 2014: Synthesis Report. Contribution of Working Groups I, II and III to the Fifth Assessment Report of the Intergovernmental Panel on Climate Change. Geneva: IPCC; 2014. 151 p

[2] United Nations Framework Convention on Climate Change. The Paris Agreement [Internet]. 2021. Available from: https://unfccc.int/process-and-meetings/the-paris-agreement/the-paris-agreement [Accessed: 2021-07-18]

[3] Lucon O et al. Buildings. In: Edenhofer O, et al., editors. Climate Change 2014: Mitigation of Climate Change. Contribution of Working Group III to the Fifth Assessment Report of the Intergovernmental Panel on Climate Change. Cambridge. New York: Cambridge University Press; 2014. p. 671-738

[4] United Nations, Department of Economic and Social Affairs, Population Division. World Population Prospects 2019, Volume I: Comprehensive Tables (ST/ESA/SER.A/426). New York: United Nations; 2019. 395 p

[5] Smith S, Rayer S, Smith E. Aging and disability: Implications for the housing industry and housing policy in the United States. Journal of the American Planning Association. 2008;74(3):289-306. DOI: 10.1080/01944360802197132

[6] Fujihira K. Requirements for sustainable housing design. In: Fujihira K, editor. Sustainable Home Design by Applying Control Science. Rijeka: IntechOpen; 2017. DOI: 10.5772/intechopen.71322 Available from: https://www.intechopen.com/chapters/57469

[7] Global Workplace Analytics. Latest Work-at-Home/Telecommuting/Mobile Work/Remote Work Statistics [Internet]. 2021. Available from: https://globalworkplaceanalytics.com/telecommuting-statistics [Accessed: 2021-08-06]

[8] Great Work Life. 78 Work From Home Statistic & Trends in 2021 [Internet]. 2021. Available from: https://www.greatworklife.com/work-from-home-statistics/ [Accessed: 2021-08-06]

[9] Fujihira K. Comprehensive strategy for sustainable housing design. In: Cakmakli A, editor. Different Strategies of Housing Design. London: IntechOpen; 2019. DOI: 10.5772/intechopen.86278 Available from: https://www.intechopen.com/chapters/67084

[10] Osuka K, Adachi S. Approach to Systems Control (in Japanese). Tokyo: Corona Publishing; 1999. 177 p

[11] Denton F, Wilbanks T, Abeysinghe A, Burton I, Gao Q, Lemos M, Masui T, O'Brien K, Warner K. Climate-resilient pathways: Adaptation, mitigation, and sustainable development. In: Field C, et al., editors. Climate Change 2014: Impacts, Adaptation, and Vulnerability. Part a: Global and Sectoral Aspects. Contribution of Working Group II to the Fifth Assessment Report of the Intergovernmental Panel on Climate Change. Cambridge, New York: Cambridge University Press; 2014. p. 1101-1131

[12] Fujihira K. Basic schemes: Preparations for applying control science to sustainable design. In: Fujihira K, editor. Sustainable Home Design by Applying Control Science. Rijeka: IntechOpen; 2017. DOI: 10.5772/intechopen.71325 Available from: https://www.intechopen.com/chapters/57401

[13] Fujihira K. System control for sustainability: Application to building design. In: Thomas C, editor. Complex Systems, Sustainability and Innovation, Rijeka: IntechOpen: 2016. DOI: 10.5772/65875 Available from: https://www.intechopen.com/chapters/52736

[14] Fujihira K. Methodology of applying control science to sustainable housing design. In: Fujihira K, editor. Sustainable Home Design by Applying Control Science. Rijeka: IntechOpen; 2017. DOI: 10.5772/intechopen.71324 Available from: https://www.intechopen.com/chapters/57412

[15] Fujihira K. Discussion and conclusion: Effectiveness, characteristics and future prospects of the methodology. In: Fujihira K, editor. Sustainable Home Design by Applying Control Science. Rijeka: IntechOpen; 2017. DOI: 10.5772/intechopen.71321 Available from: https://www.intechopen.com/chapters/57437

[16] Fujihira K. How to design sustainable structures. In: Sarvajayakesavalu S, Charoensudjai P, editors. Environmental Issues and Sustainable Development. London: IntechOpen, 2020. DOI: 10.5772/intechopen.95012 Available from: https://www.intechopen.com/chapters/74360

[17] NOAA Climate.gov. Climate change: Atmospheric carbon dioxide. [Internet]. 2020. Available from: https://www.climate.gov/news-features/understanding-climate/climate-change-atmospheric-carbon-dioxide [Accessed: 2021-07-18]

[18] Bergman D. Sustainable Design: A Critical Guide. New York: Princeton Architectural Press; 2012. 144 p

[19] Snow M, Prasad D. Climate Change Adaptation for Building Designers: An Introduction [Internet]. 2011. Available from: https://apo.org.au/sites/default/files/resource-files/2011-02/apo-nid72346.pdf [Accessed: 2021-08-06]

[20] Australian Government, Department of climate change and energy efficiency. Your Home: Australia's Guide to Environmentally Sustainable Homes, Adapting to Climate Change [Internet]. 2013. Available from: http://www.yourhome.gov.au/housing/adapting-climate-change [Accessed: 2021-08-06]

[21] European Union. Working Group on Ageing Populations and Sustainability [Internet]. Available from: http://europa.eu/epc/working-group-ageing-populations-and-sustainability_en [Accessed: 2021-08-06]

[22] Harding E. Sustainable Planning for Housing in an Ageing Population: A Guide for Regional-level Strategies. London: International Longevity Centre UK; 2008. 29 p

[23] Ozanne-Smith J, Guy J, Kelly M, Clapperton A. The Relationship between Slips, Trips and Falls and the Design and Construction of Buildings [Internet]. 2008. Available from: http://www.monash.edu/__data/assets/pdf_file/0019/217342/muarc281.pdf [Accessed: 2021-08-06]

[24] United Nations. Convention on the Rights of Persons with Disabilities [Internet]. Available from: http://www.un.org/disabilities/documents/convention/convoptprot-e.pdf [Accessed: 2021-08-06]

[25] Australian Agency for International Development. Accessibility Design Guide: Universal Design Principles for Australia's Aid Program. A Companion Volume to Development for All: Towards a Disability-Inclusive Australian Aid Program 2009-2014 [Internet]. 2013. Available from: https://www.dfat.gov.au/sites/default/files/accessibility-design-guide.pdf [Accessed: 2021-08-06]

[26] Australian Government Department of Social Services. National Dialogue on Universal Housing Design - Strategic Plan [Internet]. 2010. Available from: https://www.dss.gov.au/sites/default/files/documents/05_2012/national_dialogue_strategic_plan.pdf [Accessed: 2021-08-06]

[27] Victorian Government of Australia, Department of Planning and Community Development. Visitable and Adaptable Features in Housing: Regulatory Impact Statement. Victorian Government of Australia: Department of Planning and Community Development; 2009. 190 p

[28] Gratton L, Scott A. The 100-Year Life: Living and Working in an Age of Longevity. London: Bloomsbury Business; 2017. p. 407

[29] Recruit Sumai Company Ltd. Questionnaire about teleworking and housing under the situation of the COVID-19 pandemic (in Japanese). [Internet]. 2020. Available from: https://www.recruit.co.jp/newsroom/recruit-sumai/data/upload/07a49f312ad3ef3f6eb08d2e4dac6f08.pdf [Accessed: 2021-07-22]

[30] D'Amico B, Pomponi F. A compactness measure of sustainable building forms. 2019; R. Soc. open sci. 6: 181265. DOI: http://dx.doi.org/10.1098/rsos.181265

[31] Fujihira K. Case study: Detached house designed by following the control system. In: Fujihira K, editor. Sustainable Home Design by Applying Control Science. Rijeka: IntechOpen; 2017. DOI: 10.5772/intechopen.71323 Available from: https://www.intechopen.com/chapters/57400

[32] National Institute of Building Sciences. Whole Building Design Guide: Windows and Grazing [Internet]. 2016. Available from: https://www.wbdg.org/resources/windows-and-glazing [Accessed: 2021-07-22]

[33] State of New York, New York State Energy Research and Development Authority, University at Buffalo School of Architecture and Planning. Climate Resilience Strategies for Buildings in New York State. [Internet]. 2020. Available from: http://archplan.buffalo.edu/content/dam/ap/PDFs/NYSERDA/Climate-Resilience-Strategies-for-Buildings.pdf [Accessed: 2021-07-22]

[34] Interstate Mercury Education and Reduction Clearinghouse (IMERC). IMERC Fact Sheet: Mercury Use in Lighting [Internet]. 2015. Available from: http://www.newmoa.org/prevention/mercury/imerc/factsheets/lighting_2015.pdf [Accessed: 2021-08-06]

[35] Gooley J, Chamberlain K, Smith K, Khalsa SB, Rajaratnam S, et al. Exposure to room light before bedtime suppresses melatonin onset and shortens melatonin duration in humans. Journal of Clinical Endocrinology & Metabolism. 2011; 96(3):E463-E472. DOI: https://doi.org/10.1210/jc.2010-2098

[36] Harvard Health Publishing. Blue Light has a Dark Side: What is Blue Light? The Effect Blue Light has on Your Sleep and More [Internet]. 2020. Available from: https://www.health.harvard.edu/staying-healthy/blue-light-has-a-dark-side [Accessed: 2021-08-03]

Chapter 4

Lessons from the World Sustainable Housing (Past Experiences, Current Trends, and Future Strategies)

Amjad Almusaed and Asaad Almssad

Abstract

The term vernacular architecture is widely accepted by architects and derives from the Latin "VERNACULUS", which means "domestic, native, indigenous", local. So vernacular architecture designates the entire culture built from a particular place. In modern housing design, the inheritance and development of traditional architectural culture is reflected in the inheritance of classic architectural forms and craftsmanship and the rational application of traditional building systems. With the progress of social civilization and the improvement of technological level, various innovative building systems emerge in an endless stream and are widely used in modern housing design. In today's rapid social and economic development, housing changes are coming quickly, sometimes even seeming a bit rough. At present, more attention is paid to the construction of new residential areas. As far as the field of urban architecture is concerned, the human settlement environment can be understood explicitly as people's living and living environment. Today architects need to design a settlement that balances all social functions between meeting current needs and future development, designing energy and material-saving buildings, so that it is in harmony with the environment, and is conducive to the physical and mental health of the human body. In other words, the planning process requires attention to human behavior, psychology, emotions, and interpersonal relationships.

Keywords: Sustainable housing, human settlement, vernacular architecture, housing strategy

1. Introduction

A house represents one of the primary material conditions of human existence. It is created to protect people from the effects of unfavorable meteorological factors (cold, heat, wind, atmospheric precipitation) and take care of leisure, work, and living needs [1]. The human settlement perspective must outline a precise vision for sustainable human settlements "everyone enjoys adequate housing, a healthy and safe environment, basic services, and productive and freely chosen work." Sustainable development is essential to the development of human settlements [2]. It must include the following two aspects:

- Everyone must have a suitable house; It refers to an appropriate place where individuals are not disturbed; appropriate space; proper security; legal guarantees during the land use period; proper lighting, heating, and ventilation; right infrastructure such as water supply, sanitation, and garbage management facilities; good quality of the environment and health-related factors.

- Residential areas must be supported by the relevant infrastructure and services; It refers to "safe water supply, environmental sanitation, waste management, social welfare, mass transportation and communication facilities, energy, health and emergency services, schools and public safety, and green space management, etc. Adequate essential services are one of the critical factors for housing [3].

"Sustainable Human Settlements in an Urbanization Process" and "Sustainable Human Settlements in an Appropriate Housing" (Sustainable Human Settlements in an Urbanizing World, Adequate Shelter for All) [4]. In the transformation, renewal, reconstruction and new construction of public housing, excessive savings and blindly lowering the cost of Inside once occupied a dominant position. This tendency appeared in the 1950s and 1970s of the former Soviet Union in the late 1970s and Sweden to end the housing shortage once and for all, the Swedish parliament decided that a million new dwellings should be built in the period 1965 to 1974 and this was achieved [5]. Many suggestions for blindly pursuing cost reduction It needs attention. That affected on the essential quality of housing and hit households with insufficient spending power particularly hard.

In recent years, sustainability as a concept in housing has taken on an increasingly prominent role. This role is made clear in all layers of housing components, from the ministerial to the institutions and to the individual consultant, contractor, and manufacturer [6]. As the concept in the sustainable housing context is still relatively new, the discipline of designing sustainable housing is correspondingly new. In the past, in many contexts, there have been similarities between efficient housing and sustainable housing [7].

The term "quality of life" is often used by city planners to reflect all aspects of the physical environment that are closely related to the productivity, satisfaction, and happiness of residents. Improving residents' quality of life is essential for regions to meet the needs of existing residents and attract and retain new businesses, employees, and other talents [8]. Cities are critical to people. Those who live, work or visit them, and those who depend on the growth that cities generate for both the city and the surrounding area [9]. However, the realities of each country form a specific local perception of what social housing means. The solution is not always offered in the form of a physical structure, sometimes more favorable conditions can be created for having a home. At the same time, in most parts of the world, the term is directly related to a problem called the affordable housing crisis. In our reality, most people are facing a "housing problem" [10]. In the modern world, housing is the most significant asset most individuals or families will ever have. Given the scale of this global problem, when the study debates social housing, it required to understand the housing stock, which will be protected by the layers of price regulation, property security, quality, and stability policies. Environmental sustainability, which helps reduce utility costs, is not the only aspect of sustainability that should be considered in such projects. The housing types are conditioned by the level of development of the country's productive forces, social relations, forms of family life, cultural and household traditions, and geographical environment. In recent years, the term "social housing" has become popular. Many texts, projects, programs have been produced under the name of this idea.

2. Reading in the world vernacular houses (visions, conception, implementation, and operation)

In architecture, "vernacular" is the term used to refer to famous constructions, made by people whose main activity is not necessarily in the field of construction [11]. It is based on empirical knowledge of materials, gained over time, through repeated trials (and failures). Knowledge is passed down from generation to generation, orally. Today, when the need for authenticity is so great, the word "traditional" seems worn out and abused [12]. It is increasingly difficult to distinguish between authentic and inauthentic. If for food there is a clear definition of what is and what is not traditional, things are not as clear and regulated in the case of architecture. The rural environment is full of boarding houses that call themselves traditional but have nothing to do with the architecture of the place [13].

An architect will say that the traditional is related to vernacular architecture - that is, the place itself. It has been developed and passed down from generation to generation, which does not mean that it has not evolved over time [14]. On the contrary, the traditional architecture has developed and refined, adapting to the times and needs, but permanently reflecting the environmental, cultural, technological, economic, and historical conditions of the local context. Although the phrases of vernacular architecture, and traditional architecture are considered synonymous, there are differences in nuance between them [15]. While vernacular architecture is created without the contribution of construction professionals, ie without architects, vernacular architecture can use craftsmen specialized in the construction process (but not architects) and is also based on local techniques and materials [16]. Traditional architecture denotes, first, the mode of transmission, from generation to generation and orally, but this is a valuable feature in the case of the first two forms.

2.1 Experience of housing in India

Houses in India are rooted in its history, culture, and religion. Tamil Nadu: The traditional architecture of South India is sometimes considered synonymous with the Agrahara-style of Tamil Nadu. The traditional house of Tamil Nadu Agrahara or Agraham reflects the primary Hindu roots of the state [17]. The Brahmins' home is considered a perfect example of this architecture. The name itself derives from the way it was placed in a village, which was like a garland. It is included in the houses leading to the primary temple of the village which are either dedicated to one deity or to different gods.

Mohenjo-Daro, city is the best-preserved and most extensive city, estimated to have had a population of 40,000. Mohenjo-Daro has a planned layout with rectilinear buildings arranged on a grid plan [18]. In contrast to both Mesopotamia and Egypt, the Indus settlements seem to have been relatively egalitarian societies. There are neither palaces nor royal tombs, and no great temple complexes to indicate a concentration of power and wealth (**Figure 1**) [19].

Buildings were durable, being constructed of fired bricks of uniform size throughout the region, and houses were provided with underground drains connected to a well planned sewer system (see **Figure 2**) [20]. The houses organized around intern al courtyards that were open to the sky for light and air.

The plans vary, but all houses presented virtually complete facades to the street. Most were built of fired and mortared brick, some combined sun-dried mudbrick, and wooden structures [21]. While the buildings do not seem elegant in terms of architectural refinement, the clear urban layout, careful provision of a water supply [22]. Rooms were small, perhaps because there was a scarcity of wood to serve as beams for second floors and roof framing (**Figure 3**).

Figure 1.
Mohenjo-Daro city.

Figure 2.
A traditional house from Mohenjo Daro city.

Ancient cultures in Greece, Ancient Egypt, Babylonia, Japan, and India had all similar lists, sometimes referring in local languages to "air" as "wind" and the fifth element as "void" the classification of the material world, these five are earth, water, fire, wind/air, and void. These came from Indian Vastu shastra philosophy and Buddhist beliefs. The system of "five great elements", of Hinduism are Bhūmi – earth, Ap – water, Tejas – fire, Pavan - air or wind, Shunya (space or zero) void.

2.2 Experience of housing in China

Chinese dwellings are folk dwellings designed and built by residents with a certain degree of representativeness and rich local characteristics [23]. Among the houses in China, the most characteristic residences include Beijing courtyard houses, cave dwellings on the Loess Plateau in Northwest China, ancient dwellings in Anhui, Hakka earth buildings in Fujian and Guangdong, and Mongolian yurts [24]. The Hakka Weilong House, Beijing's "siheyuan", Shaanxi's "cave dwelling",

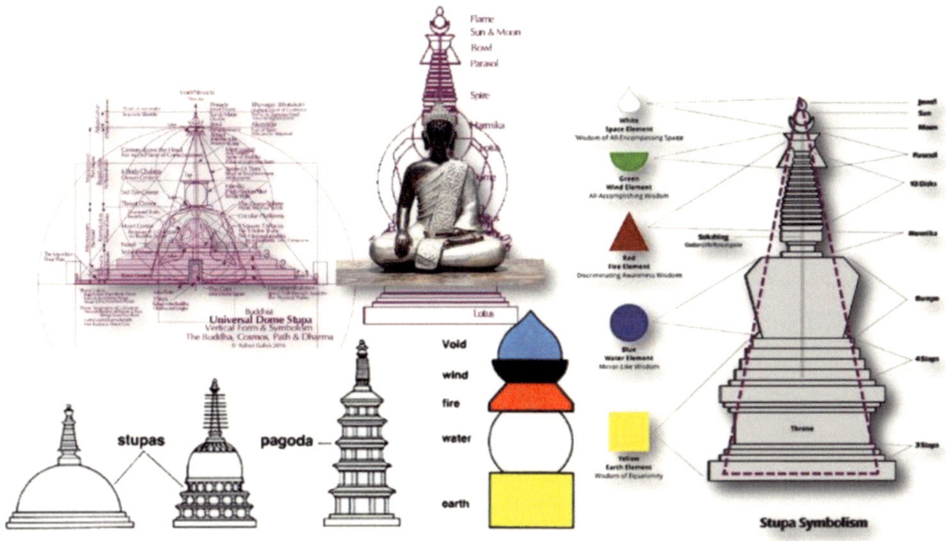

Figure 3.
Semiotics in Indian traditional building.

Guangxi's "bar-style" and Yunnan's "one seal", are collectively referred to as the five most rural traditional residential architectural forms in China [25]. It is called one of the five major characteristics of China's residential architecture in the field of architecture.

Houses in China reflects the most essential and representative things in the ethnic area in historical practice, especially the characteristics closely related to the life and production methods, customs, and esthetic concepts of the people of all ethnic groups [26]. Typical Chinese houses used "JIAN" – a structured bay as a standard unit to construct all buildings. The "Jian" represents the basic unit for wooden construction. "JIAN" was a rectangular space marked by adjacent structural frame. A modular unit called the Jian (about 3 m - 4.5 m) was defined as the primary measure in construction. "Jian", as the primary interior unit, can be expanded or repeated along the architectural plan axis to join to create a hall, then a building [27]. For example, two rows of four columns make three "JIAN". The "JIAN" is like room, in that it is a restricted space, but various a room, a "JIAN" does not necessarily have walls on all sides. Most dwellings are three to five "JIAN" in size. See **Figure 4**.

One of the great religious beliefs that influenced the design of the classical Chinese city and Chinese architecture is Confucianism [28]. To create a stable social order. Confucianism established the strict principles putting the society in order with rules and filial piety.

The experience of the nation mainly refers to the experience of how the residential houses meet the needs of life and production and the struggle against the natural environment under the social conditions at the time [29, 30]. For example, the experience of combining the use of the terrain, the experience of adapting to the climate, the experience of using local materials, and the experience of adapting to the environment. And so on, this is what is commonly referred to as the experience of adapting measures to local conditions and adapting to materials. The folk houses are distributed all over the country [31]. Due to the differences in national historical traditions, living customs, humanistic conditions, and esthetic concepts, as well as the different natural conditions and geographical environments of various places, the plane layout, structural methods, modeling, and detailed characteristics of traditional houses are also different. Different, showing simplicity and nature,

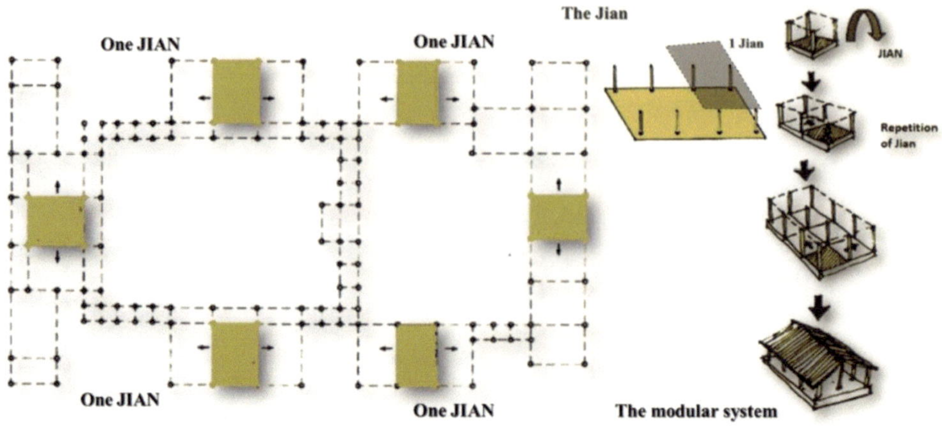

Figure 4.
The diagram of a typical Chinese house configuration.

but with their own characteristics. Especially in residential buildings, people of all ethnic groups often reflect their wishes, beliefs, and esthetic concepts, and use natural or symbolic methods to reflect their wishes, beliefs, and esthetic ideas to the decoration, patterns, colors and styles of residential buildings. Wait for the structure to go. Such as cranes, deer, bats, magpies, plums, bamboo, lilies, Ganoderma, Wanzi pattern, Hui pattern of Han nationality, lotus of Yunnan Bai nationality, elephant, peacock, betel nut tree pattern of Dai nationality, etc. In this way, the dwellings of various ethnic groups in multiple regions show a colorful and colorful ethnic characteristic. The mainstream of traditional houses in the various areas of China is the structured houses, which is represented by the "BEIJING SIHEYUAN", which adopts a symmetrical layout in the central axis [32]. The "BEIJING SIHEYUAN" is divided into two courtyards. The main house system in the center is the most respected. It is a place for family etiquette and meeting distinguished guests. Each building faces the courtyard and is connected by a verandah. Although "BEIJING SIHEYUAN" is a concrete manifestation of Chinese feudal society's patriarchal concept and family system in residential buildings, the courtyard is expansive, suitable in scale, quiet, friendly, and well-organized, making it an ideal outdoor living space. Residential houses in North China and Northeast China are mostly such spacious courtyards [33]. A SIHEYUAN is a historical type of residence that was commonly found in Beijing and rural Shanxi. The SIHEYUAN composition was the basic pattern used for homes, palaces, temples, monasteries, family businesses, and government offices (**Figure 5**).

Residential buildings do not have a set of procedural rules and practices like official buildings. They can build houses according to local natural conditions, their own economic level, and the characteristics of building materials [34].

2.3 Experience of housing in Japan

Houses in Japan have a short lifespan, so even if they are remodeled, they are structurally strong and will not get tired of the design, which can be inherited by the second and third generations [35]. The floor plan can be flexibly changed according to changes in the family structure and lifestyle of the residents. Barrier-free so that you can live with peace of mind even as you grow older. By devising the color and finish of the outer wall and roof, the appearance is unified and continuous in a group of housing complexes, or the traditional landscape that is transmitted to the land is preserved and conveyed. When building, it protects the

DOI: http://dx.doi.org/10.5772/intechopen.100533

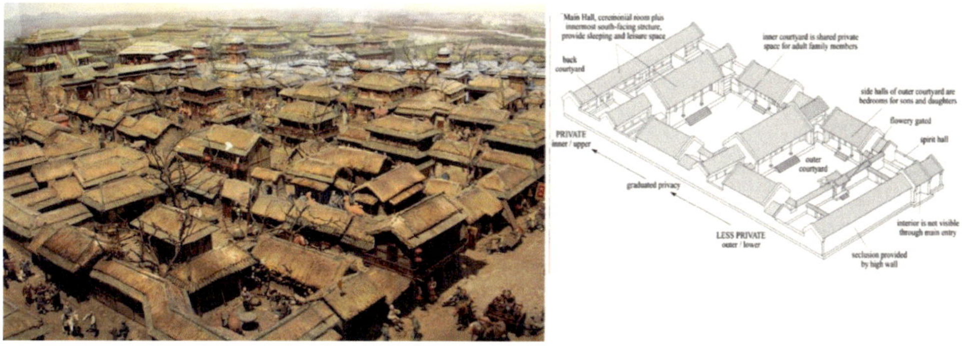

Figure 5.
The "BEIJING SIHEYUAN" house (courtyard house).

environment by minimizing the impact on the surrounding natural environment. The houses in Japan have been influenced by the climate. They were derived from China but maintained its own unique characteristics of lightness as fragility. The "KEN" is a traditional Japanese unit of length, equal to six Japanese feet "SHAKU", and equal to 1.8 m. 'KEN' is known as standard measurement of inter spaces [36] (see **Figure 6**).

Using materials produced in the land or in Japan, designing on the premise of living longer, making the structure easy to repair, remodeling, partial demolition, etc., and using the same material repeatedly will enrich the local industry. In addition, production, consumption, and waste material disposal can be circulated in the area. There are many types of house roofs, IRIMOYA, KIRIZUMA, YASEMUNÈ, and HOGYO as shown in **Figure** 7 [37].

In traditional city forming many architectural elements, represent an essential symbol in city configurations and composition [38], for example, A torii is a traditional Japanese gate most found at the entrance of or within a Shinto monument, where it symbolically marks the transition from the normal to the sacred (see **Figure 8**) [39].

A vital house form and composition in architecture from Japan. Yoshimura House is one of the traditional houses located in Saga Prefecture. It was constructed in 1620 [40]. This house was used to be the prosperous farmer because it has bigger musts and beams to make this house substantial and significant, see **Figure 9**.

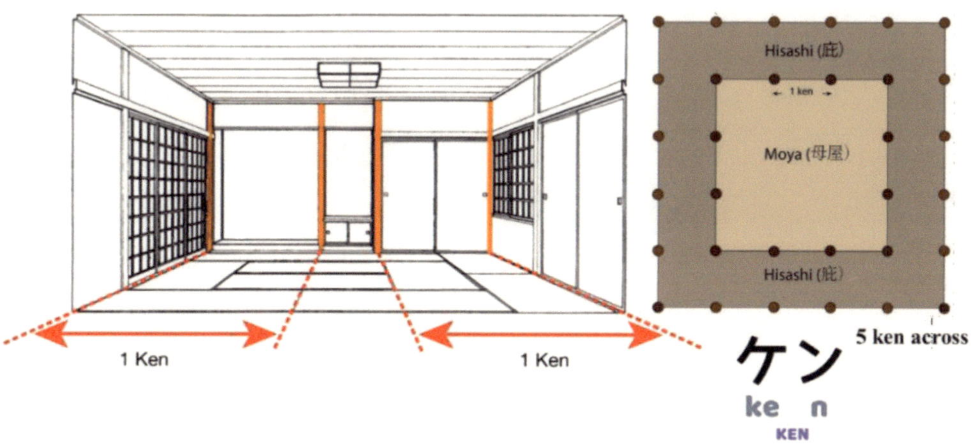

Figure 6.
"KEN" in Japan standards measurement.

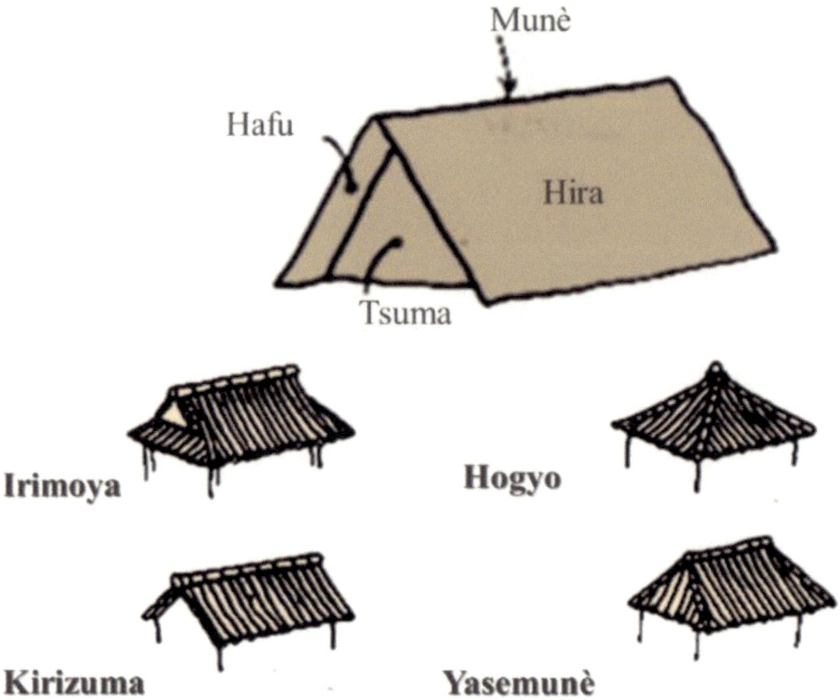

Figure 7.
Japan roof types in traditional houses.

Figure 8.
Torii in traditional Japan city elements symbols.

A Palace – mansion was a house style from the Heian period, which was a very similar to the Western European's Italian Renaissance, for this was a time for art and poetry to flourish in Japan. A "SHINDEN-ZUKURI" was an architectural style house that flourished during the Heian period [41] (see **Figure 10**).

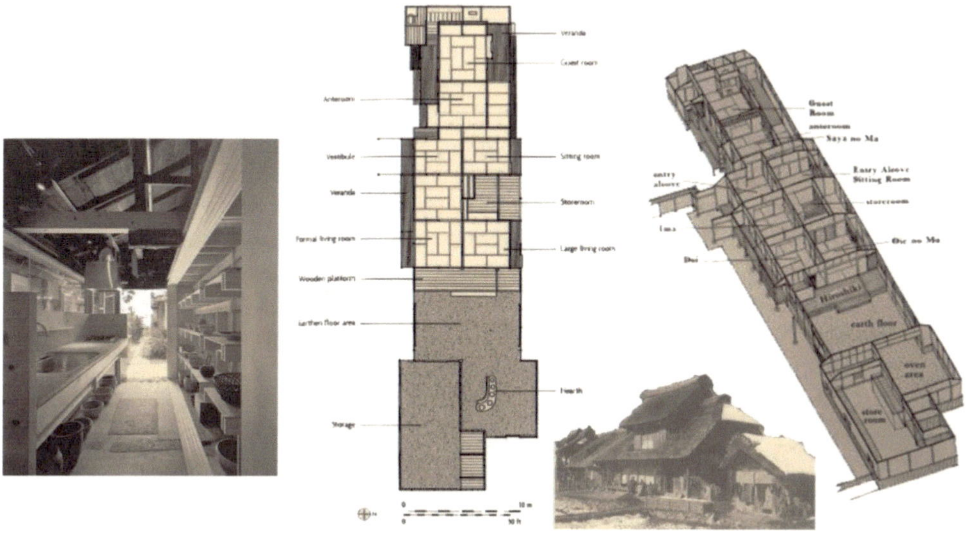

Figure 9.
"YUSHIMURA" house near Osaka – 1620.

Figure 10.
A layout of a "HEIAN MANSION".

2.4 Experience of housing in Korea "HANOK"

A "HANOK" is a traditional house using traditional Korean architecture. It is a traditional house of the Joseon Dynasty that reflects the ideal of building with the mountain facing the back and facing the water in the south. It is a traditional house of the "JOSEON DYNASTY".The origin of "HANOK" is a hut in the early Neolithic period around 6,000 BC, and it is considered that traditional "HANOK" was completed in the late Joseon period [42]. During this period, the ondol, floor, and kitchen, which are the basic units of space composition, were wholly combined to form a close relationship with each yard, and "HANOK" was differentiated into various regional types.

It has various characteristics that have been developed according to the environment of the Korean Peninsula and the traditional food, clothing, and shelter patterns of Koreans, and although the wooden structure tiled house in the photos is often thought of, thatched houses made of rice straw and ocher also fall within the scope of "HANOK". In modern Korea, the number has decreased due to Western-style buildings, but it continues to exist through the construction of temples. There is a theory that the word "HANOK" itself was derived from the opening of the door in the late Han Dynasty and modernization after liberation and the rapid spread of 'western-style houses', which is a contrasting meaning to traditional houses called 'HANOK' [43]. From a foreigner's point of view, it refers to a house in which the Koreans living on the Korean Peninsula live (see **Figure 11**). Originally, "HANOK" itself was a form of residence, so today, it was called "JUJU" and "JETAEK" as if it were just a house. It would have been divided into tile-tiled houses and thatched-roof houses as if they were divided into houses. The essential materials are the window and square that connect the front and rear columns in a straight line, the beam that connects the front and rear columns back and forth, and the rafters and the ribs that support them. When you think of "HANOK". that you can see often, you think of a "HANOK" house with an octagonal roof.

2.5 Experience of housing in Iraq

Houses in Iraq were, compact with interior courtyard. The streets are sinuous and pass-through building volumes. In the meantime, between yard and street at least a wall or a building is constantly interrupted (see **Figure 12**) [44]. This isolation from the road indicates concerns for defense. The architectural elements are intensely decorated, reproducing typologies and traditional houses [45].

Figure 11.
An example of "HANOK" house in Korea.

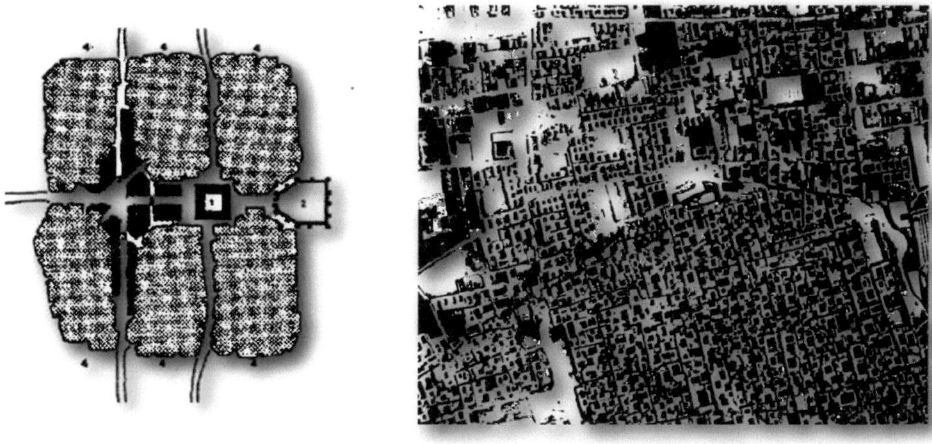

Figure 12.
A specific urban texture.

The patio is for the traditional dwelling the outside space that creates a microclimate and the most efficient form of using the inside space [46]. The shady interior courtyard has the effect that the rooms do not communicate directly with the overheated air outside, but through intermediate buffer spaces. The windows are small sized, located in the upper part and wooden framed (**Figure 13**).

"SHANASHIL" is a wooden decorative element piece or made from tiny wooden fragments allowing the inside [47]. ventilation and lighting and preventing the penetration of the outside excessive heat because wood. The thermal role of those elements is also a reflection of the sunlight and changing the current of air direction (**Figure 14**) [48].

Outside decoration are profiled elements of large volumes under various forms play the role of creating pronounced shadows on the sun-warmed facades [47]. Ventilation gaps, this element is opening located at the upper part of the houses, which is decorated with a grid network under the form of a drilled screen wall and used for ventilation and lighting [47, 48].

2.6 Experience of housing in Turkey, Iran, Central Asia

In this region a part of the old Islamic civilization developed, in which the architecture of the residential buildings is distinguished by the ingenious way in which it offers protection against excessive horses. Although the predominant type of housing is the one with an inner courtyard, an inner courtyard closed on all sides, either with buildings or with high walls. Here several variants coexist. The streets and buildings that border them form an organic unit, with the aim of thermal protection. There are different specific spaces with a thermal role. TAKURE, the area of Iran and Central Asia, RAWAQUL SHURFA, etc. [49]. All these spaces being enriched with interior and exterior decorations. The windows are replaced with sanasil, so that the partitions on the ground floor and first floor are permanently cooled by the shadow left by them Courtyards are the center of the plans of houses in south turkey. Courtyard houses, which represent the cultural layers of the Middle East, also characterize the traditional dwellings in some neighboring countries, such as Syria, Iraq, and Iran (see **Figure 15**). However, it is impossible to separate this plan type from the Anatolian cultural layers, like house plans [50].

Figure 13.
A traditional Iraqi dwelling [46].

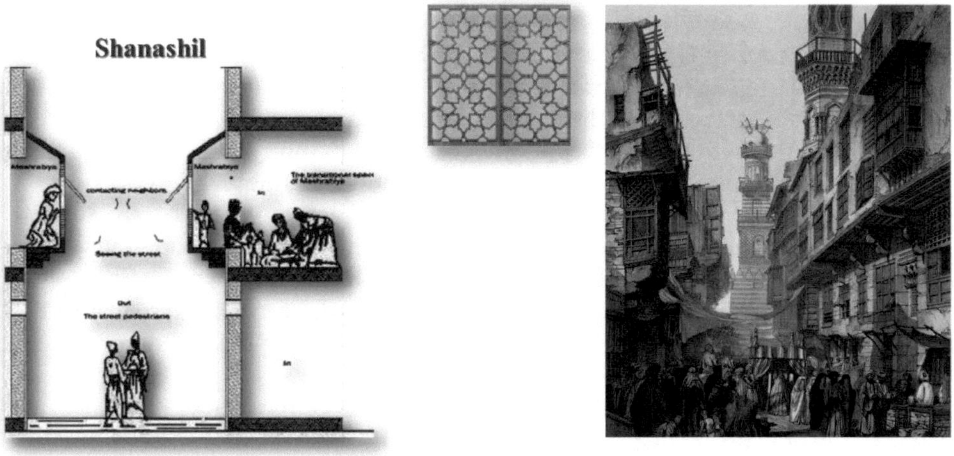

Figure 14.
The "SHANASHIL" in traditional Iraqi dwelling.

2.7 Experience of housing in Mexico

Houses in Mexico have currently been influenced by various styles, from the traditional pre-Columbian style, with its intense colors, rough textures, and thick walls, to functional modern architecture with simple straight lines. Land and housing are part of one thing [51]. The traditional vernacular dwelling becomes one more element of the same territory. The houses were made of wood and marl, the roofs were made of reeds, although the pyramids, temples and palaces were generally made of stone. The homes have a dwelling function, a cellar, and a stable, which are separated by thick structural walls and few openings (see **Figure 16**). Inside the divisions are few, some wall to isolate the kitchen, which sometimes moves towards the corridor in the simplest homes.

Most of the Mexican house style is inspired by the old famous Spanish architecture of the 18th and 19th centuries, better known as "Mission" or "Colonial" style houses, where the typical stone or white walls stucco, with red-tiled roofs and triangular ceilings but low, since it does not rain, or with small vaults. Plants, especially *aloe vera* and the like, in addition to cacti or those that need little water, usually populate the garden [52, 53].

Figure 15.
The house types from Iran, Syria, and Central Asia.

Figure 16.
Different traditional Mexican houses.

2.8 Experience of housing in in Morocco, Tunisia, Libya, and North Africa

Rooms in houses from this area are not communicate directly with the overheated air from outside, but through a buffer space. At the same time, between the yard and the street there is at least one wall, or a building, often with several cores [54]. This isolation from the road indicates defense concerns (see **Figure 17**).

The cities of North Africa are compact with inner courtyards, winding streets between the built volumes. In this way the surrounding streets have the role of a cold air reserve and similarly the air in the inner courtyard will replace the air in the surrounding spaces [55]. This phenomenon occurs in the following cases:

- The surrounding streets are narrow, planted and mostly shaded with irregular profiles.

- In the shaded and chained inner courtyards, a colder microclimate is created compared to the surrounding areas from the same perimeter. This fact determines a reciprocal conditioning between the type of traditional construction and its neighborhoods on the one hand and the inner courtyard and the spaces that surround it on the other.

2.9 Experience of housing in Balkan

Houses in Balkan has been traditionally labeled either an "Ottoman house" or a "Balkan house." The focus here is on constructing national interpretations of the vernacular residential architecture in question, meaning symbolic appropriations or "nationalizations" of a shared cultural heritage from the Ottoman era 8 see **Figure 18**).

Figure 17.
Typical house from the north of Africa.

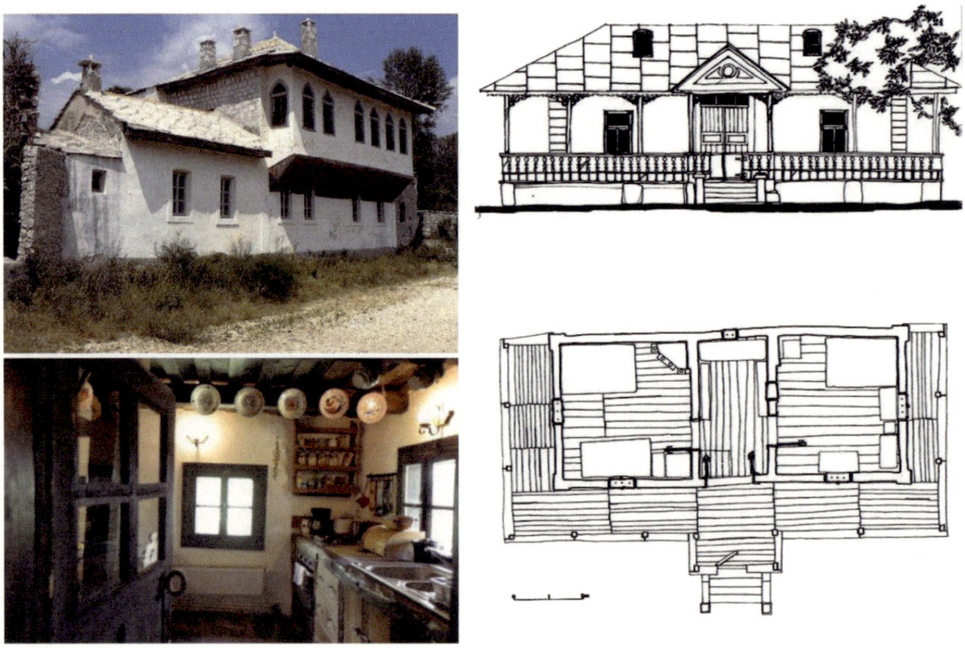

Figure 18.
The characteristic houses in Balkan.

Although dedicated to a particular form of traditional architecture, this text does not present the outcome of research in the field of art/architectural history or ethnography [56]. The house area dominates by simplicity and modesty, functional and natural elements, fabrics made with care and attention to detail, and solid furniture made to last. Each component of the traditional houses is justified by the owner's lifestyle, by the traditions he respects, and by the relationship with nature.

The traditional Balconies house is not unitary but differs depending on the area. Specialists classify these as distinct ethnographic areas. Among the most critical ethnographic regions, which have preserved mainly their local specificity today. In other areas, traditional architecture can only be found in the village museums. Even if these areas have a distinct character of their own, there are still some common features. The peasant houses had few rooms, and the oldest were single rooms. The house usually consisted of a porch, a space, and a pantry. The access to the house was made through the porch, after which one entered the hall, and from the hall one entered the room. The porch, sometimes called porch or porch, is the open and covered space that mediates the interior-exterior relationship and is also the most ornate area of the house. The decorative elements usually covered the porch pillars, also called slippers or forks; their role was both esthetic and magical, protecting the home from evil. The porch was not just a passageway but sometimes served as a place to store snacks or household utensils. At other times, there was a hearth in the hall and thus, this space became the kitchen and living room of the house. Most of the time, however, the room was the place where the hearth was and the place where it was cooked, eaten, and slept. This is not accidental, because the fire was also used to prepare food and heat the building.

Therefore, the oldest houses had a single room, also called "house". In the summer, on hot days when the fire was not lit inside, the hearth in the yard or a summer kitchen was used for cooking. or the "cellar" or "file" (so named because it was in the back of the house) was the storage space for food, tools, or small tools. The main food storage space was either the attic of the house or the cellar. The attic was used to store grain and smoke meat, and in the cellar were kept fruits, vegetables, pickled pots, and barrels of brandy or wine. The roof of the house was usually high and in four glasses of water. Multi-room houses appeared a little later, in the 19th and 20th centuries, and in this case, one of the rooms was kept for special days. In the "beautiful room", the "good house" or the "house before" guests were received at the big holidays, important family events were held. Here were placed the most precious furniture and decorative objects and here were kept the holiday clothes. Because it was used only occasionally, the beautiful room rarely had a heating system. Today there is so much talk about vernacular architecture because it is the condition for the sustainable development of rural communities. Vernacular architecture has the quality of being ecologically sustainable by adapting to the climate and relief of a particular place and by using the natural building materials available in that area [57]. Thus, vernacular architecture has a minimal impact on the natural environment, being sustainable from a social point of view, because it is based on the science of construction (local know-how) and local labor. This stimulates the feeling of belonging to people in that space, makes them feel that their place has something unique and that they share the specifics of the area. The brick was used sporadically in some areas of the country, for example in Saxon villages from southern Transylvania or Banat, and this happens under the influence of culture urban. Brick became a material favorite construction with the era of industrial. Clay or adobe was used in older times, being easy to find and by the process. They are still found today, mainly in the houses on the plain, where it has been kept construction system with clay reinforced with twigs woven, specific to Neolithic architecture [58]. The predominant decorations in most houses in the country have religious themes or depict flora, fauna, or anthropomorphic elements. The fabrics are adorned with decorative motifs with definite meanings, and the colors are usually bright.

2.10 Eţxperience of housing in Southern Netherlands, Belgium, and Germany

Many cities and villages in South-West Friesland are built on what is called terpen. It is an artificial dwelling mound found on the North European Plain that has been created to provide safe ground during storm surges, high tides, and sea or river flooding [59]. There are incredibly many terpenes in Friesland, and now there are about 400 of them. The front door was in a narrow pediment and led to the utility part [60]. The hearth was still almost in the center of the room, but its transition was already outlined closer to the rear wall and the residential part's allocation. The stalls located behind the pillars were divided by low wicker partitions (see **Figure 19**).

Vernacular Houses from the Southern Netherlands and Belgium were of a different type. Although here and there were also rectangular semi-dugouts like the northern ones, all the same for these areas, starting from the Neolithic, round or oval-shaped huts, deepened 1.5–2 m into the ground, were characteristic. The walls were very low or completely absent, only the roof towered above the ground. It was built of long poles, had a cone shape, and was covered with straw and leaves [61]. Such a dugout is characterized by a roof structure with a DAKZUl - one pillar supporting the roof. There was an open hearth near the pillar. These houses bear a resemblance to the places of the ancient Celts in the British Isles. Apparently, in the Belgian and Dutch lands, the evolution of the house proceeded in the same way as in Britain - from a round hut through an oval to a rectangular single-chamber dwelling with an open hearth in the middle. An interesting type of dwelling, often found at the same time as round huts, in the eastern regions of Belgium: elongated buildings with an open canopy "VOORHALLE" over the main entrance from the narrow pediment. Two pillars supported the roof of the shed. The ridge beam of the gable roof was reinforced on several posts located on the centerline of the house. The area of homes varies from 20 to 84 m. Some of these dwellings have a three-chamber plan: for example, in North Limburg, dating back to the 1st century. n. e. the "VORHALLE" building was an elongated building made up of three rooms of the same size. Unfortunately, it is difficult to judge their purpose since not even the remains of the hearth have survived development of rural dwelling in the Netherlands and Belgium.

Stable house (HALLEHUIs) is one of the most ancient houses of the Netherlands, which also existed everywhere in Germany in the early Middle Ages, had only a narrow passage between two rows of pillars separating the stalls. With the development of agriculture, and especially when agriculture began to play an increasing role, the central longitudinal aisle expands and is used as a vast threshing floor. Such a transformation occurred because the high humidity of the air, frequent rains, and fogs made it challenging to thresh grain in an open room. Large houses, combining a residential part, a barn, and a barn under one roof, have been found since the 12th century, but they have become widespread since the 16th century. The residential part is divided into several rooms, more often into three: the central place is occupied by the main living room with a frontal fireplace and niche beds

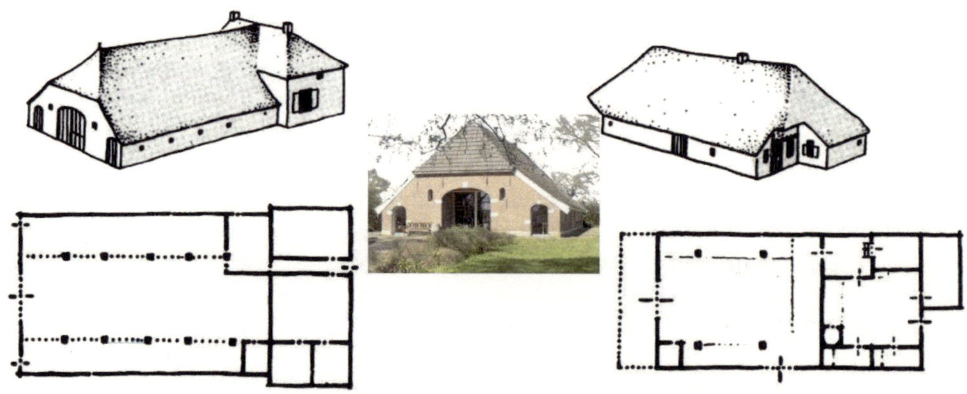

Figure 19.
The characteristic of the Southern Netherlands, Belgium, and Germany houses.

on both sides of it. To the left and right of this room are small bedrooms. One of the distinctive features of the stall house is also a very steep and high four- or two-pitched roof, which is apparently caused by a large amount of precipitation in the Netherlands. Such shelters were also convenient because an ample attic space was formed under them, where hay and grain supplies were stored. The roofs were covered with thatch, later with tiles. The walls of ancient houses were made up of vertical pillars connected by wattle fences, in the Middle Ages and up to the 19th century, frame walls predominated everywhere.

Frisian house (GULFHUIS) is, typical in the southern regions of the Netherlands and almost everywhere in Belgium, was often called Franconian or even Celtic by romantic ethnographers. Archeological materials indicate that its origin was somewhat different from the two previous types. It differed in the internal layout, the kind of connection between residential and utility premises, and design. The house's core was formed from a single chamber building with a central hearth, from which very early with a transverse partition at one end stalls were fenced off. In the early Middle Ages, another living room was fenced off behind the hearth on the other side. Frisian house (GULFHUIS) is characterized by a transverse division of the entire room, which is why in the Netherlands it is also often called "DVARSHHOIS". A small front or vestibule is usually separated by a light partition near the entrance in the main living room. The fireplace is located at the left or right of the wall from the front door.

2.11 Experience of housing in North Europa (Scandinavia)

The Longhouse is the traditional primary habitat in the Scandinavian region, dating back to the Iron Age, 2000 BC. This study examines the influence of climate on the conformation of habitats. Climate had a substantial impact on the conceptions of habitat form and internal space [62]. The climate's role in the conformation of the vernacular houses of the Scandinavian region was notable and can be observed clearly in the urban texture in the following:

- Roads, ways and moving arteries were east and west.
- The long facades of houses were designed to be north and south.
- The courtyard is essential in forming habitat units, combatting the adverse effects of high winds.

The walls of habitats in vast areas of Scandinavia define the edge between public spaces and private spaces and physically and perceptually explain the public roads and squares, the spaces in which communal life occurs (see **Figure 20**) [40].

Figure 20.
Environmental answer to vernacular habitat conformation from Scandinavia.

3. Future reading on a global sustainability in housing policy

Establishing a sustainable housing concept will point out the purpose, direction, and means for future housing planning, architectural design, improvement, and innovation of human settlements. It is essential to establish proper architectural and planning guidelines that conform to the laws of objective development. it required reforming and innovating to create a good and comfortable living environment for residents, benefit the people, and enable them to live and work in peace and contentment [63]. Housing is a crucial issue for any sustainable development. In recent years, under the strong organization and promotion of the government, excellent pilot projects, demonstration projects, and housing projects have emerged one after another, enabling people to obtain comfortable, convenient, safe, and sanitary housing. These pilots and demonstrations have already had a lot of good experience in implementing sustainable development strategies, such as energy-saving, land saving, water saving, new technology, new material application, etc. A sustainable framework for housing policy should focus on the future and have a strong contemporary focus, as otherwise it quickly becomes intangible in everyday inhabitants' lives. Therefore, the vision for a new framework for sustainable housing policy can be about methods by which sustainable solutions are made an attractive and advantageous alternative for all [64]. The intention is that sustainable solutions must be based on the city inhabitants' daily needs. For example, it is not required to cycle solely because it is environmentally friendly and healthy, but perhaps rather because it is easy, fast, cheap, and accessible. This principle can also be advantageously transferred to sustainable housing development. Everyday life must not become more complex with a sustainable change, then there is a risk that the broad popular support will be lacking. The shift in sustainable residential areas forbids an individual and collective project that can strongly support the city strategy in sustainable development. Housing strategies given the complexity of the challenges considered, there is a need for an overall housing strategy policy framework for how they are addressed, and the potentials exploited. By thinking about social-cultural conditions, environment, and economy together, the probability of being able to implement coherent solutions is improved. Sustainable housing strategy must be social - culturally viable, while at the same time considering climate adaptation, energy and resource efficiency, environment, architectural quality, and social security. In this way, the traditions and the environment can generate added value when they are considered together in holistic considerations [65]. A healthy environment should have indispensable intrinsic value. Humans and the entire natural environment are ecologically interdependent, and the persistence of this interdependence requires all aspects to establish a partnership, equal and balanced relationship. Urbanization based on residences' dreams can contribute to a more sustainable society by linking different urban functions in housing development strategies. Architectural design requirements are increased. To the standard of sustainable design, sustainable design should comprehensively consider the use of resources and energy, the use of healthy buildings and materials, land that is sensitive to ecology and society, and an esthetic that can inspire, affirm, and cultivate Sensitivity.

On the one hand, continuous design can significantly reduce the negative impact of humans on the natural environment, and on the other hand, it can improve the quality of life and improve living standards. Therefore, architectural design should pay more attention to ecological and environmental issues. Architecture has entered the era of "ecological architecture" (or "green building"). Sustainable development is proposed from the relationship between environmental pollution and human survival and growth. Still, sustainable development should also pay attention to the relationship between economy, population, society, and resources. Coordinated development with the five aspects of the environment [66].

What is a sustainable house in the context of the climate change era?

Since the Rio Declaration, which is also the origin of the Environmental Summit, the term sustainable has become used to refer to economic activities that consider the global environment. Sustainable housing is a long-lived housing that is easy to live in and will be passed down to the next generation of children. Human beings will create homes that use natural clean energy such as solar heat and wind power, without using petroleum energy that causes global warming. In addition, because it is a residence that considers the cycle of tree growth and regeneration, it also leads to the effective use of recycled materials such as demolished old folk houses. In addition to making the house last longer, sustainable housing is also characterized by consideration for building a home that can reduce waste when dismantled and reuse building materials. Structure and performance with ruggedness and comfort. The critical point in the system is robustness. It is assumed that the suitable material is used in the right place and that a sturdy frame is assembled. In the case of a wooden house, not only is it a sturdy house with a structure, but it also has the advantage of being easy to remodel, such as extension and renovation, so you can continue to live without rebuilding. The interpretation of housing in construction, economy, society, and politics is also different. He summarized the meaning of housing: sheltered places, private Space, a product of location, a combination of buildings and neighborhood facilities, investment tools, a symbol of wealth and socio-economic status; also summarized. It has the characteristics of immovability, indivisibility, longevity, heterogeneity, expensiveness, investment products, and consumer products. Harsman and Quigley [67] also pointed out the characteristics of housing which is different from other commodities.

- Housing is a complex commodity, for It is difficult for both parties to trade effectively.

- Housing is fixed in space. Choosing accommodation means choosing a neighborhood environment.

- Housing is costly, so it is common to rent houses, mortgage loans are generally required to purchase homes, housing accounts for a large proportion of expenditure, and new house construction is a large part of the new investment every year.

- The life cycle of housing is extended, new housing only accounts for a small part of the housing service supply, and small changes in housing demand have a major impact on housing construction activities.

- Housing is a necessity. They emphasized that it is these characteristics that together determine the high transaction costs of the housing market.

In the construction idea, design, and design phase the framework for large parts of the building is laid out future, both in relation to the architectural and functional, but equally so in regard to the building's environmental, social and economic footprint. The personal impressions depend in height degree of what choices and opt-outs are made in these and later phases [68]. This release is aimed at to make the parties to the construction aware of the dilemmas often encountered and the considerations one must therefore do in connection with design and design of sustainable building.

4. Conclusion

Residential buildings are the most basic type of architecture, appearing the earliest in the history of world architecture, the most widespread, and the most significant number. Due to world vast territory, many ethnic groups, and a long history, the geographical and climatic conditions and lifestyles vary from place to place. Therefore, specific residential buildings' architectural styles and styles are relatively rare in the history of world vernacular architecture. The vernacular architecture had a rich and beautiful element symbol and a solid philosophical charm. It fully complies with the laws of nature and cleverly integrates the natural scene. That can be a solid basis to modern architecture which focuses on the pursuit of humanized characteristics, while traditional architectural culture advocates the harmony and unity of man and nature. Traditional architectural culture can provide connotative reference materials and broad thinking space for modern architectural design and further highlight the individual characteristics of architectural design and enhance the connotation of art and culture. The integration of traditional architectural culture into the field of modern architectural design can inherit national culture, highlight modern scientific features, demonstrate characteristic humanistic feelings, and reflect the new style of the development of the times. A sustainable house where people can live comfortably forever while being a friendly house to the global environment in future home building, the idea of sustainable housing will be strongly required. In sustainable housing, it is considered to create a house that is friendly to people and the earth everywhere, such as the structure, floor plan, equipment, and building materials used. Combining modern architectural design theories with traditional culture and creating a series of works that were in line with national conditions, adapted to nature, and recognized by the public, making modern architecture famous It is a model of new vernacular architecture. Citizen involvement is a central element in the housing development of the future and helps maintain the vision for human and diverse cities. It provides vibrant and sustainable housing strategies where everyone has a place and can have a say. It places different and new demands on both the individual, architects, and companies, but it also provides a wide range of opportunities for new collaborations across traditional structures. Housing strategy should include three aspects: first, it has changed in the size, density, and design of the population in different regions; second, it includes There are fundamental changes in the socio- economic structure; finally, it is the changes in people's behavior that need to be pointed out. Architects should pay more attention to the application of traditional housing systems, deepen the cognition and understanding of national culture, and enhance modern architecture's cultural connotation and artistic value. In modern architectural design, to seek the integration of cultural characteristics of different times, traditional cultural symbols will be summarized, refined, and refined. Local features will be added based on retaining essential values and then reshaped to achieve the inheritance and spread of traditional culture. At the same time, it can also improve the modern architectural design and highlight the connotation. Today, the worldwide urbanization process has reached a turning point. Its main manifestations are as follows: the population is highly concentrated in the cities, and the rural areas are highly concentrated. A good housing strategy should include a good connection with the local traditional housing system with a concordance of a metropolitan and global city, an attractive and inclusive opportunity, a green area, and a livable and resilient residential area, have a significant grade to regenerate, and reflect the fragrant history and cultural heritage of the local areas. That is the first step to sustainable housing.

Author details

Amjad Almusaed* and Asaad Almssad
Department of Architectural Engineering, Jönköping University,
Jönköping, Sweden

*Address all correspondence to: amjad.al-musaed@ju.se

References

[1] Amjad Almusaed and Asaad Almssad (November 5th 2018). Introductory Chapter: Housing Policy Matters, Housing, Amjad Almusaed and Asaad Almssad, IntechOpen, DOI: 10.5772/intechopen.81622. Available from: https://www.intechopen.com/books/housing/introductory-chapter-housing-policy-matters

[2] Trovato, M.R. An Axiology of Residual Green Urban Areas. Environments 2021, 8, 53. https://doi.org/10.3390/environments8060053

[3] Amjad Almusaed, Asaad Almssad, Asaad Alasadi, Analytical interpretation of energy efficiency concepts in the housing design process from hot climate, Journal of Building Engineering, Volume 21, January 2019, Pages 254-266, https://doi.org/10.1016/j.jobe.2018.10.026

[4] Raad Z.Homod, Almssad, A., Almusaed, A, Et al (2021), Effect of different building envelope materials on thermal comfort and air-conditioning energy savings: A case study in Basra city, Iraq, Elsevier Journal of Energy Storage,volume 34, February 2021, 101975, https://doi.org/10.1016/j.est.2020.101975

[5] Thomas Hall & Sonja VidenThe Million Homes Programme: a review of the great Swedish planning project, Planning Perspectives, 20:3, 301-328, (2005). http://doi.org/10.1080/02665430500130233

[6] Jian Z., Zhen-Y. Z., Green building research–current status and future agenda: A review, Renewable and Sustainable Energy Reviews, Volume 30, February 2014, Pages 271-281, https://doi.org/10.1016/j.rser.2013.10.021

[7] Cascone, S.M.; Cascone, S.; Vitale, M. Building Insulating Materials from Agricultural By-Products: A Review. In Sustainability in Energy and Buildings. Smart Innovation, Systems and Technologies; Littlewood, J., Howlett, R., Capozzoli, A., Jain, L., Eds.; Springer: Singapore, 2020; Volume 163

[8] Amjad Almusaed, et al., Biophilic architecture, the concept of healthy sustainable architecture, PLEA2006 - The 23rd Conference on Passive and Low Energy Architecture, Geneva, Switzerland, 6-8 September 2006.

[9] Muzaffer U., M. Joseph S., Eunju W. Hyelin K.,Progress in Tourism Management; Quality of life (QOL) and well-being research in tourism, Tourism Management, Volume 53, April 2016, Pages 244-261, https://doi.org/10.1016/j.tourman.2015.07.013

[10] Amjad Almusaed and Asaad Almssad (December 13th 2018). Introductory Chapter: Overview of Sustainable Cities, Theory and Practices, Sustainable Cities - Authenticity, Ambition and Dream, Amjad Almusaed and Asaad Almssad, IntechOpen, DOI: 10.5772/intechopen.82632. Available from: https://www.intechopen.com/books/sustainable-cities-authenticity-ambition-and-dream/introductory-chapter-overview-of-sustainable-cities-theory-and-practices

[11] Kevin D., Menelaos G., Architecture and rural planning: 'Claiming the vernacular, Volume 41, November 2014, Pages 334-343, Land Use Policy, https://doi.org/10.1016/j.landusepol.2014.06.013

[12] Senanayake, S.G.J.N.,. Indigenous knowledge as a key to sustainable development. Journal of Agricultural Sciences – Sri Lanka, 2(1), pp.87-94. 2006. DOI: http://doi.org/10.4038/jas.v2i1.8117

[13] Almusaed A., Almssad, A. & Alasadi, A. (2019). Analytical

interpretation of energy efficiency concepts in the housing design, Journal of Building Engineering 21, ss. ISSN: 254-266, 2352-7102/ Published by Elsevier Ltd. https://doi.org/10.1016/j.jobe.2018.10.026.

[14] Almusaed, A.; Almssad, A.; Homod, R.Z.; Yitmen, I. Environmental Profile on Building Material Passports for Hot Climates. Sustainability 2020, 12, 3720. https://doi.org/10.3390/su12093720

[15] Bogdan V. C., Alexandru L. P., Traditional Architecture in Romanian Philately (IV): the Households from Ieud, Jurilovca, Mastacăn, Moișeni and Năruja, Asian Journal of Education and Social Studies, 17(1): 29- 38, 2021; Article no.AJESS.67445, http://doi.org/10.9734/AJESS/2021/v17i130412

[16] Christian Illies, Nicholas Ra, y Philosophy of Architecture, Philosophy of Technology and Engineering Sciences, Handbook of the Philosophy of Science, 2009, Pages 1199-1256, https://doi.org/10.1016/B978-0- 444-51667-1.50047-1

[17] J.Safrin R. D., C. Ezhil M. Sustainable quotient of traditional houses - a panoramic view of agraharams in ayanavaram, international journal of engineering sciences & research technology (IJESRT), Dulcie * et al., 7(3): March, 2018, http://doi.org/10.5281/zenodo.1189052.

[18] Mohan Pant & Shuji Funo The Grid and Modular Measures in The Town Planning of Mohenjodaro and Kathmandu Valley, Journal of Asian Architecture and Building Engineering, 4:1, 51-59, (2005) https://doi.org/10.3130/jaabe.4.51

[19] Thomas P. Leppard, Social Complexity and Social Inequality in the Prehistoric Mediterranean, Current Anthropology, 2019, (3) Volume 60, Number 3, https://doi.org/10.1086/703174

[20] De Feo, G.; Antoniou, G.; Fardin, H.F.; El-Gohary, F.; Zheng, X.Y.; Reklaityte, I.; Butler, D.; Yannopoulos, S.; Angelakis, A.N. The Historical Development of Sewers Worldwide. Sustainability 2014, 6, 3936-3974. https://doi.org/10.3390/su6063936

[21] Robert S. Homsher, Mud Bricks and the Process of Construction in the Middle Bronze Age Southern Levant, Bulletin of the American Schools of Oriental Research, No. 368 (November 2012), pp. 1-27, The University of Chicago Press, https://doi.org/10.5615/bullamerschoorie.368.0001

[22] Ali, M.M.; Al-Kodmany, K. Tall Buildings and Urban Habitat of the 21st Century: A Global Perspective. Buildings 2012, 2, 384-423. https://doi.org/10.3390/buildings2040384

[23] Hyun Bang Shin, Urban conservation and revalorisation of dilapidated historic quarters: The case of Nanluoguxiang in Beijing, Cities, Volume 27, Supplement 1, June 2010, Pages S43-S54, https://doi.org/10.1016/j.cities.2010.03.006

[24] Shaoqing G. et al. Climate responsive strategies of traditional dwellings located in an ancient village in hot summer and cold winter region of China. Building and Environment, Volume 86, April 2015, Pages 151- 165, https://doi.org/10.1016/j.buildenv.2014.12.003

[25] Lowe, K.D. Heaven and Earth—Sustaining Elements in Hakka Tulou. Sustainability 2012, 4, 2795-2802. https://doi.org/10.3390/su4112795

[26] Almusaed, A., Yitmen, I, Almssad, A., & Homod, R. Z.(2020). Environmental profile on building material passports for hot climates. Sustainability (Switzerland), 12(9). https://doi.org/10.3390/su12093720

[27] Donia Zhang, Classical Courtyard Houses of Beijing: Architecture as

Cultural Artifact, Space and Communication 2015; 1(1): 47-68, https://doi.org/10.15340/2148172511881

[28] Low K., Ang S., Confucian Leadership and Corporate Social Responsibility (CSR) the Way Forward, Asian Journal of Business Research, Volume 2 Number 1 2012, https://doi.org/10.14707/ajbr.120006

[29] Amjad Almusaed, Asaad Almssad, Building materials in eco-energy houses from Iraq and Iran, Case Studies in Construction Materials, Volume 2, June 2015, Pages 42-54, https://doi.org/10.1016/j.cscm.2015.02.001

[30] Ji Youn Kim, Cultural entrepreneurs and urban regeneration in Itaewon, Seoul, Cities, Volume 56, July 2016, Pages 132-140, https://doi.org/10.1016/j.cities.2015.11.021

[31] Katarina Rus, Vojko Kilar David Koren, Resilience assessment of complex urban systems to natural disasters: A new literature review, International Journal of Disaster Risk Reduction, Volume 31, October 2018, Pages 311-330, https://doi.org/10.1016/j.ijdrr.2018.05.015

[32] Jiaran Zhanga,Yingrui Chib, Mengdi Shi, On the Transition from Ming Dynasty Ceramic Courtyard to North China Residence Take the Ming Dynasty Pottery Courtyard of Henan Museum as an Example, International Journal of Frontiers in Sociology, Vol. 3, Issue 4: 93-98, https://doi.org/10.25236/IJFS.2021.030420.

[33] Zhang Fengjie, The Architectural Typology Analysis of Han Traditional Courtyard in Northeast China, International Journal of Structural and Civil Engineering Research Vol. 10, No. 1, February 2021, http//doi.org/10.18178/ijscer.10.1.9-16

[34] Akadiri, P.O.; Chinyio, E.A.; Olomolaiye, P.O. Design of A Sustainable Building: A Conceptual Framework for Implementing Sustainability in the Building Sector. Buildings 2012, 2, 126-152. https://doi.org/10.3390/buildings2020126

[35] Ken Tadashi Oshima, Rediscovering Japanese Urban Space in a World Context, Volume: 42 issue: 3, page(s): 623-633, https://doi.org/10.1177/0096144216635165

[36] Gozde Çakır, the effects of natural environmental data in the traditional japanese house design, International Journal of Advanced Research 6(7):361-371, http://dx.doi.org/10.21474/IJAR01/7370

[37] Amjad Almusaed and Asaad Almssad (December 22nd 2019). City Phenomenon between Urban Structure and Composition, Sustainability in Urban Planning and Design, Amjad Almusaed, Asaad Almssad and Linh Truong-Hong, IntechOpen, DOI: 10.5772/intechopen.90443. Available from: https://www.intechopen.com/books/sustainability-in-urban-planning-and-design/city-phenomenon- between-urban-structure-and-composition

[38] Yan Liu, A full moon in another land: The Moon Bridge in the Japanese garden of the Huntington Library, Frontiers of Architectural Research Volume 9, Issue 3, September 2020, Pages 556-567, https://doi.org/10.1016/j.foar.2020.02.004

[39] Nobuyuki MIYAZAKI, A study on the theory for floor planning of the central living-room type house in modern ages of japan; In the case of the pamphlet presented by the union for the improvement of living conditions in 1921, Journal of Architecture and Planning (Transactions of AIJ), Volume 75 (2010) Issue 649, https://doi.org/10.3130/aija.75.551

[40] Almusaed, A., Almssad, A. (2015). Environmental reply to vernacular

habitat conformation from vast areas of Scandinavia. Renewable & Sustainable Energy Reviews Journal, Elsevier, ISSN: 1364-0321, Volume 48, ss. 825-834, August 2015. https://doi.org/10.1016/j.rser.2015.04.013

[41] Shigeo Kawamoto, On the origin of the shinden-zukuri residence, Journal of Architecture and Planning (Transactions of AIJ), 2016 Volume 81 Issue 729 Pages 2497-2506, https://doi.org/10.3130/aija.81.2497

[42] Nuri Seo, Yunsub Lee, Yeheun Jeong and Youngsoo Jung, Evaluation Framework for Korean Traditional Wooden Building (Hanok) through Analyzing Historical Data, Pages 614-621 (2018 Proceedings of the 35th ISARC, Berlin, Germany, https://doi.org/10.22260/ISARC2018/0084

[43] Kim Bok-rae, Past, Present and Future of Hallyu (Korean Wave), American International Journal of Contemporary Research Vol. 5, No. 5; October 2015.

[44] Yitmen, Ibrahim, Amjad Almusaed, et al (2021), ANP Model for Evaluating the Performance of Adaptive Façade Systems in Complex Commercial Buildings, Engineering, Construction and Architectural Management Journal, https://doi.org/10.1108/ECAM-07-2020-0559

[45] Amjad Almusaed and Asaad Almssad (March 14th, 2012). Passive and Low Energy Housing by Optimization, Effective Thermal Insulation - The Operative Factor of a Passive Building Model, Amjad Almusaed, IntechOpen, DOI: 10.5772/36922. Available from: https://www.intechopen.com/books/effective- thermal-insulation-the-operative-factor-of-a-passive-building-model/passive-and-low-energy-housing-concept-

[46] Almusaed, A. Biophilic and Bioclimatic architecture, Analytical Therapy for the Next Generation of Passive Sustainable Architecture (2011), Springer-Verlag London Limited, p 406, London, UK. https://www.springer.com/gp/book/9781849965330

[47] Mofidi, S. Majid, R. Vakilinezhad, and Mehdizadeh Mehdizadeh. 2013. "Shanashil: A Sustainable Element to Balance Light, View and Thermal Comfort." The International Journal of Environmental Sustainability 8 (4): 101-110., https://doi.org/10.18848/2325-1077/CGP/v08i04/55066

[48] Amjad Almusaed and Asaad Almssad (March 14th 2012). Improvement of Thermal Insulation by Environmental Means, Effective Thermal Insulation - The Operative Factor of a Passive Building Model, Amjad Almusaed, IntechOpen, DOI: 10.5772/35744. Available from: https://www.intechopen.com/books/effective-thermal-insulation-the-operative-factor-of-a-passive-building-model/improvement-of-thermal-insulation-by-environmental-means-

[49] Hossein Sarhaddi-Dadian, Zohre Oveisi-Keikha, Vahid Purzarghan, Introducing Troglodyte Architecture at Chabahar City in South-east of Iran, International Journal of Archaeology. Vol. 5, No. 1, 2017, pp. 1-5. doi: 10.11648/j.ija.20170501.11

[50] Nimet Öztank, An Investigation of Traditional Turkish Wooden Houses, Journal of Asian Architecture and Building Engineering, 9:2, 267-274(2010), DOI: 10.3130/jaabe.9.267

[51] Paavo Monkkonen, The Housing Transition in Mexico: Expanding Access to Housing Finance, SAGE journals, Volume: 47 issue: 5, page(s): 672-695, https://doi.org/10.1177/1078087411400381

[52] Amjad Almusaed (September 6th 2017). Introductory Chapter: Overview

on Grass Topic, Grasses - Benefits, Diversities and Functional Roles, Amjad Almusaed and Sammera Mohamed Salih Al-Samaraee, IntechOpen, DOI: 10.5772/intechopen.70407. Available from: https://www.intechopen.com/books/grasses-benefits- diversities-and-functional-roles/introductory-chapter-overview-on-grass-topic

[53] Amjad Almusaed (September 19th 2018). Introductory Chapter: A General Reading Process on Landscape Architecture, Landscape Architecture - The Sense of Places, Models and Applications, Amjad Almusaed, IntechOpen, DOI: 10.5772/intechopen.77971. Available from:https://www.intechopen.com/books/landscape-architecture-the-sense-of-places-models-and-applications/introductory-chapter-a-general-reading-process-on-landscape-architecture

[54] Moghadam, Valentine M. "democratization and women's political leadership in north Africa." Journal of International Affairs, vol. 68, no. 1, 2014, pp. 59-78. JSTOR, www.jstor.org/stable/24461706. Accessed 8 July 2021.

[55] Richard Stren, Urban Service Delivery in Africa and the Role of International Assistance, Development Policy Review, 2014, 32 (S1): s19–s37, https://doi.org/10.1111/dpr.12067

[56] Velika Ivkovska, Aiming Towards the Sky: The Vernacular „skyscrapers" of the South-West Balkans, ISVS e-journal, Vol. 6, no.3, Special Issue on Ottoman Vernacular, June, 2019, DOI: 10.5194/isprs-archives- XLIV-M-1-2020-1005-2020.

[57] Amjad Almusaed and Asaad Almssad (July 18th 2018). Introductory Chapter: Overview of a Competent Sustainable Building, Sustainable Buildings - Interaction Between a Holistic Conceptual Act and Materials Properties, Amjad Almusaed and Asaad Almssad, IntechOpen, DOI: 10.5772/intechopen.77176. Available from: https://www.intechopen.com/books/sustainable-buildings-interaction-between-a-holistic-conceptual- act-and-materials-properties/introductory-chapter-overview-of-a-competent-sustainable-building

[58] E. B. Banning, The Neolithic Period: Triumphs of Architecture, Agriculture, and Art, A journal of the American Schools of Oriental Research, Volume 61, Number 4, December 1998, https://doi.org/10.2307/3210656

[59] J.Bazelmans D. Meiere A. Nieuwho T. Spek P. Vos, Understanding the cultural historical value of the Wadden Sea region. The co-evolution of environment and society in the Wadden Sea area in the Holocene up until early modern times (11,700 BC–1800 AD): An outline, Ocean & Coastal Management, Volume 68, November 2012, Pages 114-126, https://doi.org/10.1016/j.ocecoaman.2012.05.014

[60] Appolloni L, D'Alessandro D. Housing Spaces in Nine European Countries: A Comparison of Dimensional Requirements. International Journal of Environmental Research and Public Health. 2021; 18(8):4278. https://doi.org/10.3390/ijerph18084278

[61] Tawayha FA, Braganca L, Mateus R. Contribution of the Vernacular Architecture to the Sustainability: A Comparative Study between the Contemporary Areas and the Old Quarter of a Mediterranean City. Sustainability. 2019; 11(3):896. https://doi.org/10.3390/su11030896

[62] Amjad Almusaed and Asaad Almssad (July 16th 2020). Urban Social Sustainability - Case Study; Gellerupparken–Denmark, Sustainability in Urban Planning and Design, Amjad Almusaed, Asaad Almssad and Linh Truong-Hong,

IntechOpen, DOI: 10.5772/intechopen.93124. Available from: https://www.intechopen.com/books/sustainability-in-urban-planning-and-design/urban-social- sustainability-case-study-gellerupparken-denmark

[63] Ibrahim, I.A. Sustainable housing development: role and significance of satisfaction aspect. City Territ Archit 7, 21 (2020). https://doi.org/10.1186/s40410-020-00130-x

[64] Sipos N, Pap N, Gonda T, Jarjabka Á. Feasibility and Sustainability Challenges of the Süleyman's Türbe Cultural-Tourism Centre Project in Szigetvár, Hungary. Sustainability. 2021; 13(10):5337. https://doi.org/10.3390/su13105337

[65] Silvia Mete, Jin Xue, Integrating environmental sustainability and social justice in housing development: two contrasting scenarios, Progress in Planning, Available online 10 July 2020, 100504, https://doi.org/10.1016/j.progress.2020.100504

[66] Matheus F.A.Goosen, Environmental management and sustainable development, Procedia Engineering, Volume 33, 2012, Pages 6-13, https://doi.org/10.1016/j.proeng.2012.01.1171

[67] Bjoern Harsman, John M Quigley, Housing Markets and Housing Institutions: An International Comparison, 1991, Springer

[68] Almusaed A., Almssad, A., et al, Coherent Investigation on a Smart Kinetic Wooden Façade Based on Material Passport Concepts and Environmental Profile Inquiry, Published by MDPI Energy in Construction and Building Materials (Switzerland). https://doi.org/10.3390/ma14143771

Chapter 5

The Impact of Contemporary Housing Functions on Its Social Sustainability

Andreas Koch

Abstract

Social housing functions are interrelated in manifold ways, expressing different needs and preferences of heterogeneous and socially unequal modern societies. The home as a place of individual shelter and privacy and as a node of interaction in social networks interferes with activities that had been spatially outsourced in the past and reintegrated again in recent times, such as productive labor, care or supply. In addition, social housing functions compete with economic functions of capital accumulation and profitmaking, transforming the dwelling into a tradeable commodity. Likewise, ecological functions of saving land and resources and reducing greenhouse gas emissions have to be satisfied. These interdependencies challenge sustainable housing politics, most prominently signified in the UN's Sustainable Development Goals 1, 10, and 11. The contribution captures this network of housing functions by advocating to strengthen social housing functions against economic functions. Political and philosophical justification of this position refers to theories of social capital and relational justice. Political measures feasible of being applied within the neoliberal system will be delineated, aiming to sustain social housing functions.

Keywords: housing needs, housing preferences, social capital, equity of relationships, housing commodification, capital accumulation, local housing governance

1. Introduction

Sustainable Housing is a significant challenge of contemporary societies worldwide, addressed explicitly by the Sustainable Development Goals 1, 10 and 11 of the United Nation's Agenda 2030 [1]. By taking the triangle model of sustainability as a reference, the economic sustainability of housing rests upon the idea to push back a dominant profit-making strategy of capital accumulation and claiming a social infrastructure approach of housing instead. Ecological sustainability deals with eco-sensitive products and property-protecting housing construction without losing the practical needs of availability and affordability of income-poor households. While ecological sustainability is considered here rather implicitly, the contribution focuses on the conflicting interlinkages between the social sustainability of housing and the growing meaning of economic housing commodification.

Housing in its social dimension does not only provide a place for individual shelter and privacy that help humans survive physically. Moreover, it offers opportunities for community engagement and social inclusion. The neighborhood represents a significant institution to embed residents in social networks, enabling social capital creation such as trust, solidarity, and mutual support. From a theoretical perspective, the paper primarily though not exclusively refers to Bourdieu's concept of social capital and Rosanvallon's relational equity approach, with its ingredients of 'singularity', 'reciprocity' and 'community'.

However, the mentioned social functions of housing are challenged by a significant transformation and pluralization of lifestyles, labor conditions, family structures, and cultural aspirations of living, affecting sustainable housing policies. Housing has become more flexible, fragmented, multi-located and biographically dependent. More single (parent) households of younger and older people ask for social co-housing and intergenerational living solutions. Increasing housing prices are a big problem for low- and middle-income households, and the current COVID-19 pandemic crisis contributes to a massive relocation of functions to the home (working, consumption, schooling, and leisure). These multiple and conflicting housing functions induce a search for tailored political solutions to socially sustainable housing.

The paper delineates the outlined challenges by, firstly, discussing the numerous and different housing and neighborhood functions of contemporary modern societies. This section is meant to be a plea for considering the social functions of housing primarily. Secondly, a political and philosophical justification for this plea is presented, with reference to the theories of social capital and relational equity. Social functions of housing will, thirdly, be confronted with neoliberal economic housing functions to argue for political interventions that encourage social sustainability of housing. Several political measures will then be presented, which deemed suitable and necessary in this regard. Examples from European countries and cities are used to illustrate and elucidate the arguments raised throughout the contribution.

2. Housing and neighborhood functions

Housing functions are diverse and assessed differently by people and social milieus of varied socio-economic and socio-demographic status. Housing functions' appraisal depends, on the one hand, on individual needs, expectations and aspirations, which, in turn, are influenced by biographical and cultural circumstances as well as economic conditions. Housing biographies change in phases of commencing work or study, growing or shrinking households, new job or family opportunities, or after retirement. These phases often affect residential movements, along with shifting expectations towards housing needs. Also, changing earning capacities (growing or falling) affect a household's autonomy concerning the housing situation. The housing culture shapes different imaginations of how people live – in a detached house or an apartment in a multi-story housing block – and thus how housing contributes to the production of personal identity.

Beyond individual needs, housing functions' appraisal is subjected to social and cultural aspects. Until today, prevailing housing architecture is designed for the traditional two-generational family, represented in separated housing units without or few commonly shared spaces. Moreover, an increase of single-person households, both younger (voluntarily intended) and older (involuntarily accepted) persons, is to be considered. Alternative forms of housing such as social co-housing [2], or housing associations that include jointly used spaces (e.g., kitchen, gym, library, co-working spaces), are growing but still marginal compared to mainstream architecture. While standard housing architecture designs rooms to be used more or less mono-functionally, co-housing concepts promote ideas of rooms that allow to integrate different functions in one room or to provide spaces that are used by multiple households (e.g., a room for taking care of children or a kitchen that is used by several families). Social housing functions are also influenced by a transformation of how living spaces are utilized. The change of (global) labor markets and the digitization of almost every aspect of our lives contributes to a spatio-temporal intersection and decentralization of activities of work, provision, education and recreation at home, which manifests in a functional mix of living spaces (e.g., working in the kitchen).

Historically, the spatial interference of domestic work had been the norm; farming, handcraft, and trade had been tied locally to the home [3]. The system of the estates of the realm, political sovereignty of aristocracy and the church inhibited social mobility by and large. Industrialization provoked an increasing spatial separation of productive and reproductive labor, which led to an extensive functional fragmentation and spatial specialization of activities to date. Today, we face spatial-functional re-unification processes, which results in a complex amalgamation of housing needs. Ongoing COVID-19 pandemic measures (homeschooling, home office, online delivery services) boost and exacerbate this development, affecting the home and the neighborhood. It is assumed that this development will become a sustainable housing trend, not least due to requirements to mitigate climate change effects [4]. However, not all households will be able to cope with these housing functions' transformations.

The neighborhood with its opportunities for social interaction is likewise relevant to satisfy the requirements of accommodation. Public spaces provide chances to establish and strengthen mutual recognition to enhance social inclusion opportunities to both new and long-established residents [5]. A well-functioning neighborhood also promotes social integration of all residents but mostly of minorities and marginalized groups like migrants, income-poor, or single-parent households. In addition, local communities can serve as a source for local social and political engagement. Even though all these social functions cannot be reduced to the neighborhood alone, it remains an essential promoter to residents being embedded into social networks and able to generate social capital. The next section will refer to this topic in more detail.

Social integration and inclusion do not rest upon a general and external natural law. Still, they are the temporary results of negotiation processes within and between social collectives whose participants are equipped with different amounts of social power, which, in turn, result in uneven social relations, local knowledge and social positions. Housing and neighborhood functions are therefore linked to issues of social inequality social (in)justice, which must be raised in the context of the prevailing economic model of neoliberal capitalism. This model places an individual's liberty and autonomy and self-reliance at the core of economic and societal action. All types of equal opportunities, which, among other things, affects access to housing, are to be valued and criticized against a neoliberal understanding of the equal opportunity. An alternative approach to the concepts of equal opportunity is given with Rosanvallon's [6] equity of relationships, delineated in the next section.

Housing space is geographically not evenly distributed nor available but depends on market regimes influenced by numerous factors, such as economic development, local and regional labor markets, infrastructure, return expectation of real estate markets, and local and regional planning. Housing preferences reflect these factors with spatio-temporal variations. Urban agglomerations are characterized by periods of urbanization, suburbanization and re-urbanization, while rural regions are affected by out- and in-migration, depending on location and infrastructural facilities. The last three or so decades represent periods of novel forms of residential segregation such as gentrification or gated communities [7]. They represent a growing social fragmentation whose spatial mechanisms of exclusion reveal different aspirations of housing [8].

Transnational and seasonal labor migration, forced migration due to civil war or changing climate conditions, lifestyles of multi-locality or long-distance commuting are some manifestations of globally mobile societies which affect the individual and social assessment of housing functions too. An example of the highly dynamic residential mobility processes at the local scale is given with statistical data of the twenty largest cities in Germany, illustrating residential movement of households: on average, 8.5 percent of households move each year, with a range of six to eleven percent [9]. Mobile households compete with the needs of the long-established population with regard to needed shops and infrastructures, cultural and sports facilities or the availability of outsourcing housing services (cleaning, food provision).

Individual and social functions of housing and neighborhood compete with economic functions of the home in several ways. With the advent of neoliberal market principles in the 1980s, the commodification of the house became an essential strategy of capital accumulation [10]. In fact, housing commodification has turned out to be the predominant investment strategy to date since many alternative investments yield lower rates of profit. Besides expectations on return of capital investment, it represents additional income during one's old age (because pensions are expected to decline). In this case, the residence is not always intended to be used by the buyer but serves as a second or touristic home (Airbnb) for others. These housing units are then divested of the regular housing market. Social housing strategies – represented by public housing, housing associations, or private housing assemblies – are seen as an attempt to mitigate sky-rocketing housing prices due to housing commodification. Housing policies dedicated to all types of settlements (urban, suburban, rural) are confronted with a fundamental dilemma. On the one hand, (attractive) new construction is meant to be a measure to grow demographically (and by tax revenues). On the other hand, competing with other cities or municipalities on residents is prone to risks of vacant housing stocks, which, in turn, impacts ecological sustainability and endamages a city's image.

Three overall domains can be extracted from the discussion of a multi-layered, interdependent and competitive amalgamation of housing functions, which are inevitably relevant to sustainable housing and correspond with the three dimensions of sustainability. The first domain represents the home as part of the social infrastructure, which is one precondition for residents being embedded into social networks (social sustainability). The second domain emphasizes the preservation of the building. Investment of capital is seen as a strategy to maintain the building substance without pursuing the goal of maximizing profit (economic sustainability). The third domain accounts for the fact that land and construction materials are finite resources. New building is neither excluded nor ranked first (ecological sustainability). The next section delineates theories of social capital and equity of relationships to make a plea for the social sustainability of housing, followed by a critical discussion of the capitalistic commodification of housing to shed some light on the problems of economic sustainability. Both sections result in reflections on measures taken to ultimately strengthen *local* political power in order to promote ecological, social, and economic sustainability as a function of political- and social-ecological transformation [11]. This kind of promotion aims to take a particular perspective on the social sustainability of housing that carries further the idea of transformation as a bottom-up process. **Figure 1** illustrates the functional

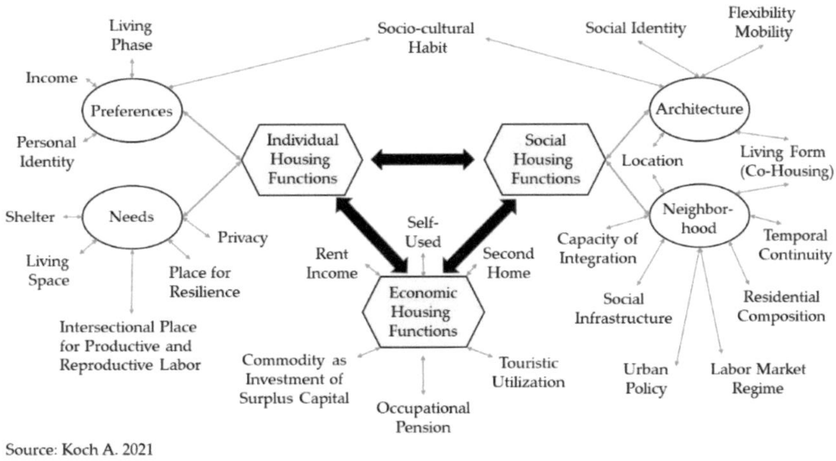

Figure 1.
Overview of interdependencies of housing functions.

interdependencies of housing (arrows do not represent exclusive relations, as there are many more relationships between the mentioned objectives; all the other connections are excluded deliberately to keep the illustration clear).

3. Social capital and equity of relationships

Theories of social capital contribute significantly to assess the relevance of housing as being a part of the social infrastructure, even though the idea of capitalizing on social relations is questionable [12]. Definitions of social capital are different in scale and incorporated values, but they all emphasize its meaning in social network relations. Bourdieu [13, p. 248] defines social capital as "[...] the aggregate of the actual or potential resources which are linked to possession of a durable network of more or less institutionalized relationships of mutual acquaintance and recognition". Since resources are incorporated into actors, his utilitarian perspective primarily focuses on the nodes of the networks. The values are given by solidarity, reciprocity, and trust expressed as network connections.

Coleman [14, p. 302], in contrast, accounts for assigning value to the edges, thus relations, of a social network: "Social capital is defined by its function. It is not a single entity, but a variety of different entities having two characteristics in common: They all consist of some aspect of a social structure, and they facilitate certain actions of individuals who are within the structure". Though different in their focus on social networks, both definitions share a common understanding of the preconditions – resources and social structure – necessary to trade social capital and of the selective mechanisms of exclusion and inclusion that emerge with the quantity and quality of social capital recognized and appreciated in particular social networks. In order to explicate those different forces within and between social networks that are responsible for social inequalities, it is worth referring to Woolcock's [15, p. 10] distinction of "bonding", "bridging", and "linking" social capital [16].

Even though we take into consideration that the value units of social capital derive from societal sources that lay outside social networks – "trust, reciprocity, moral, or attitudes will be reproduced and modified in social networks, but not exclusively produced" [12, p. 11] – we are able to recognize the relevance of social capital in community building and participation, notwithstanding. The home serves as one, albeit not a single spatial node of their residents' social networks. The home functions as the material source for creating and exchanging the ingredients social capital is made of, irrespective of the intensity of its use and the social status, ethnic background, or any other discrimination of its inhabitants. Any social interaction needs to be embedded into a local context to be recognized and accumulated as a potentially valuable contribution of belongingness and inclusion. Social capital is, therefore, one core concept to strengthen the social housing function against other concurrent functions.

A second core concept is given with the equity or justice of relationships [6]. The requirement for equal relationships arises from the insufficient applicability of equal opportunities in capitalistic societies. The relational equity principle does not claim to represent a utopian counter-program against the free-market liberal paradigm but to propose a realistic attempt of prioritizing social prosperity over individual justice of achievement. None of the concepts of equal opportunities – be it legal, institutional, instrumental or radical – aim for a comprehensive reduction of social inequality, nor do they fairly value individual success (or failure) against societal achievements. Furthermore, they largely fail to offer satisfactory suggestions to the range of legitimate inequality, a just redistribution of wealth or a commonly accepted poverty level [6, p. 286ff., 17].

According to Rosanvallon, social inequality rests, at least, upon three causes: (i) the privilege of social origin, that correlates with unevenly distributed capabilities among people; (ii) the growing commodification of life, that jeopardizes social and ecological well-being; (iii) the social-spatial segregation, that threatens social cohesion due to limited possibilities of mutual recognition and communication across these segregated areas.

Individual and social functions of housing suffer from economically preferred individual justice of achievement and social inequalities. Against these problems, Rosanvallon proposes a relational justice that satisfies three conditions: firstly, 'singularity' appreciates the mutual recognition of human beings. It does not qualify individual autonomy and liberty as a state but as a relational property. Social rights derive from unconditional access to societal institutions, including housing markets. Mutual recognition is particularly committed to less privileged and/or less affluent people struggling to articulate their rights in economic relationships. Secondly, it claims equal participatory rights of co-produced goods and services, as well as commonly shared values, referred to as 'reciprocity'. This relates to commons, not to trade and exchange. Needs are anticipated and shared democratically. In our context of housing functions, reciprocity accounts for the globally claimed human 'right to housing' and the targets addressed by the Sustainable Development Goals, such as affordability, basic protection of tenants against disproportionately rising rents or displacement. The third condition of relational justice refers to 'communality', which avoids the creation and maintenance of communities to be realized by socio-cultural ascriptions. These discriminations imply a commodification of the person, with which inclusion and exclusion – and asymmetrical power relations that derive from these processes – are established. Residential segregation is one of the results, reproduced and seemingly legitimated by those discriminations. Communality urges all people to strengthen local democratic relationships and local social spaces [6]. It can be understood as a synthesis of singularity and reciprocity criticizes contemporary processes of gentrification because they discriminate against humans socially and reward economic Darwinism.

3.1 In a nutshell

Theories of social capital and relational equity/justice are two approaches that delineate common principles of establishing and preserving social communities.

Other approaches not explicitly considered in this contribution are sociological and social geographical theories, such as actor-network theory [18], system theory [19], theory of structuration [20], and theory of action [21], as well as social- and political-philosophical theories like, for example, the theory of justice [22]. Social capital and relational equity theories take a relational perspective on communities and highlight the functions of relational structures. In so doing, they enable a critique of current housing policies from a sustainability's perspective. Social capital theories exhibit the principles of social relationships, the mechanisms on which social networks of different size, quality, and hierarchy develop and differentiate, and the roles, positions and power relations that emerge due to capital allocation. Theories of relational justice set out alternative norms of collective participation and engagement. Priority is entitled to an equitable status of community members against individually justifiable inequalities.

4. The capitalistic exploitation of dwellings

Industrialization implicated a comprehensive division of labor and, consequently, the functional and spatial division of living and working. While productive activities had been relocated to firm sites, the home mainly served for tasks of reproduction. Further outsourcing of housing functions followed through with the proliferation of health, education, and cultural institutions. With the growing availability of leisure time, recreational and sports activities were outsourced increasingly as well. This intense spatial differentiation of functions has changed with the development of modern information and communications technologies over the last 20 or so years, as both productive and reproductive tasks can now be relocated to the home. This development, however, happens to take place selectively, voluntarily, and involuntarily. The measures to combat the COVID-19 pandemic are currently the most obvious signifier of a complex mix of functions at home. All these processes will influence – and partly even determine – the economic meaning of the dwelling as a commodity.

With the implementation of the neoliberal regime in market societies since the 1980s, economical housing functions have also altered. Globalization and deregulation of financial markets led to significant demand for properties as capital investment. Commercial and housing real estates are valued profitably, primarily in large urban agglomerations such as global cities, not least due to other capital investments are assessed less attractive because of low interests. Also, the privatization of public commodities like social housing complexes increased the revenue expectations of investors. For example, Germany's number of social housing units declined from almost 4.0 million (1987, only West Germany) to 1.2 in 2020 [23]. The economic and financial crises of 2008 further boosted housing commodification and reinforced housing market structures – with private interests outperforming social needs. Accumulating capital clearly prevails [24], though personal risk management is increasing likewise [25].

Land and real estates are finite and scarce commodities whose utilization depends on diverse interests and needs – agriculture, commerce, housing, transportation, recreation or natural livelihoods' preservation. Scarcity and competition ask for high demand, which also varies due to locational parameters. Globalization, deregulation, and privatization of commodity and financial markets made numerous and flexible possibilities of capital reinvestment available simultaneously. Overproduction of commodities and falling prices thereof produce a surplus of capital in the "primary circuit of capital" [26]. This capital seeks valorization in the "second circuit of capital", to which real estate markets belong too. Urban and suburban housing markets are predominantly affected by increased exploitation of capital [27].

The influence of economic geographical parameters manifests, among other places, in processes of gentrification. Gentrification, defined as a process of economic, architectural, cultural and social upgrading of mostly attractive urban areas (concerning location), often accompanied by a displacement of less affluent households which have lived in these areas for long [28–31], is theoretically analyzed by different approaches. Rent gap and value gap theories focus on the investment behavior of financial companies (real estate agencies, insurances, mortgage lenders, private housing companies, etc.) and government agencies. Driven by suburbanization processes with rising land prices in these regions, "[...] the relative price of inner-city land falls. Smaller and smaller quantities of capital are funneled into the maintenance and repair of the inner-city building stock. This results in what we have called a *rent* gap in the inner city between the actual ground rent capitalized from the present (depressed) land use and the potential rent that could be capitalized from the "highest and best" use [...], given the central location" [30, p. 133]. **Figure 2** illustrates the rent gap theorem. Though similar in its principle explanatory objective, value gap theories consider rent-housing markets of European cities mainly [32].

Unlike top-down approaches of capital accumulation, theories such as the invasion-succession-cycle theory primarily take the investment behavior of households into account. According to this theory, gentrification is distinguished into phases of invasion (firstly of pioneers who are engaged in activities of cultural upgrading, followed by gentrifiers whose primary interest lies in profit-making) and succession (an increase of firstly pioneers and then gentrifiers lead to a transformation of the residential population due to rising rent and property prices, referred to as 'social' upgrading). Despite these analytical differences, both approaches highlight the function of housing as a commodity.

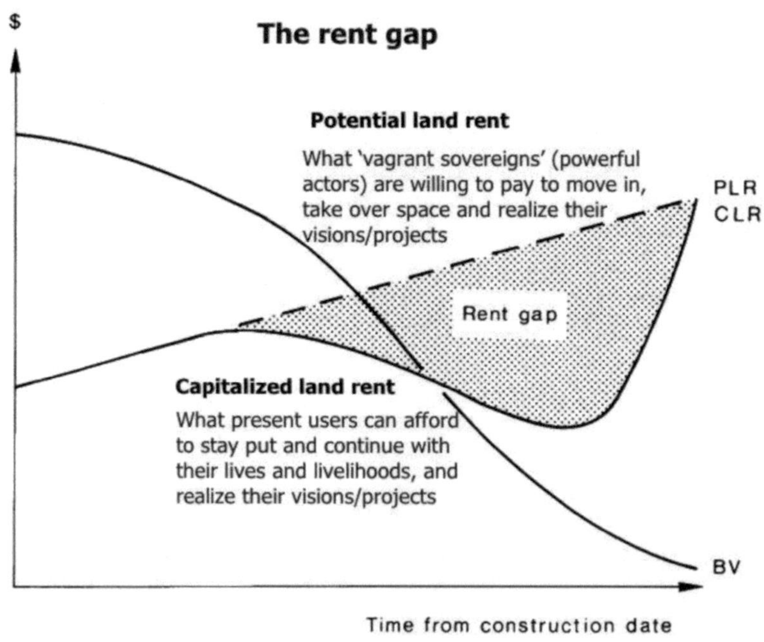

Figure 2.
The rent gap theory by Neil Smith (idealized). Source: Clark E (2010). https://www.slideshare.net/environmentalconflicts/eric-clark. BV = building value; PLR = potential land rent; CLR = capitalized land rent.

Furthermore, the strong position of the economic functions of housing can be explained by the heavy meaning of the competition law and private property within the European Union. One conclusion of this meaning is that social housing policies must not distort competition. In other words, social housing is exclusively restricted to low-income households unable to afford homes offered on the private market [33, p. 20]. Private property is explicitly legally protected as a basic liberal right, while social housing needs are not. The human right to housing is interpreted as a right to "housing assistance" that obliges the state to issue adequate political conditions and not accomplish an individual right to housing. The consequences of this politics are far-reaching: Sweden, for example, had to adapt its previous praxis of fixing rents when it joined the EU in 1995. The "reasonable rent" was related to the dwellings of the public housing companies; private rent prices were not allowed to exceed five percent. Today, rents of public housing companies have to comply with free-market conditions. In addition, Sweden has reduced public funds in housing to zero [33, p. 24].

It is not only private companies but also cities and municipalities which share a common interest in responsibly taking care of the economic functions of housing. In contrast to the private housing economy, public institutions are required to gauge economic, social and ecological concerns against each other, for which they have to adjust their local programs and plans to regional and national ones. Given their sovereign right to designate land to be utilized for construction, they govern the settlement development of their territory mostly independently, however. Income and commercial tax revenues force cities and municipalities to promote the economic development within their territory selfishly and independently, under which a trans-locally coordinated settlement development suffers. Planning rules (e.g., building density and height) as hard facts and cultural aspirations (e.g., preference of detached houses in suburban and rural regions) determine the housing forms sustainably. Middle- and high-income residents are supposed to be most attractive since they pay more taxes and spend more money which, in turn, increases municipalities' scope of freedom for future local development – a self-enforcing process has been initiated, which ultimately is economically risky and ecologically unstainable.

Even a balanced local housing market development is prompted to take the private interests of landlords into account, even though it affects low-income households adversely. Social housing associations increasingly struggle to supply affordable housing for all income groups (income that relates to the regional labor market with its salary structure). For example, the income and living expenses of Salzburg's residents grew by 22 percent, while rent prices did so by 30 percent and property prices by 70–100 percent [34]. Current rent prices for tenants of social housing are almost 11 Euro per m^2, compared to private rent prices of 15 Euro per m^2 on average [35]. Although a significant difference is given, for a growing number of low-income households, it gets more and more difficult to afford these rent prices.

A related problem is that housing markets are spatially and temporarily volatile due to socio-demographically varying attitudes of housing preferences. Suburbanization, reurbanization, counter-urbanization or living in the countryside are all trends of changing popularity that affects the supply and demand of housing. While vacancy in demographically shrinking regions is accompanied by a dereliction of buildings and a decay of prices, it is a signifier of speculation in regions of growth. Economic sustainability of such processes is at least questionable, since individual profit-seeking is local and costs of housing conservation are socially outsourced. Ecological and social sustainability are disastrous because the emigration of people is followed by the emigration of businesses and services.

4.1 In a nutshell

The growing dominance of housing market functions against their personal and social functions jeopardizes the social sustainability of housing. It neglects the justifiable needs of those persons, being unable to rent or purchase dwellings offered to private (and increasingly social) market conditions. Solidarity across social strata is threatened, and novel mechanisms of exclusion provoke intersectional fragmentation and marginalization, introducing room for encapsulation and resistance. Contemporary capitalistic housing functions are, furthermore, in sharp contrast to challenges of ecological sustainability (land and resources consumption, housing forms, housing locations). Economic sustainability is challenged by short-term profit-seeking versus long-term income security (rent revenue). Housing markets are socially and ecologically blind [36].

5. Strengthening social functions of housing sustainably

The increased evidence of economic housing functions is, to a large degree, the result of global developments, particularly the financial economy. Although housing commodification manifests locally and regionally differently, a dominance of global interests versus local needs is empirically evident [37]. Therefore, local countermeasures are required, aiming to strengthen the social objectives of sustainable housing. The following suggestions are subdivided into three domains, have been discussed in the literature with different intensity, and are anything but complete. The first domain refers to measures meant to directly strengthen individual and social functions of housing, while the second domain is seen as a framing condition that helps promote personal and social housing functions by forcing back economic housing functions of capital accumulation. The third domain appreciates the local level of political activism by its regional counterpart.

5.1 Strengthening individual and social functions of housing

Measures that strengthen individual and social functions of housing justify their purpose by the fundamental meaning of housing as being part of the social infrastructure. One of these measures is to prioritize permanent housing function over touristic usage (e.g., Airbnb) or as second homes. In a study of the city of Salzburg, Van-Hametner et al. [38] uncovered 17 percent of housing units being used alienated; political measures like registration, fixing a maximum number of permitted housing units to be used for touristic purposes, or introducing penalty fees, may alleviate this.

Another measure that helps foster social justice in housing politics is seen in a transformation of funding principles. A politics of social housing is doing better if public funds are dedicated to the construction of housing instead of supporting households. Although the latter is tied to income thresholds, political governance restricts itself to the economic strength of households instead of social need.

Because societies and communities are diverse in their lifestyles, cultural attitudes, demographic structure, and capital composition (economic and social capital), housing politics should account for this diversity by expanding the forms of housing. Up to now, the standardized model of the traditional two-generational family has prevailed. Changing household structures (beyond the biological family) call for new co-living forms, such as social co-housing or intergenerational housing, which provide spaces for collective usage [39]. Ideas comprise spaces such as the kitchen or a room for co-working, co-caring or co-recreation. They promote social interaction and safe resources.

A measure to cope with increased biographical mobility of people (due to increased social mobility) and thus residential mobility is to offer more forms of mobile housing, not only as caravans or campers but also as creative forms of tiny houses [40], for which dedicated public spaces are made available by cities and municipalities. 'Mobile' also refers to a growing supply of furnished dwellings, which simplify residential mobility within a city, a region or beyond.

Residential mobility, furthermore, implies an ongoing change of the household composition in urban (as well as rural) neighborhoods. Establishing relational justice among residents and promoting social interaction to accumulate social capital between residents, cannot be expected to be taken for granted under these dynamic conditions. Instead of hoping for an ideal situation of self-organized, bottom-up-driven communication spaces, it would be more appropriate to support interaction among residents by introducing facilitators and intermediaries as appreciated nodes in local social networks. This mode of communication, referred to as "organized urbanity" [41], can be realized through particular (elected) residents themselves or through representatives of public social institutions. Urban quarter management thus strives for solving conflicts (not necessarily resolving them) and creating a social environment that enables the reconciliation of interests.

All these measures are dedicated to enhancing the social infrastructure's role of housing against its growing and powerful capital accumulation function.

5.2 Framing individual and social functions of housing

The construction of new dwellings is seen as an adequate strategy to mitigate rising housing prices in mainstream social politics, preferably in urban and suburban regions. Increasing the supply of housing units appears to be a strategy that satisfies an urgent demand due to an increasing population. This, however, is only partly true. Besides the quantitative growth of an urban population, habits of housing are changing: single-person households, multi-local forms of living ("living apart together"), and second homes in attractive regions do apparently increase the demand for dwellings. This seems to result in the paradox of growing property prices due to increasing housing supply. In fact, profit-seeking strategies are the true cause for this contradiction.

Countermeasures that help mitigate those developments are temporarily and regionally adapted moratoria of housing construction [42]. These are socially sustainable because they stem the causes of maximizing capital accumulation as commercial exploitation of dwellings becomes less attractive, even though a further increase of housing prices during the initial phase of transformation cannot be strictly prevented. A moratorium enables affected cities and municipalities to think about the potential future utilization of the existing stock of buildings. No further land will be sealed, no further resources to construct and open up building sites will be needed, and no further energy needs for construction and maintenance will be consumed. With this measure, the daily consumption of land – in Germany, it is 190 square kilometers per year, in Austria it is 44 [43, 44] – can be reduced considerably. Also, building materials can be saved (for example, cement production contributes to two percent of CO_2 emissions in Germany; it is six to eight percent worldwide [45]).

A large part of the existing building stock that can be incorporated into the regular housing market are vacant housing units. 2.8 percent of all housing units in Germany (2019) are estimated to be vacant, i.e. would be available immediately or within six months after renovation [46]. For the city of Salzburg, a conservative estimation is four percent (3.500 housing units) [47] but goes up to nine percent. Accompanying measures of public administration are needed to reactivate vacant homes. Measurements can be restrictive (e.g., a penalty charge) by expressing the political will to prohibit vacancy as a violation of the human right to housing, though they might likely be less effective because reactivation cannot be guaranteed. Constructive measures, such as private-public-partnerships to offer vacant dwellings of private homeowners, would be another approach that would be beneficial as they involve the local knowledge of public authorities about housing needs of households of different income constraints.

All these measures and programs can be easily transferred to vacant office buildings in order to further increase the availability of living spaces. Estimates of 3.5 percent vacancy rate in Germany [48] and 4.7 percent in Vienna [49] (no Austrian data available) likewise illustrate the considerable potential of transformation. Converting office space to living space has primarily to contribute to increase the availability of affordable housing and to restrain strategies of capital accumulation.

Cities and municipalities in Germany and Austria are not only invited to collaborate with private landlords or housing companies more tightly but also to take their general right of pre-emption of housing units more seriously into account. A few cities like Berlin apply this right more intensely in recent years [50]. In order to increase the stock of public housing units by this measure, a more comprehensive financial subsidy and political encouragement of national governments or the European Union is needed. Recalling the above mentioned EU's competition law and property rights illustrates how complex this operation will be, however.

5.3 Complementing individual and social functions of housing

Cities and municipalities do have a considerable sway to develop, govern and shape their territories, which implies both numerous policy spaces and risks. To maintain or enhance their attractiveness to residents, businessmen or tourists, they are liable to the growth paradigm. New building programs or the renovation of expensive housing blocks is seen as an indicator of economic prosperity in local settlement development. By generalizing this idea, a selfish competition of potential residents takes place, producing winners and losers even if the national population is growing. In order to oppose this thinking and doing, which is anything but sustainable, mandatory regional planning is necessary, which prioritizes regional needs democratically. Designating land to be built is then not any longer subjected to local individual interests. Strengthening regional planning this way has to cope with a reform of local financial planning that enables cities and municipalities to fulfill their local duties in the future as they do today. Merging municipalities would be an alternative which, however, is less likely realizable though not impossible.

Another measure to account for the social and ecological sustainability of housing without causing severe economic disadvantages for local communities is introducing a fee or a tax to redesignated land (farmland that is designated to be built). With this fee, a comprehensive intervention into private property rights will be avoided, and municipalities' pressure to designate land will be reduced. Revenues can be used ring-fenced for public duties as, for example, health and education infrastructure. This way, compensation between private and public interests is feasible.

6. Conclusion

The proposed solutions of the previous section entail a prospective movement towards a more substantial commitment of considering social sustainability of housing a severe issue. Simultaneously, they indicate unfolding the weaknesses of the current circumstances and the obstacles that hinder a progressive socio-ecological transformation of societies. The critical observation that the socio-ecological transformation does not happen [51] results, among other things, in a defensive attitude of liberty, values, and lifestyles of the middle-classes, including the seemingly progressive cultural-left. Ecological behavior in one respect (e.g., buying organic food) serves as to justify a behavior that jeopardizes ecological or social sustainability in another respect (e.g., living in a detached house out in the green).

While the United Nations' Sustainable Development Goals frame the global political requirement of a comprehensive transformation of predominantly neo-liberal societies (by leaving the question unanswered whether their goal strategies are good suggestions compared with the problems a global population is facing), a complementing local perspective of strategies' implementation is seen necessary to comply with the requirements of sustainability. Strengthening social functions of housing with the proposed instruments appears to be an essential step in this direction.

Relational equity and justice are not meant to be a "nice to have" ingredient to appreciate social relations but an essential component to reducing social inequality remarkably. Localizing social needs and wants in general and social functions of housing, in particular, argues for a relational understanding of geographical space that takes local relationships of people but also of land, commodities, services and capital circulation primarily into account. The presented competition between social and economic housing functions is only one though highly relevant facet in the political and social arena that determines the wellbeing of the present and future generations. Similar efforts are undertaken that refer to ideas of the commons, degrowth, sharing or circular economies. They all have in common a different understanding of co-habitation – co-habitation that rests upon local and thus trans-local governance and shared responsibility. It eludes comparability for profit-seeking accumulation strategies and favors subsistence over global production and consumption. Promoting localizing strategies can help amplify a comprehensive understanding of sustainability.

Author details

Andreas Koch
University of Salzburg, Salzburg, Austria

Address all correspondence to: andreas.koch@sbg.ac.at

References

[1] United Nations. The 17 Goals. 2021. Available from: https://sdgs.un.org/goals [Accessed 2021-06-18]

[2] Wang J, Hadjri K, Bennett S, Morris D, The role of cohousing in social communication and sustainable living environments. WIT Transactions on the Built Environment, 2020, Vol. 193: 247-258. DOI: 10.2495/GDI170201

[3] Sennett R. The Craftsman. London: Penguin Books; 2009. 336.

[4] Krau I. Corona und die Städte. München: Oekom; 2021. 120.

[5] Hallman H.W. Neighborhoods. Their Place in Urban Life. Beverly Hills, CA: Sage Publications; 1984. 320.

[6] Rosanvallon P. Die Gesellschaft der Gleichen. Hamburg: Hamburger Edition HIS; 2013. 384.

[7] Lees L, Shin HB, Lopez-Morales E, editors. Global gentrification: Uneven Development and Displacement. Bristol, Chicago: Policy Press; 2015. 416.

[8] Sennett R. The Uses of Disorder. Personal Identity and City Life. New York, London: W. W. Norton; 1970. 198.

[9] Statista. Umzugsquote in den größten Städten Deutschlands im Jahr 2017. 2021b. Available from: https://de.statista.com/statistik/daten/studie/731899/umfrage/umzugsquote-in-den-groessten-staedten-deutschlands/ [Accessed 2021-06-18]

[10] Harvey D. Seventeen Contradictions and the End of Capitalism. London: Profile Books; 2014. 338.

[11] Berger H. Entgrenzte Städte. Zur politischen Ökologie des Urbanen. Münster: Westfälisches Dampfboot; 2003. 181.

[12] Koch A. The Capital, The Social, The Institution – Bourdieu's Theory of Social Capital Revisited. In: Kapferer E, Gstach I, Koch A, Sedmak C, editors. Rethinking Social Capital: Global Contributions from Theory and Practice. Newcastle upon Tyne: Cambridge Scholar Publishing; 2017. p. 3-17

[13] Bourdieu P. The Forms of Capital. In: Richardson J, editor: Handbook of Theory and Research for the Sociology of Education. New York: Greenwood; 1986. p. 241-258. Available from: https://www.marxists.org/reference/subject/philosophy/works/fr/bourdieu-forms-capital.htm [Accessed 2021-06-15]

[14] Coleman J.S. Social Capital in the Creation of Human Capital. The American Journal of Sociology. 1988; 94: 95-120.

[15] Woolcock M. The Place of Social Capital in Understanding Social and Economic Outcomes. 2000. Available from: https://www.oecd.org/innovation/research/1824913.pdf [Accessed 2021-06-15]

[16] Putnam R. Bowling Alone: The Collapse and Revival of American Community. New York: Simon & Schuster; 2001. 544.

[17] Lessenich S. Grenzen der Demokratie. Ditzingen: Reclam; 2019. 119.

[18] Latour B. Reassembling the Social. An Introduction to Actor-Network-Theory. Oxford: Oxford University Press; 2005. 301.

[19] Luhmann N. Soziale Systeme. Frankfurt a.M.: Suhrkamp; 1993. 675.

[20] Giddens A. The Consequences of Modernity. Oxford: Polity Press; 1990. 200

[21] Werlen B. Society, Action and Space. An Alternative Human Geography. London: Routledge; 1992. 272.

[22] Rawls J. A Theory of Justice. Harvard: Harvard University Press; 1971. 560.

[23] Deutscher Mieterbund. Eklatanter Mangel an Sozialwohnungen in Deutschland. 2020. Available from: https://www.mieterbund.de/startseite/news/article/57520-eklatanter-mangel-an-sozialwohnungen-in-deutschland.html [Accessed 2021-06-15]

[24] Harvey D. Rebel Cities. London, New York: Verso; 2013. 187

[25] Heeg S. Wohnungen als Finanzanlage. Auswirkungen von Responsibilisierung und Finanzialisierung im Bereich des Wohnens. sub\urban. zeitschrift für kritische stadtforschung. 2013; 75-99. DOI: https://doi.org/10.36900/suburban.v1i1.71

[26] Harvey D. The Urban Process under Capitalism: A Framework for Analysis. In: Fyfe N.R, Kenny J.T. editors. The Urban Geography Reader. New York: Routledge; 2005. p. 109-120.

[27] Harvey D. Social Justice and the City. Revised Edition. Athens, London: The University of Georgia Press; 2009 . 354.

[28] Helbrecht I, editor. Gentrifizierung in Berlin. Bielefeld: Transcript; 2016. 322.

[29] Holm A. Wiederkehr der Wohnungsfrage. Bundeszentrale für politische Bildung: Gesucht! Gefunden? Alte und neue Wohnungsfragen. 2019; 98-111.

[30] Smith N. Gentrification, the Frontier, and the Restructuring of Urban Space. In: Fyfe NR, Kenny J.T. editors. The Urban Geography Reader. New York: Routledge; 2005. p. 128-137.

[31] Vollmer L. Strategien gegen Gentrifizierung. Stuttgart: Schmetterling Verlag; 2018. 163.

[32] Dangschat J. Gentrifizierung: Kein 'back to the city' – weder von Menschen, noch des Kapitals. sub\urban. zeitschrift für kritische stadtforschung. 2019; 7(3): 101-108. DOI: https://doi.org/1036900/suburban.v7i3.516.

[33] Shah N. Europäische Union und der soziale Wohnungsbau – ein Spannungsverhältnis?! In: Club of Vienna, Editor. Brennpunkt Wohnbau. Vienna: Mandelbaum; 2019. p. 9-28

[34] Stadt Salzburg. Gebäude, Wohnungen & Grundstückspreise 2017. 2019. Available from: https://www.stadt-salzburg.at/fileadmin/landingpages/statistik/gebaeude_und_wohnungen/gebaeude_wohnungen_und_grundstueckspreis_00477006.pdf [Accessed 2021-06-12]

[35] Hölzl & Hubner. Der Salzburger Immobilienmarkt 2019. 2019. Available from: https://www.hh-immo.at/presse.html [Accessed 2021-06-16]

[36] van Vliet W, editor. International Handbook of Housing Policies and Practices. Westport, CT: Greenwood Press; 1990.

[37] Ley A, Ur Rahman Md A, Fokdal J, editors. Housing and Human Settlements in a World of Change. Bielefeld: Transcript; 2020. 275.

[38] Van-Hametner A, Smigiel C, Kautzschmann K, Zeller, C. Die Wohnungsfrage abseits der Metropolen: Wohnen in Salzburg zwischen touristischer Nachfrage und Finanzanlagen. Geographica Helvetica. 2019; No. 74: 235-248.

[39] Sørvoll J, Bengtsson B. Autonomy, democracy and solidarity. The defining principles of collaborative civil society housing and some mechanisms that may

challenge them. Urban Research & Practice. 2020; 13: 390-410. https://doi.org/10.1080/17535069.2019.1573267

[40] Wotton J, Skates H, Shutter L. Tiny House – when size matters. Australian Planner. 2019; 3-4: 209-220. https://doi.org/10.1080/07293682.2019.1634112

[41] Smigiel C, Koch A. Formen von Urbanität in der kapitalistischen Moderne und ihre Effekte. Polylog. 2018; 39: 19-33.

[42] Fuhrhop D. Verbietet das Bauen! München: Oekom; 2015189.

[43] BMU Bundesministerium für Umwelt, Naturschutz und nukleare Sicherheit. Flächenverbrauch – worum geht es? 2020. Available from: https://www.bmu.de/themen/europa-internationales-nachhaltigkeit-digitalisierung/nachhaltige-entwicklung/strategie-und-umsetzung/reduzierung-des-flaechenverbrauchs/ [Accessed 2021-06-14]

[44] Umweltbundesamt. Flächeninanspruchnahme. 2021. Available from: https://www.umweltbundesamt.at/umweltthemen/boden/flaecheninanspruchnahme [Accessed 2021-06-14]

[45] WWF Deutschland. Klimaschutz in der Betonindustrie. Hintergrund und Handlungsoptionen. 2019. Available from: https://www.wwf.de/fileadmin/fm-wwf/Publikationen-PDF/WWF_Klimaschutz_in_der_Beton-_und_Zementindustrie_WEB.pdf [Accessed 2021-06-14]

[46] Statista. Entwicklung der Leerstandsquote auf dem deutschen Wohnungsmarkt in den Jahren von 2001 bis 2019. 2021. Available from: https://de.statista.com/statistik/daten/studie/74463/umfrage/wohnungsleerstand-in-deutschland-seit-2001/#:~:text=Die%20Leerstandsquote%20misst%20die%20Zahl,733.000%20Wohneinheiten [Accessed 2021-06-14]

[47] SIR Salzburger Institut für Raumordnung & Wohnen. Wohnungsleerstand in der Stadt Salzburg. 2015. Available from: https://www.salzburg.gv.at/bauenwohnen_/Documents/endbericht_wohnungsleerstand_final.pdf [Accessed 2021-06-15]

[48] Statista. Leerstandsquote für Büroimmobilien in Deutschland 2019. 2021a. Available from: https://de.statista.com/statistik/daten/studie/912966/umfrage/leerstandsquote-fuer-bueroimmobilien-in-deutschland/#:~:text=Diese%20Statistik%20zeigt%20die%20Entwicklung,der%20Büroflächen%20in%20Deutschland%20leer [Accessed 2021-06-15]

[49] EHL. Büromarktbericht Wien, Herbst 2020. 2020. Available from: https://www.ehl.at/fileadmin/user_upload/EHL_BMB_20201007_ohne_VRF-zahlen_v2.pdf [Accessed 2021-06-17]

[50] Senatsverwaltung für Stadtentwicklung und Wohnen Soziale Erhaltungsgebiete. Vorkaufsrecht. 2021. Available from: https://www.stadtentwicklung.berlin.de/staedtebau/foerderprogramme/stadterneuerung/soziale_erhaltungsgebiete/vorkaufsrecht.shtml [Accessed 2021-06-14]

[51] Blühdorn I. The Politics of Unsustainability: COP15, Post-Ecologism, and the Ecological Paradox. Organization & Environment. 2011; 24(1): 34-53.

Chapter 6

Sustainable Housing in Developing Countries: A Reality or a Mirage

Ibiwumi Saliu and Evangelisca Akiomon

Abstract

Efficient houses built in a way that respect resources and could last long in quality systems are said to be the way forward in achieving a low carbon footprint and a sustainable environment. These houses are constructed from high performance, energy saving materials with an energy maximizing building orientations. Findings have shown that as much as housing is a basic human need, in developing countries, around 40–75% of the population in fast growing cities is housed in squatter settlements without basic amenities and services. In sub-Saharan Africa, 59% of the populations in urban regions live in slums, about 30% in the Latin and Caribbean, 28% in Asia and Pacific region. Population migrate massively to big cities in search of green pastures, which has invariably turned green pastures into 'brown' if not 'red' pastures due to overcrowding and other social menace, turning houses to mere shelters. Energy efficiency or environmental friendly housing is far from the thoughts of dwellers which still crave to have or maintain a roof over their heads. Whereas government policies are majorly jeered towards constructing houses, if sustainable, would only be affordable for a few well to do population neglecting the homeless masses. Therefore, this chapter aims to expound on the situation of housing in developing countries as well as the possibility of achieving sustainable housing.

Keywords: carbon footprint, environmental sustainability, sub-Saharan Africa

1. Introduction

Worldwide, there has been large scale proliferation in construction of houses due to population growth, economic development, urbanization and migration, which has in turn had a ripple effect on sustainability [1]. According to the National strategy for ecologically sustainable development, sustainability is referred to as the development that improves the total quality of life, both now and in the future, in a way that maintains the ecological processes on which life depends [2]. It is about meeting human needs and improving their quality of living by minimizing negative impact on the environment. According to Queensland Department of public works [3], sustainability could either be social, which include safety, security and universal design; or environmental, for example, water, waste, energy efficiency; or even economic, which entails cost efficiency, peace of mind and resale value [4].

Meanwhile, housing in itself is referred to as the central of sustainable development [5]. It is one of those social conditions that determine the quality of life and welfare of people and places; a social necessity of life recognized worldwide as one of the most important needs of man, after food [6]. It is a basic necessity that holds

Stage of house life cycle	Examples of environmental sustainability considerations
Planning stage	Impact of planned site on the local environment; relationship with the city; quality of the local built environment; mixed-use and density; poly-centricity; infrastructure; public transport; green areas; environmental hazards.
Building Design	Considering embodied energy and resource utilization; enabling energy and water efficiency by design; integrating district heating and micro-generation; sustainable waste management green roofs; robustness and resilience; future-proofing; possibility of upgrading; shaping of lifestyles.
Construction	Safe, environmentally-friendly, local affordable material; minimization of environmental impact from building activity.
Operation	Energy performance; air conditioning; air quality; pollution by residents and impact of the local pollution on residents, water use and water management; water recovery comfort and hygiene of home; quality and energy efficiency of the local infrastructure and transportation property maintenance and management, waste management and recycling; greening the area; natural hazards.
Refurbishment	Choice of refurbishment material; energy efficient design; disturbance of the environment; management of construction waste.
End of life	Demolishing or reusing; recycling of building components; management of construction of waste.

Source: www.UNHabitat.org.

Table 1.
Guiding matrix for assessing sustainable housing.

a place of singular importance in the general strategy of development [7]. The daily lives of people, their health, security and wellbeing are affected and influenced by locations of their homes, its construction and design and how well they are weaved into the environmental, social, cultural and economic fabric of communities [8].

Sustainable housing offers rare opportunities to promote not only environmental conservation and economic development but also quality of life and social equality while mitigating numerous precarious problems relating to population growth, urbanization, slums, poverty, climate change, lack of access to sustainable energy and economic uncertainty [9–11]. Its goal is to reduce the impact of consumption of natural systems by keeping it within natural limits while simultaneously enabling human system to be optimized without impairing the quality of life [8]. It is to integrate both green agenda, which involves maintaining the natural environment and brown agenda, i.e. ensuring a well built environment [6]. However, this has always being a dilemma for many countries, especially developing countries, which are still way behind in attaining their targets of the sustainable development goal 11 of making cities and human settlements inclusive, safe, resilient, and sustainable, and the 2030 target, which is to ensure access to adequate, safe, and affordable housing and basic services, upgrade slums and support least developed countries [4]. Sustainable housing development could be successfully achieved, if an optimal balance is ensured between sustainable housing and residents' satisfaction [12]. It should be environmentally safe, socially inclusive and economically productive [12] according to **Table 1**. Below is the table that shows the guiding matrix for the assessment of sustainable housing, **Table 1**.

2. Situations in developing countries compared to other countries

Developed and developing world are facing sustainable housing and urbanization challenges in different ways but the developing countries are moving slowly or even on a negative direction towards adopting feasible sustainable strategies [13]. Developing countries are those countries whose standard of living, income, economic and industrial development remains more or less below average. According to the IMF, there are 52 developing countries with current population of 6.53 billion, which is a considerable proportion (85.09%) of the world's population [14]. It includes the whole of Africa, central and South America, almost all of Asia countries and numerous other island states [14]. Due to low standard of living, income, among others, decent and safe housing remains a dream for a majority of the population in most developing countries, yet housing development in itself create amplified carbon footprint and negative impact on the environment if it's not sustainable [8]. The government even considers sustainable housing as merely social burden, while the so called pro-poor housing programs provide accommodation of poor standards in remote locations with little consideration to the residents' lifestyle and livelihood strategies [8]. Neglecting the fact that, it is through sustainable solutions that the tensions between the sporadic urban growth, climate change, access to clean energy and environmental conditions alongside other issues can be alleviated, while the potential of improved economic prosperity and social development can be further unlocked [8].

There is indeed no doubt in the magnitude of housing problems in developing countries compared to developed countries with good standard of living and industrial development [15]. Urbanization in developing countries is in sharp contrast to that of western industrial urbanization. According to the UN-Habitat 2008b, in developing countries, 1 out of 4 households live in poverty, 40% in African cities. 25–50% of people in developing cities live in informal settlement, while not further than 35% of cities in the developing countries have their waste water treated; 2.5billion people have no sanitation and 1.2 billion do not have access to clean water [16, 17]. Half of the urban population in Africa, Asia, Latin America and the Caribbean suffer from one or more diseases associated with adequate water and sanitation [18]. Whereas between one third and one half of the solid waste generated within most cities are not collected, and less than half of the cities have urban environmental plan [16]. This has in turn made slum dwellers constitute 36.5% of the urban population in developing countries, with percentage being as high as 62% in Sub-Saharan and 43% in Southern Asia [17]. This has led to increasing urban poverty from rapid rural–urban rush, alongside inability to access sustainable and affordable housing. In developed countries there is high emission of CO_2, in terms of energy consumption, 60% of the world's electricity is consumed by residential and commercial building. Space heating accounts for 60% of residential energy consumption and water heating for 18%, but different strategies are relentlessly developed to alleviate these issues [13].

3. Lessons from world experiences

Balancing the equation between societal, economical and ecological issues could be quite difficult but it's not unachievable [19]. Different countries, more of developed countries, are adopting different approaches for sustainable construction and have set different priorities depending on their economic conditions. Nations with high economic growth are developing sustainable buildings making use of latest technologies and innovations [13].

In the order of countries with best sustainability measures to the poorest, Finland among many other countries was voted best in the quality of natural environment, with very innovative and eco-friendly built environment [20]. Finland was a pioneering country of energy efficiency, after the energy crisis over three decades ago. Her goal of sustainable building was to build a house with as low energy use as possible and by utilizing the best available technology [21]. For building designs, solar heat from solar collector, geothermal heat from under building drilled borehole, solar electricity with high insulation level as well as low energy windows were adopted. Some houses adopted the use of locally available biomass to generate electricity and heat with a very low carbon footprint. Cutting edge LED application, developed through interdisciplinary research is very useful in energy saving as shown in **Figures 1** and **2**. Local waste is managed with interconnected conduits that carry them away for proper disposal and recycling [21].

India, a developing country, on the other hand, has 12.5% of all deaths caused environmental issues such as air pollution [20]. Most of research experts do not support government's approach towards environmental protection policies,

Figure 1.
Ecofriendly house in Finland. Source: The culture trip.com.

Figure 2.
House built with energy efficient materials. Source: Skimbaco lifestyle.com.

meanwhile more than half of the experts are of the opinion that the total population is not interested in environmental issues [20] as shown in **Figures 3** and **4**. Energy resources used in India for manufacturing and transporting building materials has green high gas emission and related environmental issues. The sustainable use of these materials is also a call for concern as about 300metres depth of fertile top soil of the entire country have been said to be consumed for burnt brick clay bricks production in the next six decades if growth rate is assumed as 5% [22].

However there are improvements in some countries that were time past taking a down toll in sustainability. In United Arab Emirate, ABU-Dhabi as a case study, innovative technologies that are consistent with overall state expansion and growth over the years have been adopted in the creation of sustainable cities in Dubai and Abu-Dhabi [23]. A pearl rating system was established to achieve sustainability of housing throughout its life cycle from design to construction and operation. This

Figure 3.
House in Mumbai, India. Source: India ink, New York times.

Figure 4.
Aerial view of slums in India. Source: Hindustan.com.

has introduced green building norms and regulation, minimizing water and energy consumption; improving waste recycling and using local, environmentally friendly materials for construction [23].

Moreover, despite the housing crisis in the UK, environmental stewardship and long-term sustainability is the foremost in the minds of individuals, governments and businesses [13]. Environmental impact is factored in to safeguard the environment; therefore environmental sustainability is always embedded in every level of construction. Environmental stewardship is seen as not just creating great places and improving local environment, but having regard for the global climate as well by reducing energy and material consumption [13].

Since sustainability is all about meeting today's needs without compromising the needs of future generations, sustainable housing must target economic, social and environmental sustainability ensuring that houses are affordable, accessible and posing no harm to the environment [24].

4. Challenges and way forward to sustainable housing in developing countries

Challenges plaguing sustainable housing in developing countries are numerous and diverse and will be discussed based on four dimensions of sustainability.

4.1 Social challenges

4.1.1 Urbanization

One of the greatest problems facing sustainable housing is rapid urbanization which is common among developing countries. Urbanization takes a huge toll on the environmental resources which are mostly non-renewable resulting in shortage and limited access to basic amenities such as potable water, roads, waste disposal facilities, sanitation facilities and electricity [25–27].

4.1.2 Slum development

The movement of growth towards metro cities and mega cities poses a greater challenge to provide housing in the already saturated urban areas, transforming them into areas of crowded habitations without basic amenities, thus giving rise to urban slums [28].

4.1.3 Policy and legislations set back

A study by Ibem and Azuh supported the fact that weak sociopolitical structure and institutional frameworks are the banes of failed housing policies and its implementation mechanisms. As a result, growth in the quantity and quality of the housing stock in the country remains poor due to non-existent or non-functional standard and norm system for accreditation of green buildings [29].

4.2 Environmental challenges

4.2.1 Climate change

Some of the resulting effects of urbanization are an overwhelming increase in the number of high energy consumption buildings, increased vehicular density, leading to more consumption and burning of fossil fuels and release of greenhouse gases. These activities are major culprits of climate change.

4.2.2 Land resource and vegetation degradation and destruction

Due to excessive pressure on land resource and vegetation arising from the need to construct more buildings, there is a huge loss of green covers leading to decrease in green belts, biodiversity loss, and loss of other green infrastructure [30].

4.2.3 Climatic conditions

Conventional sustainable housing methods may not be applicable in places with harsh climatic conditions and water scarcity.

4.2.4 Topography

The terrain of an environment can be challenging for sustainable housing. Landslides is a risk associated with mountainous and hilly areas, sudden occurrence can be very fatal for residents.

4.3 Economic challenges

Economic instability implies higher cost of living in developing countries like Nigeria, financing of housing development is a major challenge. One of the numerous reasons is that access to housing credit; mortgage or loans are associated with to high interest rates and collateral guarantees which are beyond the reach of majority of the citizens. Empirical data around the globe highlights Nigeria's homeownership rate in urban areas is around 10% compared to 97% for Romania; 74% for Brazil; and 62% for South Africa [31].

4.3.1 Affordability

The major housing challenge in developing countries is decent affordable houses this is due to rapid urbanization caused by increasing population and rural–urban migration [32]. A major consequence of this influx in population is an intense demand for affordable housing not necessarily sustainable ones [31].

Implementation of sustainable housing in developing countries will require a relatively high initial investment, which makes it quite difficult to take off [2].

4.4 Cultural challenges

Cultural sustainability takes into consideration cultural worldviews and values, norms and traditions, as well as lifestyles and behaviors of occupants, communities and society, thus supporting the dignity of communal life. Culturally appropriate and responsive built environments are an important dimension of sustainable housing. Communities may resist sustainable housing projects that interfere with their culture and traditions.

5. Benefits of sustainable housing in developing countries

Considering the benefits of sustainable housing over the challenges could help in charting a new course towards sustainability in most developing countries.

Asides the fact that sustainable housing is a means of achieving several global goals, there are multiple benefits of sustainable housing and they can be divided into three groups.

1. environmental benefits

2. Economic benefits

3. social benefits

5.1 Environmental benefits

Just as the housing sector plays a crucial role in the global environmental crisis, sustainable housing also provides great possibilities of mitigating global climate change through:

Reduced environmental footprints from housing, in terms of energy and associated GHG emissions, water, land and material use [33]. Entec in his study for Defra (2004) reported that with excellent standard sustainable houses about 200 million tonnes of CO2 emissions per year could be avoided, eliminating external costs of the order of £2.9 billion. Also, with improved housing, the pressure on environmental components such as water, air and soil in slums are considerably reduced to minimal levels [33].

Water Efficiency: In regions where water availability is a limiting factor to development, the scale and pace of development could increase if less water is used. There is therefore a greater likelihood of more homes being constructed more quickly and easily if they are water efficient [18, 25]. Sustainable buildings manage water in a more effective and environmentally friendly manner. Such as systems that recycle water e.g. harvesting rainwater for toilet cleaning. Rain water harvesting associated with aspirational standards would have additional environmental benefits of controlling storm-water run-off and reducing flood risk.

Reduced waste going to landfill or being incinerated: The use of renewable sources and materials employed in the construction of sustainable housing minimizes waste generation. In a study for Defra, it was reported that £19 million can be saved from waste going to landfills, if standard sustainable homes are built. Products such as demolition debris, sand and burnt coal can be used with excellent environmental and esthetic results [25].

Mitigate environmental hazards and promote biodiversity by improving green spaces: If developments incorporate more green spaces, there will be minimal effect on biodiversity thus, promoting diversification. Green areas also help in carbon sequestration, mitigate heat waves through their cooling effect, prevent soil erosion and the need for piped drainages by acting as soil covers [34].

5.2 Economic benefits

Employment generation and improved standards: New employment can be created through the housing sector, which is especially important in the context of developing countries. Jobs can be created through new construction and retrofitting, production of energy efficient or recycled materials and though renewable energy and technologies related to it [35]. In 2014, Canada's green building industry generated $23.45 billion in GDP and represented nearly 300,000 full-time jobs [36].

Through energy and water efficiency, household can save costs on utility bills. Energy efficient buildings are the most cost-effective way to battle fuel poverty in households [37].

5.3 Social benefits

Reduce medical bills: use of sustainable materials, safer building materials, design and components increases the quality of life of individuals and the community and reduction in cases arising from sick buildings [19].

Increase homeowners'/users' productivity: Qualities such as better indoor air quality, effective noise control mechanisms can improve performance and enhance productivity. Various studies have shown that sustainable buildings increase occupants' performance and wellbeing [38–40]. A recent study in Australia also showed that occupiers' cognitive scores were 61% higher for green buildings compared to standard buildings [38].

More sustainable and socially inclusive urban growth: sustainable environment promotes cultural and neighborhood integration. Communities create a sense of place, neglected or abandoned locations will result in neighborhood instability and a loss of economic activities [40].

6. Recommendation and conclusion

The best approach to achieve sustainable and affordable housing is to embark on a comprehensive approach spearheaded by appropriate standards and regulations and capacity building schemes which will oversee and ensure Strategic planning, cooperation and participation of stakeholders, supportive institutions, economic instruments and financial incentives.

6.1 Strategic planning

Strategic planning is important for ensuring efficiency and effectiveness in policy design and implementation. It enables the aspirations of different stakeholder groups to be included in a common vision that gets translated into objectives [1]. A comprehensive and clear cut plant would determine the success of a sustainability project. Goals, targets, key performance indicators and deadlines are indices that will help measure level of progress. Also, it is imperative that pilot studies are carried out which will be scaled up following their success.

6.2 Participation and cooperation of stakeholders

It is important to educate and thoroughly sensitize all stakeholders on the necessity of sustainable housing and involve them in every stage from the planning phase to scaling up. A city based approach which is a combination of "bottom up" approach which borders on self reliance and "top down" approach bordering on support will be ideal because it encourages learning and knowledge sharing platforms between stakeholders and communities, motivates communities to take "ownership" of finished products and reduce conflict between groups [1]. Sustainable housing implementation requires strong support from the government (leaders), communities and the housing industry.

6.2.1 Leadership

It is imperative that sustainable housing initiatives are backed by a clear and strong leadership and political will, as they are essential components to successful public interventions. Strong leadership is required to bring the various groups and stakeholders together also to initiate a process for collaborative decision-making, Review and adapt existing planning legislation and regulatory planning controls.

6.2.2 Communities

It is very important to involve the communities from the planning phase to the implementation phase. Such approach will build learning and knowledge sharing platforms between stakeholders and communities, encourage communities to take "ownership" of finished products and ensure that the ideas, beliefs and traditions of the communities is taken into account. All these enhance occupants' feelings of belonging. Other actors such as built environment professionals and manufactures are responsible for ensuring that housing design criteria, materials or product specifications are environmentally responsive [41].

6.3 Regulation and standards

Sustainable housing should be long term, requiring a healthy and clear institutional setting that allows all stakeholders to play their part without fail [35]. To achieve this, the government which is the key stake holder by virtue of the crucial position which they occupy in the country has to introduce a national housing strategy and a strong legislative framework. Policies and governance structures as it pertains to various developing countries has to be reformed, strategic investment, research and training programmes launched. All these will help full institutionalization of sustainable housing policy in both governmental and non-governmental structures that is not subject to changes in government [34].

6.4 Financial incentives and economic instruments

Mobilization of financial resources by advocacy with government institution, involving the private and public-private partnership for the implementation of sustainable housing projects.

Provide funding to support emerging businesses and innovative technologies.

Low-income households, especially in developing countries, often do not have the initial capital needed for building sustainable housing or can face problems of paying back loans [10], therefore it is important to make available financial support that provide cheap credit [42].

Creating jobs for locales through sustainable construction projects.

6.5 Capacity building

Building capacities of institutions and actors is crucial for scaling up sustainable housing practices. Capacity development refers to the development of the whole housing sector whereas capacity building is targeted at improving skills of stakeholders through education, skill acquisition programs, collecting and sharing data bank of best practices. This is important for recognizing crucial needs, develop capacities to implement housing that takes care of these needs and scale up sustainable practices [34, 35].

6.6 Pilot studies and scaling up

Embarking on pilot projects addressing sustainable housing is very important to test the viability of the project before executing it at full scale. It helps manage risk and reveal serious deficiencies or errors in the plan before committing major resources to the project. However, it is crucial to scale up sustainable housing practices to meet the massive housing demand that exists, and that will be in demand over the coming decades as expected of developing countries. Scaling up requires three key ingredients; a supportive institutional and regulatory environment, timely monitoring and evaluation mechanisms, and appropriate capacity development of the housing sector and capacity building of housing sector actors [35, 43]. Achieving sustainable housing in developing countries could still remain a mirage if necessary measures such as the recommendations listed above still remain out of place.

Author details

Ibiwumi Saliu[1,2]* and Evangelisca Akiomon[1]

1 Department of Environmental Health Sciences, Faculty of Public Health, University of Ibadan, Ibadan, Oyo State, Nigeria

2 Blue-gate Public Health Initiative, Ibadan, Oyo State, Nigeria

*Address all correspondence to: saliuibiwumi@yahoo.com

References

[1] UN-Habitat (2012). Urban Patterns for a Green Economy Leveraging Density. UNON, Publishing Services Section, Nairobi.

[2] Nguyen T Lang. (2013). Cultural and Social Elements in the Development of Green Architecture in Vietnam. Procedia - Social and Behavioral Sciences 85:16-26.

[3] Queensland Government Department of Public Works. (2008). Smart and sustainable homes design objectives. Brisbane: State of Queensland.

[4] Oyediran, O.S. (2013). Institution and housing development: mirage, magic and miracle of low cost housing in Nigeria. Vol 57.

[5] Okpala, D.C. (1994). Financing housing in developing countries: A review of the pitfalls in the development of formal housing formal housing finance systems. Urban studies, Vol 31. No 9.

[6] Okeyinka, Y.R. (2014). Housing in the third world cities and sustainable urban development. Developing countries studies. Vol 4; 8:112-120.

[7] Environment Agency (2005) Sustainable homes – the financial and environmental benefits.

[8] UN-Habitat (2012). Sustainable housing for sustainable cities: A policy framework for developing countries. ISBN: 978-98-1-132488-4.

[9] Ingrao, C. Messineo, A. Beltraino, R. Yigiticanlar, T. Ioppolo, G. (2018). How can lifecycle thinking support assessment applications for energy efficiency and environmental performance. J clean. Prod. 201: 556-569.

[10] Ioppolo, G. Cucurachi, S. Salovione, R. Shi, L. Yigitcanlar, T. (2019). Integrating strategic environmental assessment and material flow accounting A novel approach for moving towards sustainable urban futures. Int. J. Life cycl Asess. 24 (7): 1269-1284.

[11] Mulanga M (2011). Presentation in the UN-Habitat International Expert group meeting in Nairobi, Kenya: Constrains and Opportunities of Sustainable Housing in Low-income Urban Settings. International Institute for Environment and development.

[12] Chan, A.P., Adanbre. M.A. (2019). Bridging the gap between sustainable housing and affordable housing. The required critical succss criteria. Build Environ. 151: 112-125.

[13] Farinoosh, A. (2020). Trends in sustainable housing. Https://www. Ecomena.org sustainable.Housing/

[14] International Monetary Fund. "Frequently Asked Questions: World Economic Outlook (WEO)." Accessed May 21, 2021 https://www.investopedia. com/terms/p/percapita.asp

[15] Arnott, R. (2008). Housing policy in developing countries. The importance of the informal economy. working paper. No 13 commission on growth and development.

[16] UN-Habitat. (2008b). the state of the world: cities. UN. Habitat. Nairobi.

[17] UNEP. (2002). United Nations Environmental programme. Global environment outlook 3: past, present and future perspectives. Earth scan. London.

[18] Defra (2004) Cross-Government Review of Water Affordability Report.

[19] Spiegel, R., and Meadows, D. (2010). Green building materials: A

guide to product selection and specification. Hoboken, NJ: John Wiley and Sons.

[20] Samanda, G. 2020. The best and worst countries for sustainable living. Thestreet.com

[21] Sustainable Urban Development. (2011). World_class sustainable solutions from Finland. Https://www. Sustainable urban development.fi/en.

[22] UN-Habitat. (2017). Sustainable social housing in India. Technical Report.

[23] ADPC, Abu Dhabi planning council. (2010). Pearl building rating system; Design and construction version 1.0. https://www.solarthermalworld.org/sites/default/files/news files/2015-05-04/estidema construction_rating_ certificate pdf.

[24] Choguill CL. (1994) Sustainable housing programmes in a world of readjustment. Habitat International. 18:1-11.

[25] Environment Agency (2003) The Economics of Water Efficient Products in the Household.

[26] Fuel Poverty Advisory Group (2004) Second Annual Report. Available from www.defra.gov.uk/environment/energy/fuelpov/fpag/pdf/fpag-annrpt-03-04.pdf

[27] Gallions Ecopark in Thamesmead (2004) Ecopark Financial Analysis Report. Available from www.gallionsecopark.co.uk/home.htm

[28] Vijai Shanker Singh and Deep Narayan Pandey. (2012). Sustainable Housing: Balancing Environment with Urban Growth in India. RSPCB Occasional Paper no 6, RSPCB.

[29] Edo Oga Ojoko, Halimat Omuya Abubakar, Oga Ojoko, Elias O. Ikpe (2016). Sustainable Housing Development in Nigeria: Prospects And Challenge. Journal of Multidisciplinary Engineering Science and Technology.

[30] United Nations. (2017). New urban Agenda third world conference on housing and sustainable urban development, Ecuador.

[31] Housing Finance Information Network (HOFINET). "Home-ownership" 09.03.2016.

[32] National housing strategy. (1991). The affordability of Australian housing. Canberra. Australian Government public service.

[33] Entec (2004) Study into the Environmental Impacts of Increasing the Housing Supply in the UK. Defra. Available from http://statistics.defra.gov.uk/esg/reports/housing/default.asp.

[34] Golubchikov Oleg and Badyina Anna. (2012). Sustainable Housing for Sustainable Cities: A Policy Framework for Developing Countries. Nairobi, Kenya: UN-HABITAT, 2012, Available at SSRN: https://ssrn.com/abstract=2194204

[35] UN-Habitat. (2012). Going green. A handbook of sustainable housing practices. Nairobi. United Nations Human Settlement Programme.

[36] Canada Green Building Council/ The Delphi Group, 2016.New CaGBC market impact report demonstrates green buildings are an engine of economic growth in Canada.

[37] EST (2003) Energy Efficiency Best Practice in Housing. Energy Saving Trust.

[38] Allen, J. G., MacNaughton, P., Satish, U., Santanam, S., Vallarino, J., and Spengler, J. D. (2016). Associations of cognitive function scores with carbon dioxide, ventilation, and volatile organic compound exposures in office workers: A

controlled exposure study of green and conventional office environments. Environmental Health Perspectives, 124(6), 805-812. doi:10.1289/ehp.1510037

[39] Hamilton, A. B. (2015). USGBC Green Building Economic Impact Study. Available at https://www.boozallen.com/e/media/press-release/study-finds-green-construction-is-major-us-economic-driver.html

[40] Olanrewaju AL, Tan SY, Abdul-Aziz A-R. (2018). Housing providers' insights on the benefits of sustainable affordable housing. Sustainable Development.;1-12. doi:10.1002/sd.1854

[41] Reffat, R. (2004) Sustainable construction in developing countries, in the Proceedings of First Architectural International Conference, Cairo University, Egypt.

[42] Throsby D. (2005), On the Sustainability of Cultural Capital, Research Papers 0510, Macquarie University, Department of Economics.

[43] UN-Habitat (2012). Going Green: A Handbook of Sustainable Housing Practices, UN-Habitat: Nairobi.

Section 3

Habitation Improvement - Structure, Procedures and Legislation

Chapter 7

An Explorative Perspective on the Resilience in Neighborhoods in the Netherlands

Hendrik Marten Koolma
and Catharina Frederika van Dreven

Abstract

In this contribution we intend to shed light on the social dimension of sustainability. Thereby, we follow adhere to the goals for sustainable development as issued by the United Nations. The majority of the goals concerns social topics in the national societies. We intend to research in what way resilience lead to social progress and, through that, to sustainability. The attainment of the goals is hindered by the phenomenon of negative spillover effects between countries. In addition, we assume the presence of intra-national spillover effects which means that advancements in some areas are accompanied by deteriorations in other parts of the country. The negative effects concentrate and accumulate disadvantages into distinct neighborhoods. Inhabitants who try to overcome the disadvantages can be fostered in their resilience by beneficial interventions and conditions. However, they will be faced with impeding interventions and conditions as well. A case in the Netherlands shows that by resilience inspired policy reforms rather inflict additional problems on disadvantaged inhabitants. A set of interviews with key persons sheds light on the question of how resilience in the neighborhoods could be fostered instead. It requires support of individuals who take the initiative to improve the living conditions of themselves and others. The processes need some guidance by professional workers who do relate to everyday live in the neighborhoods, and who build relationship op the basis of trust. In their situational work, they need to be enabled and backed by their superiors.

Keywords: resilience, sustainable development goals, social ecology, trust, distrust, social comparison

1. Introduction

The chapter will set out the theoretical foundation for the concept of resilient neighborhoods as the core of social sustainability. Resilience is more than bouncing back after an experience of adversity. It paves the way for social progress. However, social progress is not only promoted but also impeded by social and institutional processes that shape interventions and create conditions. These processes can undermine the sustainability of neighborhoods and inflict losses of social and economic capital. At the same time, the neighborhood might be the level on which conflicting processes still are comprehensible and amenable.

This chapter has three sections. The next session contains an elaboration on the concept of resilience. A combination of personality research, developmental psychology, and social ecology place resilience in a person-in-environment system in which personals attributes and social institutions interact. In this adaptive system counterintuitive processes occur due to the interplay of trust and distrust and to social comparison. In this way, we point out which social and institutional processes shape interventions and conditions encompassing resilience. The section concludes with the research question and a conceptual model. In Section 3, national reports on policy reforms in the social domains will be analyzed on the relation between by resilience inspired state policies and the outcomes in terms of equality and justice. The sustainable development goals of the United Nations provide a frame of reference for the topics. In Section 4, the role of resilience in Dutch society is discussed on the basis of 10 semi-structured interviews with a selection of key persons involved in sustainable cities and initiatives in disadvantaged neighborhoods. Overall, an explorative study is conducted in order to probe and advance the approach of the social sustainability of neighborhoods and cities.

2. Theoretical framework

When one thinks about sustainability of neighborhoods, the first reflex will generate associations with technical innovations and the necessity of coping with the threats of climate change. In this contribution, the scope is on a social approach of sustainability, encompassing scientific disciplines like social ecology, social systems theory and psychology. We hold the presupposition that the ecological success of man in comparison to other animals can be attributed to a superior tool making ability, whereby the processes of social tool making (knowledge sharing, norms, institutions, and joint moral reflection) are lagging and fail to control, align, and fairly distribute the benefits and humanitarian costs of technical progress. Here, we advocate for a broader approach of sustainability in order to develop the potentialities of man is his or her environment. Thereby, a key mechanism is the resilience of inhabitants of neighborhood, cities and smaller settlements.

2.1 Sustainable development goals and spillover effects

A relation between sustainability and resilience of neighborhoods is found in the United Nations' framework of Sustainable Development Goals (SDGs). The goals were established in a general meeting of the United Nations on September. 15, 2015 and entitled 'Transforming our world: the 2030 Agenda for Sustainable Development' [1]. In this international agreement, goals are extended to social and economic dimensions of human society by putting social sustainability in the centre [2]. We attempt to follow this line of argument by taking neighborhoods as the central point of view, embodied by goals 11: 'Make cities and human settlements inclusive, safe, resilient and sustainable'. On the geographical level of urban neighborhood and rural settlements, the advancements, stagnations, or deteriorations on other goals will manifest themselves as well (**Table 1**).

Of this goals, numbers 6, 7, 12, 13, 14 and 15 are directly related to natural and cultivated environment. The targets of these goals are lagging far behind and seem impossible to achieve in due time frame [2]. The check marks in the rightmost column indicate that the results on the respective goal facilitates or hinders the achievements on goal 11.

In the trajectories of transformation, geographical spillover effects occur when countries' actions have a positive or negative effects on other countries' ability

Description Sustainable Development Goals	Social dimensions	Physical dimensions	Goals which may affect realization of Goal 11
Goal 1. End poverty in all its forms everywhere	✓		✓
Goal 2. End hunger, achieve food security and improved nutrition and promote sustainable agriculture	✓		✓
Goal 3. Ensure healthy lives and promote well-being for all at all ages	✓		✓
Goal 4. Ensure inclusive and equitable quality education and promote lifelong learning opportunities for all	✓		✓
Goal 5. Achieve gender equality and empower all women and girls	✓		✓
Goal 6. Ensure availability and sustainable management of water and sanitation for all		✓	
Goal 7. Ensure access to affordable, reliable, sustainable and modern energy for all		✓	✓
Goal 8. Promote sustained, inclusive and sustainable economic growth, full and productive employment and decent work for all	✓		✓
Goal 9. Build resilient infrastructure, promote inclusive and sustainable industrialization and foster innovation		✓	
Goal 10. Reduce inequality within and among countries	✓		✓
Goal 11. Make cities and human settlements inclusive, safe, resilient and sustainable	✓	✓	Main subject
Goal 12. Ensure sustainable consumption and production patterns		✓	✓
Goal 13. Take urgent action to combat climate change and its impacts		✓	✓
Goal 14. Conserve and sustainably use the oceans, seas and marine resources for sustainable development		✓	
Goal 15. Protect, restore and promote sustainable use of terrestrial ecosystems, sustainably manage forests, combat desertification, and halt and reverse land degradation and halt biodiversity loss		✓	
Goal 16. Promote peaceful and inclusive societies for sustainable development, provide access to justice for all and build effective, accountable and inclusive institutions at all levels	✓		✓
Goal 17. Strengthen the means of implementation and revitalize the global partnership for sustainable development	✓		

Table 1.
Sustainable development goals United Nations, Department of Economic and Social Affairs Sustainable Development [1].

to achieve the SDGs. Such international spillovers are pervasive [3], including examples in which growing wealth in some countries counteract the progress on the SDGs in other countries [3]. The concept of spillover effects is the equivalent of externalities in economics. It is based on the mechanism that agents achieve to book revenues on their own account, while passing on the costs to the private accounts of others and to the common account of society. The mechanism drives depletion of natural resources and block the way of people to prosperity, as asserted 50 years ago by Hardin and the Club of Rome [4, 5]. It is, in Hardin's words, a tragedy that we fail to manage the commons [6] and to exstablish a social order that control spillover effects. The noxious effects are driven by extra-ordinary profits of frontier agents and their opportunistic followers. The spillover effects occur in transnational trade and in the relationships between powerful and less powerful nation states [3].

Another concern is that the efforts on physical sustainability, though very urgent, will be done without consideration of the social aspects of sustainability [3]. Thus, negative spillover effects can emerge in the relationship between goals, an argument to integrally plan and implement efforts on a geographical scale on which spillover effects are communicable and manageable. In addition, we presume that on the intra-national level spillover effects occur as well. Progress on the SDG's in some regions of countries and cities are made at the 'expense'of other parts of the country. An accumulation of problems and unbalanced distribution of adversity and generation of environmental stressors may be result.

2.2 The concept of resilience and its implications

In recent years, resilience of neighborhood has become a buzz word. A Google search scores 219 million hits. In the Dutch public administration and societal sectors, resilience gets on the agenda [7, 8]. We want to take up this concept in a way that it can be assigned to social progress. This conception is proposed in motivational and development psychology by Block and Block [9], Carver [10], and Luthar et al. [11]. These authors deviate from the mainstream of research in which resilience is specifically related to overcoming events of adversity [12, 13]. Resilience does not only involves bouncing back to the original state [14, 15], but moreover, to an improved level capability and adjustment to the social and physical environment.

Carver holds that this surpassing of the original state, termed psychological thriving, reflects gains in skills, knowledge, confidence, or a sense of security in personal relationships. He assumes further, that 'personal differences in confidence and mastery are self-perpetuating and self-intensifying' ([10], p. 245). This an example of psychological reinforcement. People are more likely to surpass to an improved level, when they map and oversee the situation of the downturn and conceive a developmental challenge for themselves [10]. When unexpected problems occur during some action, one is forced to learn more than one has learned before. Hence, people develop new courses of action, they will be more flexible in coping with new, unknown events. These flexibilities can even build on each other ([10], p. 252). However, the transfer of the attainments by resilience from one domain to another is not a matter of course [16].

In his explanation, Carver emphasizes that resilience requires both processing of negative information on the situation as a search for opportunities that helps to find a way out. In a social ecological account, resilience is predominantly related to opportunities for personal growth ([13], p. 14). In contrast, another social ecologist definines resilience as 'as a relative resistance to environmental risk experiences, the overcoming of stress or adversity, or a relatively good outcome despite risk experiences ([17], p. 34). Following Carver's argument, the analysis and understanding of the risk experience is a necessary element of resilience, because it promotes learning of the interaction with the social and physical environment and prevents a falling back into similar situations of adversity and stress. In this purport, we would like to substitute 'despite risk experience' by 'due to understanding of the risk experiences' in the definition of Rutter.

Although fostering communities do contribute to resilience of individuals, we prefer to study resilience *in* neighborhoods instead of resilience of neighborhoods. Rutter states: 'It is certainly appropriate to conceptualize influences at a community level, but resilience as an outcome is still better viewed in terms of individual outcomes' ([17], p. 35).

Ungar criticizes a simple model of an individual who overcomes adversity by his personal faculties for resilience. Instead, he has developed a formal model in which social institutions, like family, schools, and commuities determine the conditions under which a person succeeds to be resilient [13]. In contrast, we hold thatIndividual differences in personality and life histories do matter as well (see [9, 17]). When variables and measures are carefully choosen, both individual psychological properties and environmental circumstances determine in interaction whether or not and to which degree adversity is successlully overcome by persons living in stressful and disadvantaged neighborhoods [18]. Still, it is necessary to distiguish between resiliency as property of personality and resilience as process [13]. Here we chose to study resilience as a process within a holistic person-in-environment system [19].

It is remarkable that a forty year old conception of personality appears to be a variable that significantly contribute to resilience in neighborhoods [9, 18], where other personality indicators failed to be discriminating in results [13]. Block and Block have based their personality inventory on two conceptsL: ego-control and ego-resiliency. Ego-control is conceived as a continuum between the two extremes undercontrol and overcontrol. In the extremes, overcontrol is associated with diseases like depression, whereas undercontrol is related to expressions of aggression [20]. Ego-resiliency refers to the dynamic capacity of an individual to modulate his or her modal level of ego-control, in either direction, as a function of the demand characteristics of the environmental context" ([9], p. 48).

Since 1980, the study of personality research and its biological foundation has made advancements. Although the terms undercontrol and overcontrol suggest a straight linear dimension, ego-control consists of two separate dimensions and independent brain systems [21]: the behavioral-activation system (BAS) and the behavioral inhibition system (BIS). The two systems regulate the approach/avoidance reflex onto attractive respectively threatening objects and subjects in in a person's psychical and social situation. People show differing proportions of BIS and BAS [22], whereby a slightly higher level of BAS presumably supports resilience. People need some optimismic inclination in the approach of new competencies [23].

When the levels diverge considerably and the environment provides impactful aversive or attractive stimuli, either BIS or BAS will be suppressed, resulting in disinhibition repsectively deactivation. When suppression of BAS or BIS occurs, people will hardly process either negative or positive information anymore. This two dimensional brain process explains the finding by Driessen and Beerenboom that dissatisfying urban living conditions like housing defects and serious disturbances around one's house like neighbor nuissance impede the appreciation of qualitative assets and novelties in the neighborhood [24].

In addition to Gray, Boyce and Ellis have found an u-curved biological sensitivity for negative (threatening) and positive (protective) environments. People gifted with a bi-directional sensitivity and grown up with a mixture of modest encouraging and disappointing experiences, are more flexibile and have higher tolerance levels for unexpected experiences [25].

Finally, ego-resiliency is more of less equivalent to self-regulation [13]. Self-regulation is a motivational, though predominantly unconsciously operating resource [26]. Self-regulation helps to downregulate negative affect during trail-and-error cycles. With each succesfull iteration, the resource expands. However, enduring stress, for instance by working on a job beyond one's competence, the resource can be depleted [27]. At any moment, people suddenly breakdown personally, in their social relations, or deviate, at the spur of the moment, from norms of their occupational environments [28].

2.3 Interventions fostering and impeding resilience

In the modeling of Ungar, social institutions shape the positive conditions for resilience. In addition, we contend that social institutions may impede as well person's faculties for finding successful responses to adverrsity and stressors. For instance, Ungar discusses a study in which accultured migrants, in their ambition to participate in society, self-report on their well-being inferioir to less accultured migrants in the same neighborhood. He contributes this paradoxal finding to social comparison processes [13].

Too protective and too neglective parenting or care-giving both deprive children from learning by success and failure after taking modest risks [29–31]. Veroff and Veroff [31] illustrate their argument with an example that can be used to understand the choice of appropriate care and subsequent interventions. A young child is reaching for an object but does not succeed and becomes frustrated. A protective parent will get it and give it to the child just the way. A resilience fostering parent waits for an inattentive moment of the child, then places the object within reach, and encourages the child to try again. Veroff and Veroff propose the concept of pacing, implying that the balance between protection and challenge is adjusted to the person's stage of development and learning speed [31].

We propose to build up an argument from a social system perspective. The institutions are elements in the person-in-environment system that shape conditions for the person and seek to intervene in his or her attempts to overcome disadvantages. Because of the complexity of a person-in-environment system, they are forced to follow strategies that reduce its complexity. An obvious strategy, is abstracting from the paradoxical fact that interventionists themselves are a part of the system as well. By doing so, the intervention is mediated as it were performed in a lower order system like a control-regulated home heating system. The outcome of this strategy are likely to be rather unpredictable because of fact taht the subjects perceive this strategy and attribute intentions to it. In his treatise on rationality, Luhmann [32] distinguishes the next complexity reducing strategies:

- Goal-means rationality
- Opportunity oriented rationality
- Value oriented or intention-guided rationality
- Problem-to-solution rationality

Goal-means rationality is based on hierarchal way of thinking, has an almost unlimited potential for expansion in detail, but is inflexible [32] and based on the antipode of resilience, namely anticipation [33]. The application of this kind of rationality is not per se conflicting with resiliece. For instance, legislation can safeguard fundamental rights, policy programs can mobilize and distribute resources, resources can be warranted by evaluation procedures, and schooling standards may offer challenges to deprived children. Opportunism seemingly associates with challenges, but pure opportunism detaches from the original situation, and all occasional side effects and long-term consequences [32]. Resilience is not served well by sole attention to assets while neglecting the exposure to stressors and risks [13, 17]. Value oriented and problem-to-solution rationality seems mostly suited for fostering of resilience in persons. However, both require an interchange with the value configurations, the problem perceptions and the opportunity preference of the addressed person. The adaption of the care intervention to the needs and hopes of a client requires an unconditional and unpremeditated exchange of information, a requirement that would be met by interpersonal and institutional trust.

Trust is a complexity-reducing interaction mechanism as well [34]. In elaboration of Luhmann, Lewicki and colleagues have proposed a two-dimensional conception of trust and distrust [35]. Trust is operent when agents are willing to approach eachother in social relationships, while distrust relects an avoidant attitude. Thr tendidencies can coexist in a relationship. In a matrix the authors render the four combination of trust (high/low) and distrust (high/low) (see **Figure 1**). High trust combined with low distrust shapes the condition for high-value congruence, interdependence, pusuit of opportunities and new initiatives. Low-trust in combination with low-distrust limites the assets of the relationship to bounded, árm-length transactions, while the combination of high-trust and high-distrust induces a fragmented and calculating exchange [35].

To our opinion, the latter two options reflect goal-means and opportunistic strategies of institutions. The combination of low trust and low distrust reflects the attidude of courteous professionals and neutral public officials. In their benevolence, they try to be objective, and do not invest in a relationship with clients or citizens and fail to adjust to personal specificities. The combination of high trust and high distrust Is found in the behavior of opportunistic agents. Like salesmen, they build relations in order to retrieve information on opportunities and risks, but preemtively take the opportunities while averting the risks to the counterpart. Neither of both strategies is well-suited to foster resilience.

In a state of low-trust and high-distrust, undesirable eventualities are expected and feared of, harmful motives are assumed, and the behavioral expressions diverge

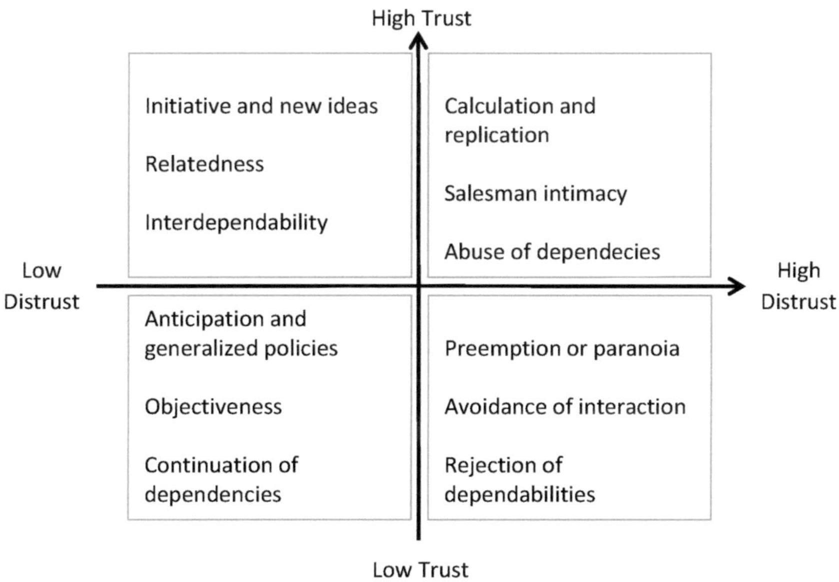

Figure 1.
Elaboration on the trust-distrust matrix of [35].

between preemption and paranoia. The mental state of a distrusting person Is characterized by fear, skepticism, cynism, wariness, and watchfulness [35].

We have added the antonyms anticipation and resilience of Douglas and Wildavsky [33] and the development from infant's dependency to interdependence of Veroff and Veroff [31].

The two-dimensional model of Lewicki and coworkers is supported by neuroimaging research [36], and can be regarded as an instance of the BIS-BAS interaction. Dimoka's study relates distrust to the fast, evolutionary older brain regions, while trust is regulated from slower parts of the modern brain. The finding is in accordance with common sayings about trust and distrust. Anyway, building of trust in a care-giving relation will often be a laborious affair.

2.4 Research question, modeling, and leveraging processes

Summarizing, we come to the next question for our research:

Which conditions and interventions foster respectively impede the resilience of people in disadvantaged neighoods in their search for outcomes that contributes to their personal development and reshape conditions and interventions to the benefit of themselves and of others? In addition, in what way do impeding conditions and interventions reinforce the disadvantages of people in the neighborhoods? (**Figure 2**).

The model for resilience in neighborhoods is conceived as a complex adaptive system. There is a public belief in leverage points. That are places in social and ecological systems were a small intervention may cause a big change [37]. Experienced adaptive systems researchers warn for the phenomenom that members within such a systems are capable of pointing out a leverage point, but choose intervention that affect change in the wrong direction. Leverage points tempt to be counterintuitive [37].

The trust/distrust matrix of Lewiecki and colleagues embodies a counterintuitive leverage point. Interventions from low distrust can be countereffective while interventions from high trust can be disadvantageous when crosswisely matched with the options of the other dimension. Second, we put to the fore the generic process of social comparison. Humans seek peers in order to compare abilities, opinions, and attributes. However, as opinions are more easily formed than abilities are improved, opinions come to substitute the evaluation of the abilities. As consequence, homogeneoused groups become indifferent to outcomes, inaccesible to newcomers with other opinions and attributes and ignorant of information from

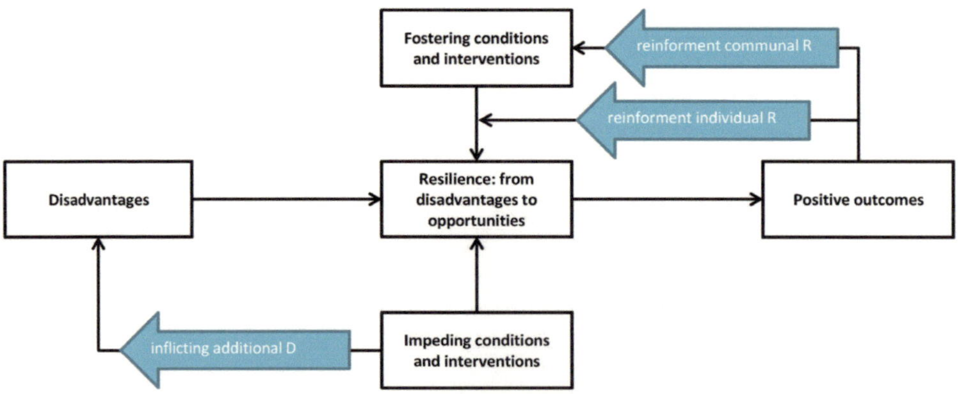

Figure 2.
Conceptual model for resilience in disadvantaged neighborhoods002E.

other groups [38]. An evolutionary process that originally would drive the learning of motor and speech acts, and of cognitive, social, and occupational skills is easily turned into a mechanism that enhances exclusion and an indifference to negative outcomes for others.

Social comparison makes communities ambivalent in their orientation as well. Granovetter has pointed at the importance of ties of members of peergroups to other groups. Within the group information is predominantly reduced to opinions and directed to conformity, wheres agents who openly connect to other groups obtain richer information [39]. In addition, Thagard [40] argues that peer-different connections are more valuable in the search for corroborative knowledge than peer-similar ones. Both contributions adocate for diversity and dynamics in social action.

3. Resilience of inhabitants of Dutch neighborhoods

In this section, we introduce the case of Dutch neighborhoods on the basis of reports of institutes related to or working in commission of the Dutch government. The aim of the case study is to understand the contribution of governmental policies on other sustainable development goals to the resilience in neighborhoods. Hence, the starting point is the reporting of the United Nations on the Sustainable Development Goals (SDG) in the Netherlands.

3.1 Negative trends regarding sustainable development goals in the Netherlands

In the Netherlands, the data collection is serviced by Statistics Netherlands [41]. In **Table 2** we show the sub-goal indicators that display deterioration in the development, while assuming that negative developments will be more manifest in disadvantage neighborhoods of the country. **Table 2** Excerpt of Sustainable Development Profile of the Netherlands ([42], p. 350), with additional national indicators [41].

Rather remarkable is Netherlands 8th score in the list of netted negative spill-over effects generating countries. An explanation can be found in the state policy of facilitating transnational tax evasion. A and for instance, in the contribution of Dutch agriculture companies to deforestation in the Amazon. Moreover, the Netherlands is at the bottom of the EU league with regard to climate action. This score is rather puzzling. Of its territory, 26% is below actual sea level and 29% is susceptible to river flooding [43]. The relationship between risk exposure and climate action in the Netherlands is giving rise to speculation: Is it short-sightedness, is the problem too large to face, or is it an expression of free riding on efforts of other states? In this chapter, we leave the questions unresolved. Rather, we follow the idea that 'the battle for sustainable development will be won or lost in the cities.'[1]

3.2 Sustainable cities and communities

The United Nations have called to make cities and human settlements inclusive, safe, resilient and sustainable. For tenants. The costs of housing relative to household income have increased since 2012 from a comparatively high level of 36.2% to 38.1%, whereas the costs for home-owners have decreased. Since 2012 as well, indicators for safety and other environmental stressors show deterioration in acknowledged disadvantaged neighborhoods, whereas other neighborhoods show a modest improvement [44]. Recently, the selective downgrading of the disadvantaged

[1] Eugenie Birch, Co-Chair of the SDSN Cities network and Professor at the University of Pennsylvania, attributing this line to Ban Ki-moon, Secretary-General of the United Nations from 2007-2016.

SD Goals	SDG Profile: deteriorating indicators in the Netherlands	Additional deteriorating SDG indicators Statistics Netherlands
1 No Poverty		Children (0–12 years) raised up in poverty People worrying about making ends meet
2 Zero Hunger	Prevalence of obesity Human Trophic Level Sustainable Nitrogen management	Intensive farming livestock Spillover of chemical crop protection Antibiotics in livestock farming Food waste
3 Good Health and Well-being	Gap in self-reported health status by income	Expenditures on healthcare Alcohol consumption Smoking Vaccination coverage (0–2 years) Mental health Self-reported health (12+ years)
4 Quality Education	Variation in science performance explained by socio-economic status Underachievers in science	Expenditures on education Use of pre-school daycare Termination without grade Mid-level education grades Reading capabilities pupils*
5 Gender Equality	Seats held by women in national parliament	Difference in income level women Physical and sexual violence against women
7 Affordable and Clean Energy	Share of renewable energy in total primary energy supply	Energy consumption Energy from fossil energy sources Energy costs per household
8 Decent Work Economic Growth		Unemployment Unused employment potential Hours worked per week* People worrying about employment
10 Reduced inequalities		Family contacts Participation in associations Contribution to voluntary sector Population share youth (0–19 years) Expenditures on healthcare Expenditures on social protection
11 Sustainable Cities and Communities		Costs of housing* Experiences of displeasure in public space* Claim on space Expenditures on protection of environment Emission of acidifying substances Exposure to particulate matter Victimization by crime Experiences of insecurity
12 Responsible consumption and production		Municipal garbage disposal People worrying about the environment
13 Climate Action	Energy-related CO_2 emissions	Reduction of CO_2 emissions*
16 Peace, Justice, and Strong Institutions		Number of policemen Population share of detainees Underaged suspects Expenditures on public service Jobs covered by collective agreements Perception of corruption

**Substantial disadvantage compared to other capabilities indicators in EU-ranking.*

Table 2.
Negative trends in realization of sustainable development goals in the Netherlands.

neighborhoods has given rise to serious concern of the Dutch government. The political attention is drawn to areas where 'combinations and cumulations of problems in schooling, employment, poverty, inclusion, safety, subversive crime, housing, and health' occur [45]. The negative trends on several SDGs and indicators prove to be concentrated in specific neighborhoods and cities. Closer look makes clear that there are 16 urban areas assigned, containing several neighborhoods. In these areas, almost a million people live ([45]: Annex). According to the Minister of Interior Affairs, residents, entrepreneurs and professionals in these neighborhoods are in the lead in the search for varying tailor-made solutions [45].

In a report titled 'Resilience in social housing', an analysis has been made of the causes of the downgrading of these neighborhoods [8, 46]:

- A nation-wide process of urban reconstruction has dimished the quantify of low-rent housing stock and reduced the stock to in specific areas.

- The Dutch government has issued legislation at the end of 2010 holding a restriction in the allocation of social housing to low-income households.

- Reforms in the mental healthcare have caused a displacement of clients from protected institutions to low-rents housing, although clients concerned were expectedly vulnarabile to stressfull environments in which the low-priced housing is located.

- A part of the disturbances of livability in the neigborborhoods can be attributed to the influx of these former clients, just like the presence of enduring unemployed people.

- When cooperation in the neighborhoods between social housing providers, municipality officials, and care employees fails, the vicious circle of cumulating problems is not reversed.

To this summary could be added that the price range of the social housing has been upleveled and narrowed just below the limites of the governmental surtax on for tenants. Households that are faced with a drop of income, sudden unemployment and personal debts can not move to cheaper homes, while at the same time they are entangled in conditions of the surtaxes on housing and other public services.

Van Gent and Horstenbach argue that the combination in the legislation of taxation of the social housing institutes and the restriction of the allocation to low-income groups has caused a residualisation of the Dutch social housing sector [47]. Nieuwenhuis and colleagues classify the Netherlands in a middle group between countries with high and low social-spatial segregation. From the middle group people manage to succeed in social mobility to less disadvantaged neighborhoods. However, the analysis regards the period from 2001 to 2011 [48]. From 2012, the social-spatial segregation in the Netherlands has increased considerably. Expectedly, the opportunities for social mobility from the worst to better neighborhoods will have been decreased proportionally.

Another addition is that the access of low-income groups to low-priced home owned houses has deteriorated as well. The market for low-priced houses is reduced by price-inflating taxation policy of the Dutch goveernment as well as by redlining of neighborhoods [49]. Redlining means that banks and other suppliers of mortgages refuse to credit applicants in marked urban areas. In a report of the

Dutch national bank, redlining is considered as a practice that undermines financial stability [50]. However, redlining is still a problem to the resilience of neighborhoods [51]. Moreover in specific neighborhoods, adjacent to social housing areas, private home owners are generally and grossly outbid by shady real estate agents who utilize the houses for extortion of unregistered migrant employees [52].

The report of Leidelmeijer et al. [8] contains an advocacy for social sustainability that would be warranted by personal and communal resilience. The plea resonates influential reports of 15 years ago [44]. First, the scientific advisory board of the Dutch government issued a call for trust cooperation within neighborhoods [53]. The main idea was that institutions would be reshaped and enabled to trustworthily deliver appropriate and connected services to the citizens. This advise is reflected in SDG goal 16 regarding accountable, inclusive, and effective institutions. Second, the former advisory board for housing, spatial planning, and environment has proposed to reconsider the physical reconstruction policy. A social-oriented policy were more apt to promote upward social mobility within the borders of the neighborhoods and would be more directly contributing to community building than gentrification would do. In this report, education is pointed out as a major carrier of social progress and a main route to paid employment [54].

Other SDG goals, like full productive employment and decent work for all (SDG 8) and reduction of inequality (SDG 10) have been subject of a policy reform in the Netherlands. However, the implementation by means of three decentralizations in the social domain is regarded as one of the circumstances that allegedly have contributed to an increase of vulnerable people in low-income neighborhoods [55]. In next section, we elaborate on these policy reforms.

3.3 Inclusive and equitable education and lifelong learning opportunities

In the Netherlands, the SDG report on education displays an increase of inequality, as variation in performance in science is increasingly explained by socio economic status. Another indicator shows that the Netherlands fall behind in excellence, implying that talent of pupils and students is decreasingly brought to full development [42]. In a international survey of learning performance of 15 year olds in OESO countries, Dutch pupils show a decline in perfomance in mathemetics and natural sciences, and to a greater extent in reading [56]. An other survey displays that less primary school pupils in their grading year (12 year olds) meet the reference level for reading. It is stated that almost a quarter of the Dutch pupils lack the reading abilities required for being articulate citizens. In the OESO survey, Dutch puplis, and more specifically boys, are to a lesser degree able to evaluate and reflect on texts. Further, there is a remarkable low score on reading motivation [56].

In addition, we expect that the reading deficits will discourage lifelong learning by upgrowing Dutch citizens. Knowledge, in a broad sense, accumulates step by step, not by hapzardly retrieving of unassessed information. Jumping to conclusions and polarization of opinions in peer groups are not precluded by reading experience. Unassessed use of information is more likely to occur if the experience is absent. Regarded in this way, the poor reading motivation and experience will negatively contribute to societal stability.

A clear analysis of the ine inequal learning opportunities is provided by the Dutch education authority [57, 58]. At the end of the primary school, pupils from low-educated and from non-western migrant families are underrated by primary school staffs, while children from backgrounds similar to the school staffs are overrated. Over a longer period recurrent studies have provided this finding. Until 2017 scores on a end-term test reduced this inequality in the admittance to higher levels of secundary education. Further, underrating and overrating was found to be corrected in the first 3 years of the secondary schools, provided the availability of transfer options at the secundary schools [58]. In the chain of cascades through the educational system to employment, children from families with little education and migrant ancestry lapse to lower levels at every cascade, resulting in low emplyment rates after educatiom [57].

The education authoriy point at neighborhood effects and sorting out between white and colored schools. Some schools manage to sustain a more diverse population. The quality of the education is not significantly related to the location of schools, however on some schools in disadvantaged neighborhoods, the turmoil in some classes and schools is so dominant that teachers can pay not enough attention to the learning by their pupils [57].

The Dutch education system is rather compartmented from an early age [59]. Differentiation is not adapted to personal learning speed and capacities. Additional need for education is arranged by parents through commisioning of private teachers or commercial providers. By this, children of high-income families have a considerable advantage, although lack of motivation is the strongest driver for additional education [58].

Recently, the national advisory board for education has advocated for a reform of the selection practice. Not at the early age of 12 year, but at the end of third year of the secundary school pupils should be sorted out for further education. To the age of 15, pupils would follow their courses in unsorted classes [59]. This design is more in accordance with international educational practices, particularly with higherly performing Scandinavian countries.

By forming of homogeneous classes and absence of individual learning trajectories, the Dutch education system is driven by social reference norms (see [31]). Comparative studies in other countries has shown that when teachers neither encourage nor challenge pupils individually, but instead, rely on social reference norms, they educate less students to a level of excellence, will have students with fear of success, will have more students with test anxiety, and generate less learning motivation [60]. Remakebly, these findings reflect the comparative profile of the Dutch education system rather recognizably. Another social comparison effect is found in the rating practice by primary-school staffs. Not the talent, motivation and the social–emotional potential of the pupil, but the socio-economic similarity between the school staff and his or her parents determines the rating. In spite of scientific evidence of the resulting inequality, these professionals did not have reconsidered their rating practice. Instead, a lobby of primary school leaders for neutralization of the end-term test results has been granted by the Dutch parliament.

3.4 Full and productive employment, decent work for all

Statistics Netherlands reports negative trends in SDGs 8 and 10. The employment potentiel of the Ducth population is increasingly underutitlized, and growing part of the jobs do not offer full employment nor the security of indefinite contracts. Further, the participation of people in social and societal activities is decreasing. Expenditures in (mental) health care and social protection display declining trends too. However, just these topics have been subject of a major policy reform in three social domains. In order to understand how this policy reform worked out since 2015, we will discuss policy evaluation reports. The aim is to understand how inititially resilience promoting reforms have turned into the countereffect of increased risk exposure of target groups and decreased opportunities to overcome their adversity.

From January 2015 the **Participation Act** is in effect, holding that people from disadvantaged positions should be guided to employment. The policy theory reflects the idea of fostering the resilience of people. The act intended to establish a uniform framework with some discretion for municipalities [61]. However, since the start of the reform these local actors were forced to realize savings on the aggregated national budget.

Due to the reform, young slightly diseabled persons are more frequently employed. However the income and job security are decreased considarbly through part-time employment and temporary contracts [61]. Decrease of full and secure emploument is a general trend in the Dutch labor market. However, these changes increases the risk of getting impairments in work [62]. Hence, the group of slightly disabled youngsters is exposed to a risk for which they are considerably more sensitive. Another target group of the act concerns persons employed in social protected workplaces. The idea was to guide these protected workers to regular employment. This transfer is not established succesfully. Persons lost their jobs and became depended on social assistance [61]. People living on social assistance found employment to a lesser degree, of an inferior quality, and with part-time cancelable contracts [61]. Inspite of legal and financial incentives, and of medical reassessments, people receiving the social assistance on the basis employment impairments appear to have almost no chance on reemployment in jobs adapted to their restricted capacities [63].

Summarizing, the government has overestimated the chances on regular and full employment for these groups. Furthermore, the policy theory was a generalization that misunderstood the individual and group-specific pacing needs. The savings on the budgets have urged municipalities to restrict their case load to clients with small problems and restrictions [61].

In 2015 as well, the **Youth Act** regarding the provision of mental health care for young peope was issued [64]. The procurement structures implemented, incent care organizations to behaviors that conflict with the purpose of care provision [64]. The compensation rules and administrative requirements promote a preference for clients with singular complaints. Clients with multiple and chronic complaints, including problems in their personal and social situations are less likely to be served. These clients and their personal supporters, if present, are faced with waitinglist and lack of timely mental health care. Their situation is considered as alarming [65].

At the start of 2015, the **Social Support Act** came into effect as well. The act aimes at a combination of formal and informal support in the housing and personal transport of people with physical and mental problems [66]. In evaluation research, there is no evidence that the personal budgets have contributed to self-reliance, participation or a higher quality of life [66]. Meanwhile, a part of the municipalities put little effort in checks on legality of applications, a condition that is associated with avoidable fraud [67]. There is a considerable likelihood to become involved into organized fraud, either as an accessory or as a vicitm [68].

3.5 Access to justice for all and build effective, accountable and inclusive institutions at all levels

This fragment of substainable development goal 16, is become a serious point of concern due to the affaire of the day-care surtax. A damning judgment by a parliamentary investigation commission has led to the resignation of the Dutch government at January 15th of 2021. The report was titled 'Unprecedented Injustice' [69]. Here, we only discuss facts and judgments in relation to the sustainable development goal. It is also an example of how a resilience promoting policy had led to the inflicting of disadvantages on citizens and the creation of institutional distrust.

The day-care surtax originates from 2005. It is introduced to promote participation of parents, particularly of the mothers in the labor market. The surtax depends on the proportion between income and costs of purchased day-care. In that way, the surtax was aimed to supports houselholds which have insufficient income for use of day-care. From the beginning, legality checks on the intake of applications were absent or, later on, provisional. The Dutch tax authorities relied on the article in the day-care surtax act, which allowed this state agency to reclaim advances for a period of five years. Arbitrarily, the tax authorities have figured out an 'everything-or-nothing' principle, meaning the response of an complete reclaim on any error in the application forms. The princinple excluded the option of a restatement of the level of surtax and the leniency of redemption arrangements. The principle is judged by the commission as being disproportional [69]. The effectuation of the reclaims has caused a cascade of household debts [70].

The title of the report is a reference to the fact that up to and including the highest national court the practice rules of the tax authorities have been confirmed. Therefore, the maltreated citizens have been excluded from access to justice as well [69]. Moreover, this crisis in the Dutch institutions is not yet revealed in full extent. It is reported and confirmed that clients with a double passport, being an indicator of a non-western migrant ancestry, is used to red flag surtax receiver on a personal attribute [71]. A following disclosure is the fact that tax authorities kept up and shared across three social domains a blacklist of 240,000 households or people of which attributes would predict fraud risks [72]. As the surtaxes serve to support the provision of basic needs of the household of which the incomes do not suffice, these people have been deliberately and multiply disadvantaged in their daily life. This state practice has, in spite of its inclusive goals, contributed to more inequality in the Dutch society.

4. Analysis of interviews

In this section, the resilience of residents in Dutch neighborhoods is analized on the basis of ten interviews with key persons. In the appendix, a short description is given of the interviewees and their involvement into the subject. In the report, randomly assigned letters refer to the ten interviewees. The reporting is in the order of the elements of the conceptual model. However, we start in the center of the model with the descriptions of the concept of resilience, as given by the interviewees.

4.1 In-practice definitions or resilience

Most interviewees relate the term resilience to social progress. C gives a rich description: 'It is a wonderful word. It includes strength and optimism. It expresses dynamics. It dents in and out and brings you further. [...] It helps to make connections. It is taking blowes as well'. According to A, resilience can be triggered by experiences of inequality. 'You have to rely on your inner anchor', says E. F adds to that: 'It has to be tensive. If you cannot lose, you cannot win.' However, some conditions have to be met in order to be resilient:. 'Have attention for the problems. [...] One needs to have a safe base. One needs to be sheltered and to have a basic income', D states.

Remarkably some interviewees address individual differences in resiliency. Some people show no tendency to be resilient (A). 'They are probably not self-reliant, nor will they become so. And (yet) we pretend that it will happen. Everybody resilient in the neighborhoods, everybody happy', A relates. Used in this way, resilience has become an empty word, H states. Furthermore, F suggests to mobilize the silent force of modest people as well: 'They are not the people who are always omnipresent in the commissions and meetings in the neighborhood.' Resilience concerns the relationship between people and their environment: 'What you inflicts on the planet, you inflict do on yourself. If you are not be aware of that, it becomes a zero sum game.' It is a referral to spillover effects and lack of environmental responsibility.

4.2 Disadvantages in neighborhoods

The subject of resilience in the neighborhood addresses a growing inequality in the Netherlands. According to I, 'The gap between *haves* and *have-nots* does not close, on the contrary, it becomes ever wider.' A variety of problems is pointed out by the interviewees, including debts, lack of employment, (mental) health problems, social isolation, inequality in schooling, substance abuse, and the attraction of criminality. Nevertheless, interviewees object to present the people in the neighborhoods as chanceless (F), vulnerable (D) or weak (F; J). They warn to choose words carefully in order to avoid patronizing (D) and disparagement (E; F). Some people have simply so much problems at the same time that they cannot help but survive (J). Rather, in carrying their worries they show personal strength, but the situation 'makes them ill of depression'(J). The interviewees agree on a concern for the problems of the inhabitants. Particularly, the accumulation of disadvantages takes away the perspectives for people. 'The [...] perspective is to live in a neighborhood where nobody is employed. That is no good [...] Children, grown up in these circumstances, are deprived of perspectives and will become frustrated, G says. 'It is anything but inspiring', D adds to that. D points at the short-term orientation of people with debts. Even in the short-term, problems can take away perspective: 'When you have every evening neighbor nuisance, you cannot make your homework. Three weeks of not making your homework, you do not pass on to the next year' (F).

4.3 Opportunities and challenges

So, people need to have perspective to overcome their disadvantaged situations. The interviewees emphasize the necessity of granting initiatives that originate from the neighborhood. The initiatives find their strength in being intrinsicly motivated (C). Working on this basis provides people the ownership of the initiatives (C), and of the positive things that hppen in the neighborhood (B). The initiatives generate stories, the narratives in the neighborhood break through negativity (B), and stimulate to undertake more initiatives (C). For instance, young people gathered in the workshops, are challenged to new initiatives (B). 'There are plenty of ideas and wishes by the people themselves', A tells. 'Of course, a small push helps', A explains, 'The small push comes from the environment. Whether it comes from the community, the municipality, the care institutions, the children, it does not matter. However, the small push is needful'. B calls it 'guiding, showing something of the way to improvement'. It is challenging to return stories about problems to whom that express them. H replies regularly: 'What can I do to help you? Then, they are going to reflect'.

In the support of young people, challenges start with attraction by activities in which they want to participate (B). For instances, sports, or debating sessions with politicians and other officials. A great attraction can be 'the step from illiteracy to university' (A). Families, and particularly mothers, show a drive to offer to the children of the neighborhood better conditions to live in, and to enable their social progress (A; B; E; J). Sisters happen to take a guiding role as well, even if it concerns the assistance of a brother at the police station (H).

4.4 Fostering interventions and conditions

Several interviewees state that fostering of resilience is established by interaction. A condition for interaction is the availabilities of places to gather and to meet. Municipalities can provide sports accommodations and community centers or more occasional facilities. Sometimes social housing corporations offer facilities as well (D; H; J). The interviewees differ in opinions on the question of whether the neighborhood as a whole facilitates interaction. For instance: 'I do not believe in mixed housing. The idea [behind it] is that than things go better by itself. That is not true' (J). This statement is substantiated by an example of a gated area with new expensive apartments. 'People simply do not meet', states J. Contrasting statements concern are explained by neighborhoods where a variety in social class has emerged over time. There, the interaction got going to the benefit of people (G; H). There is more agreement about the idea that the neighborhood is the level on which cooperation between institutions is amenable (A; C; D; G; J).

The interaction starts preferably by socializing. A direct, efficient approach can have an adverse effect: 'Accept that people first come for drinking coffee or gaming at the community center. Just join them' (A). 'Close proximity. We need to think about how to organize that', D says. 'We have made a survey of social networks in the neighborhood. It appeared to be a myriad. [...] Facilitate the networks. Avoid the reflex to take them over', G says. B and F object to this reflex as well. 'The city cannot handle this well, C states.

There is a common opinion that trust is the mechanism for fostering of resilience. 'To be trusted by them, that would help', J tells. Trust and tailor-made interactions are counteracted by accountability requirements', A states. Just like other workers, a policeman who personally has taken the initiative to invest in relationships with young people in the neighborhood need to be granted with professional autonomy and need to be warranted in these interactions by his boss, J states. 'In the frontline, it has to happen. There are a lot of people who want to participate', J explains. 'The reality is that the frontline work can be very tensive', F adds to that.

One of the interviewees tells about initiatives in which young people participate and are learned to develop a variety of skills on the basis of their interest and talents. In some schools in the neighborhoods, engaged teachers and school directors try to work in this way as well. However, they cannot get things changed (J). F refers to Sweden, where pupils get individual trajectories. The school system over there is organized around the capacities and motivation of pupils. According to F, it starts with investments in the professional skills of teachers themselves. 'There is a site that promotes their development', F adds.

4.5 Positive outcomes

Two interviewees mention explicit examples of positive outcome (A; B). Examples of positive outcomes for participants in the initiatives are the development of skills like presenting yourself well, becoming sports coaches, and knowing the procedures to resolve conflicts with authorities. More intrinsic is the development of a feelings of ownership (B). It concerns responsibility for elderly in the neighborhood. By supporting the elderly individually, the participants succeed to take away prejudices about themselves. Before the program, the elderly only judged them by their bomber jacks and images from crime shows on the TV. The young people themselves feel responsible for the neighborhood, tells A.

Remarkably, the ability of self-reflection is mentioned. 'People learn to realize that adversity and disadvantages are an unavoidable part of live', tells B. Learning outcome includes coping with differences in opinions, F states. 'How do you relate to these differences. It is not only what you thinks is right, but also what others think what is right. They can have a point', F points out.

4.6 Reinforcement of personal resilience

In the conceptual model, a distinction is made between reinforcement of personal resilience and communal resilience. However, in a part of the examples, the reinforcement jointly addresses both levels of resilience. This is due to the ideas at the basis of the initiatives as well as the consequence of participants themselves, explain A and B. The reinforcement of personal resilience is the growth of the abilities to see the meaning in what are they doing, D tells. 'The belief that you can influence the course of action in the neighborhood, the possibility of coining your own efforts, it all contributes to self-esteem', F explains. It concerns 'to be connected as individual to a meaningful whole', F adds. It is a substantiation of the joined reinforcement as well.

4.7 Reinforcement of communal resilience

A common observation is that people who have made social progress in the neighborhood, want to return of it to the benefit of the community in the neighborhood. It is partly the result of an agreement at the start of participation into projects (B). This way of social return is institutionalized by the education of experience experts. However, the drive to contribute to communal resilience gets internalized or was intrinsic from the start. D gives a salient explanation: 'I regard it as an intrinsic motivation by a group whose members have encountered hardship, a motivation to give something back. That makes not only yourself resilient, but also the community of which a part of'. However, D adds that a neighborhood needs a mix of people who serve communal resilience and people who do not succeed to overcome disadvantages by themselves.

As noted before, the women play a frontier role. Mother who have experienced and overcome disadvantages, have unionized in a part of the neighborhoods (J; B). Young women who have graded at the higher professional education and university, show responsibility for their family and the community H points at the fact that boys from disadvantages neighborhood perform less well on school. B relates it to barriers that children from non-Western migrant families encounter. Even boys who have taken the hurdle of primary school rating, often fail to grade on the top level of the secondary school', B says to her regret.

4.8 Impeding interventions and conditions

In the set of interviews, this topic is 1.7 times more scored than fostering interventions and conditions has done. Moreover, the statements of the interviewees are less ambiguous. Almost all text fragments fit in obvious sub codings.

Inequality in the distribution of opportunities is mentioned by a large majority as a cause why people fail to be resilient. The rating by staffs of primary schools is mostly pointed at. Pupils of disadvantaged neighborhood receive lower ratings whereas other pupils receive high ratings. A personal example is given of an underrating of four levels. Parents with higher incomes, pay for training their children to higher test scores (E). It is not only this rating practice. 'A part of the teachers of the secondary schools repeatedly tell pupils that they are a total loss. After some time they will lose the believe in themselves', B relates. Some teachers are demotivated and feel misacknowledged in their professional autonomy. Furthermore, schools in disadvantaged neighborhoods are understaffed, underqualified, or both (H). Some respondents mention the bad relationship between police and boys from the neighborhood (B; J). 'Why do you work in a neighborhood of which 97% of the population should not be in the Netherlands anyway?', one interviewee asks rhetorically. The interviewee refers to opinions of policemen and teachers inspired by right-wing politicians.

Both interventions and conditions are affected by an incapacity or an unwillingness (J) of local officials to understand and relate to the everyday live in disadvantaged neighborhoods. Moreover, the individual differences in the problems of people, their social situations, their life stories and traits do not fit in a policy frame. For instance, people are advised to search for support in their social network, while they lack such a safety net. 'Youth work at a community center is opened till 5 PM. Then it just starts. It is bureaucratized', H states. As a consequence, interventions are chosen and conditions are shaped which impede people in their search for solutions and opportunities. There is clear majority that contributes to this explanation. People become entangled in what is described as an impeding bureaucracy. This observation as well is shared by most interviewees. A remarkable observation is that an organization in the frontline require employees to be bureaucratic competent in order to help people in their problems with other organizations (A). It is maybe a level problem. 'To the city, the neigborhood is not a managable unit', D holds.

Six interviewees critize the way of working by tendering and contracting. 'There is more competition among welfare organizations than among private corporations', B illustrates. Organizations are forced to compete instead of to cooperate. Organizations do not gather people, but launch concepts. It is smart but not wise to compete on quantify, B explains. In the interaction with local authorities business parlance is expressed such as unrolling and upscaling and talking about neighborhoods as were they businesses. 'People are carriers of initiatives in the neighborhoods. You cannot upscale people', A argues. 'Not what is needed in the society, is leading. It saddens me', C says.

Two interviewees signalize that people feel let down by the institutions (A; C). This feeling is expressed by white people with low incomes and little education as well. 'People who feel like that, become susceptible for men who want to abuse their state of mind', G relates. When inequality in normal resilience trajectories is experienced time by time, young people are more easily seduced by the luxury showed by peers who make a career in a criminal trajectories, G states.

Interviewees indicate distrust as a strong impediment. Public services are set-ip with distrust in citizens in mind (A; G). G refers to the day-care surtax affair. People in disadvantaged neighborhood, do not trust institutions anymore (J; G). In interactions with officials they get questions which display distrust. People have fear of consequences of such interactions, like getting your children out of custody (J). In contacts with commercial suppliers, the level playing field is absent. The examples of business fraud fuels a distrust in the neo-liberal policy of the government, E states.

Finally, two interviewees (C: G) state that local and national officials do not relate to the planet either. There is no response on initiatives in neighborhoods on sustainability, whereas the big opportunities for climate action like the international airport and the sea ports are unbothered (C). It is an example of what E calls a lack of level playing field in the society. E puts to the fore an ecological argument. 'We impede resilience by creating monocultures, in housing (E; F), in agriculture (E). 'Diversity is the element that contributes to resilience of social and ecological systems. Just that element we lose sight of', E substantiates.

4.9 Inflicting more disadvantages

Through debates on public and social media, the experience of discrimination is enhanced (B; D). It induces fall back in problematic behavior by young people you are trying to keep out of the trouble, B and D say in equivocal phrases. Distrust is very contagious and reciprocal in interactions (B; D; J) The distrust showed by officials of some institutions, makes people averse of institutions in general. Through this avoidance additional problems will emerge (J). Inequality of opportunities on schools, the unequal risk of being halted or arrested (B; J), confirms boys and young men in their conviction that they will be discriminated permanently. It discourages a search for solutions of their other problems. The feeling of being let down by society expedites the step to substance abuse or delinquency, E argues. It makes problems rather unsolvable. Inflicting additional problems without any concern or empathy for the victims is in the heart a dehumanizing treatment of citizens, E adds.

4.10 Serendipities

'Think great, act small' is a slogan cited in one of the interviews. An example is displayed by a neighbor who fights for appropriate education of the talented daughter of his illiterate neighbors. Starting from the first people who succeed, a growing web of resilient citizens can activated to break through the adversity. Another insight is found in the metaphor 'In the eye of the hurricane, it is quiet'. As soon as people try to improve their situations, they are faced by conditions of regulation which bounce them back in adversity. The metaphor holds as well for public officials and professionals who want to change detrimental practices. They are faced with misunderstanding, resistance, and counteracts. They have to be resilient themselves as well, and need to be supported by superiors with an open mind. However, 'from a position of power, you cannot see resilience'. Change will require that people stand up, and emerge as leaders. The subject of change extends from social action to climate action. The inability and unwillingness of politicians and their officials to relate to citizens, to understand their needs, and to make use of their individual talent and collective initiatives perpetuates inaction.

5. Conclusion

The first question is which conditions and intervention foster resilience of people in disadvantaged neighborhoods? Interventions need to relate to the personal problems people experience and the oppportunites they see. A necessary condition is that they are supported by family, peers and, if necessary officials. The latter should be granted discreation and support by their superiors. Impeding interventions are characterized by an uncapacity or unwillingness to relate to everyday live in the neighborhoods. An impeding conditions is that professional and their organizations are forced to competition instead of cooperation. The incentives in the procurement are confllicting to fostering of resilience. The combination of misalignments generates institional distrust. Positive outcomes are found in personal development, training skills, and an increase of self-esteem. A feeling of ownership can be attained, just like the experience of being a relevant part of a meaningful whole. People who succeed to be resilient in disadvantaged neighborhoods exhibit a strong drive to return their success to the community. They contribute to less prejudice and more trust between groups. The dissemination of this success expand the positive communal outcomes. Impeding interventions and conditions inflict additional disadvantages on citizens. The lack of perspective frustrates people and discourages them in their search for positive outcomes. A generalized distrust becomes self-perpetuating, either in self-harming of other-harming strategies.

Sustainable development contribute to social progress and resilience in neighborhoods. However, fall backs in goal attainment manifest particularly in the less advantaged neighborhoods. The contrasting development of other neighborhoods and areas of the country suggests the occurance of intranational negative spillover efffects. This phenonom is observed in a country that is in the vanguard of countries that generates international negative spillover effects. A question for futher research is of whether international and intranational spillover relate and whether this relation is grounded in state policies. The presupposition that the neighborhood is a level amenable for communication and control is partly rebuted. Solutions for the disadvantages of citizens experience, will be sought for in interaction on that very level. However, the stream of disadvantages from countereffective reforms needs interruption on the national level. It is found that policy reforms that have promoted resilience have resulted in the reverse of resilience. The relevance of scrutinized ex-post policy evaluation is demonstrated in this study. Finally, we are stengthened in our conviction that social progress is a condition sine qua non for sustainability in general. It not only contributes to support for policies on climate action, it will be a sources for initiatives and cooperation across levels.

Appendix: some data on the ten interviewees

Mohamed el Achkar: Board member of Woonstad Rotterdam; Member of the Advisory board Prospect Fund.

Esther Agricola: Director of BPD (Area development). Former director of 'Space and Sustainability' department of the municipality of Amsterdam. Former director of KEI (knowledge centre for urban development).

Bouchra Dibi, Consultant and Researcher on disadvantaged neighborhoods. Former member of a municipal board. Formerly employed as social worker in Utrecht.

Ton Huiskens: Director 'Werken aan de Stad' (Social entrepreneur in social urban intiatives).

Hafida Leri: Director 2gather. (Social entrepreneur in deprived urban areas).

Arnold Molenaar: Resilience Officer at Resilient Rotterdam.

Hannan Moussaoui-el Garmouhi: Director Woonbron Delft Former manager Social work and Participation' at the municipality of the Hague.

Karin Schrederhof: Alderman at the municipality of Delft, (in portfolio Housing and Social Work).

Suzanne Wacanno: Senior associate of The Natural Step (Corporation in sustainability).

Pieter Winsemius: Volunteer in several social projects. Former Minister of Housing Spatial Planning and Environment; Former member of the scientific board of the Dutch government (WRR).

Author details

Hendrik Marten Koolma[1*] and Catharina Frederika van Dreven[2]

1 Faculty of Social Sciences, Vrije Universiteit Amsterdam, The Netherlands

2 Rotterdam School of Management, Erasmus University Rotterdam, The Netherlands

*Address all correspondence to: rik.koolma@planet.nl

References

[1] United Nations, Department of Economic and Social Affairs Sustainable Development. (2015, September 27). Transforming our world: the 2030 Agenda for Sustainable Development. Opgehaald van https://sdgs.un.org/2030agenda

[2] Arora, N., and Mishra, I. (2019). United Nations Sustainable Development Goals 2030 and environmental sustainability: race against time. Environmental Sustainability, 2, 339-342.

[3] Schmidt-Traub, G., Hoff, H., and Bernlöhr, M. (2019, July 15). International spillovers and the Sustainable Development Goals (SDGs): Measuring how a country's progress towards the SDGs is affected by actions in other countries. Opgehaald van SDSN Policy Brief: https://irp-cdn.multiscreensite.com/be6d1d56/files/uploaded/SDSN-Policy-Brief_International-spillovers-and-the-SDGs.pdf

[4] Hardin, G. (1968). The Tragedy of the Commons. Science, 162, 1243-1248.

[5] Meadows, D. H., Meadows, D., Randers, J., and Behrens III, W. W. (1972). Limits to the Groth: A Report for the Club of Rome's Project on the Predicament of Mankind. New York: Unvierse Books.

[6] Hardin, G. (2006). The Tragedy of hte Unmanged Commons. In D. J. Penn, and i. Mysterud, Evlutionary Perspectives on Environmental Problems (pp. 105-108). New York: Aldine Transaction.

[7] Commissie Derksen. (2014, March 7). Perspectief voor de steden: advies van de Commissie Derksen. Opgehaald van https://www.vnpf.nl/media/files/20150327-perspectief-voor-de-steden-advies-cie-derksen.pdf

[8] Leidelmeijer, K., van Iersel, J., and Frissen, J. (2018). Veerkracht in het corporatiebezit: Kwetsbare bewoners en leefbaarheid [Resilience in the property of housing corporations: vulnarable resiedents and livability]. Opgehaald van https://www.rigo.nl/wp-content/uploads/2018/11/RIGO_Research_en_Advies_Veerkracht_in_het_corporatiebezit_Kwetsbare_bewoners_en_leefbaarheid-november-2018.pdf

[9] Block, J. H., and Block, J. (1980). The role of ego-control and ego-resiliency in the organization of behavior. In W. A. Collins, Development of cognition, affect, and social relations the Minnesota symposia on child psychology Vol. Bd. 13 (pp. 39-101). Erlbaum, NJ: Hillsdale.

[10] Carver, C. S. (1998). Resilience and Thriving: Issues, Models, and Linkages 1. Journal of Social Issuesm 54(2), 245-266.

[11] Luthar, S. S., Cicchetti, D., and Becker, B. (2000). The Construct of Resilience: A Critical Evaluation and Guidelines for Future Work. Child Dev, 71(3), 543-562.

[12] Prince-Embury, S. (2013). The Ego-Resiliency Scale by Block and Kremen (1996) and Trait Ego-Resiliency. In S. Prince-Embury, and D. H. Sakklofske, Resilience in Children, Adolescents, and Adults: Translating Research into Praxtice (pp. 135-138). New York: Springer.

[13] Ungar, M. (2012). Social Ecologies and Their Contribution to Resilience. In M. Ungar, The Social Ecology of Resilience: A Handbook of Theory and Practice (pp. 13-32). New York: Springer.

[14] Holling, C. S. (1973). Resilience and Stability of Ecological Systems. Annual Review of Ecology and Systematics Vol. 4, 1-23.

[15] Smith, B. W., Epstein, E., Ortiz, J. A., Christopher, P. J., and Tooley, E. M. (2013). The Foundations of Resilience: What Are the Critical Resources for Bouncing Back from Stress? In S. Prince-Embury, and D. H. Sakklosfske, Resilience in Children, Adolscents, and Adults: Tranlasting Research into Practice (pp. 167-188). New York: Springer.

[16] Rutter, M. (2007). Resilience, competence, and coping. Child Abuse & Neglect, 31, 205-209.

[17] Rutter, M. (2012). Resilience: Causal Pathways and Social Ecology. In M. Ungar, The Scoial Ecology of Resilience: A Handbook of Theory and Practice (pp. 33-42). New York: Springer.

[18] Nieuwenhuis, J. (2014). Neighbourhood effects on youth's achievements: the moderating role of personality. Utrecht, the Netherlands: University of Utrecht.

[19] Lipsitt, L. P., and Demick, J. (2012). Theory and Measurement of Resilience:. In M. Ungar, The Social Ecology of Resilience: A Handbook of Theory an Practice (pp. 43-52). New York: Springer.

[20] Gray, J. A. (1970). The psychophysiological basis of introversion-extraversion. Behavioral Research and Therapy, 8, 249-266.

[21] Gray, J. A. (1987). Perspectives on Anxiety and Impuslvity: A Commentary. Journal of Research in Personality, 21, 493-509.

[22] Carver, C. S., and White, T. L. (1994). Behavioral inhibition, behavioral activation, and affective responses to impending reward and punishment. Journal of Personality and Social Psychology, 67, 319-333.

[23] Alloy, L. B., and Abramson, L. Y. (1979). Judgment of Contingency in Depressed and Nondepressed Students: Sadder but Wiser? Hournal of Experimental Psychology, 108(4), 441-485.

[24] Driessen, F. M., and Beereboom, H. J. (1983). De kwaliteit van het stedelijk leefmilieu: Bewoners en hun vooorkeuren [The quality of the urban habitat: Residents and their preferences]. Utrecht, the Netherlands: Univstiry of Utrecht.

[25] Boyce, W. T., and Ellis, B. J. (2005). Biological sensitivity to context: An evolutionary-developmental theory of the origins and functions of stress reactivity. Development and Psychopathology, 17, 271-301.

[26] Kuhl, J. (2018). Individual Differences in Self-Regulation. In J. Heckhausen, and H. Heckhausen, Motivation and Actions (third edition) (pp. 529-578). New York: Springer.

[27] Hagger, M. S., Wood, C., Stiff, C., and Chatzisarantis, N. L. (2010). Ego Depletion and the Strength Model of Self-Control: A Meta-Analysis. Psychological Bulletin, 136(4), 495-525.

[28] Kassem, R., and Higson, A. (2012). The New Fraud Traingle Model. Journal of emerging trends in Economics and Management Sciences 3(3), 91-95.

[29] Rutter, M. (1987). Psycholosocial Resilience and Protective Mechanisms. American Journal for Orthopsychiatry, 57(3), 317-331.

[30] Rutter, M., Sonuga-Barke, E. J., Beckett, C., Castle, J., Kreppner, J., Kumstra, R., ... Gunnar, M. R. (2010). Deprivation-Specific Psychological Patterns: Effects of Insitutional Deprivation. Monographs of the Society for Research in Child Development, 75(1), i-x, 1-253.

[31] Veroff, J., and Veroff, H. B. (1980). Social Incentives: a Life-span

developmental approach. New York: Academic Press.

[32] Luhmann, N. (1968). Zweckbegriff und Systemrationalität - über die Funktion vond Zwecken in sozialen Systemen. Tübbingen: J. C. B. Mohr.

[33] Douglas, M., and Wildavsky, A. (1982). Risk and Culture: An essay on the seection of technical and environmental dangers. Berkeley: University of Calafornia Press.

[34] Luhmann, N. (1973). Vertrauen - Ein Mechanismus der Reduktion sozialer Komplexität. Stuttgart: Ferdinand Enke Verlag.

[35] Lewicki, R. J., McAllister, D. J., and Bies, R. J. (1998). Trust and Distrust: New Relationships and Realities. Academy of Management Review, 23(3), 438-458.

[36] Dimoka, A. (2010). Whatdoes the brain tell us about trust and distrust? Evidence from a Neuroimaging Study. MIS Quartely, 64(2), 373-396.

[37] Meadows, D. (1999). Leverage Point: Places to Intervene in a System. Opgeroepen op january 16, 2013, van www.sustainabilityinstitute.org: www.sustainabilityinstitute.org/pub/Leverage_Points.pdf

[38] Festinger, L. (1954). Theory of Social Comparison Processes. Human Relations, Vol. 7, 114-140.

[39] Granovetter, M. S. (1973). The Strength of Weak Ties. American Journal of Sociology, 78(6), 1360-1380.

[40] Thagard, P. (1997). Collaborative Knowledge. Noûs, 31(2), 242-261.

[41] Statistics Netherlands. (2020). Monitor Brede Welvaart * Sustainable Development Goals 2020. Den Haag: Centraal Bureau voor de Statistiek.

[42] Sustainable Development Solutions Network. (2020). Sustainnable Development Report 2020, The Sustainble Development Goals and Covid-19. Cambridge: Cambridge University Press.

[43] Pieterse, N., Knoop, J., Niebelek, K., Pols, L., and Tennekes, J. (2009). Flooding risk zoning in the Netherlands. The Hague, the Netherlands: Planbureau voor de Leefomgeving.

[44] Uyterlinde, M., and Engbersen, R. (2018). Zijn vitale wijken maakbaar? Essay Kenniswerkplaats leefbare wijken Rotterdam. Opgehaald van https://www.kenniswerkplaats-leefbarewijken.nl/wp-content/uploads/Platform31_vitale-gemengde-wijken-maakbaar-1.pdf

[45] Minister of Interior Affairs. (2020, March 31). Progress program Livebility and Safety. Letter addressed to the Ducht House of Parliament.

[46] Leidelmeijer, K., Frissen, J., and van Iersel, J. (2020, January 20). Veerkracht in het corporatiebezit, de update: een jaar later, twee kaar verder /// [Resilience in the property of housing corporations. An update one year later, two years further ...]. Opgehaald van https://vng.nl/sites/default/files/2020-05/veerkracht-in-het-corporatiebezit.pdf

[47] Van Gent, W., and Horstenbach, C. (2020). The neo-liberal politics and socio-spatial implications of Dutch post-crisis social housing policies. Unternational Jpirnal of Housing Policy, 20(1), 156-172.

[48] Nieuwenhuis, J., Tammaru, T., van Ham, M., and Manley, D. (2020). Does segregation reduce socio-spatial mobility? Evidence from four European countries with different inequality and segregation contexts. Urban Studies, 57(1), 176-197.

[49] Aalbers, M. B. (2005). Who's afraid of red, yellow and green?:

Redlining in Rotterdam. Goeforum, 562-580.

[50] Houben, A., Kakes, J., and Schinasi, G. (2004). Towards a framework for financial stability. Amsterdam: DNB (Dutch National Bank.

[51] Bennink, N. (2018). Hoe Nederland zijn eigen redlining creëert, Module 3.1 Urban Development International. MCD.

[52] Aboutaleb, A. (2021, May 17). Mayor of Rotterdam interviewd in talkshow. BNN/VARA de Vooravond.

[53] WRR. (2005). Vertourwen in de Buurt. Amsterdam: Amsterdam University Press.

[54] VROM-raad. (2006). Stad en Stijging. the Hague, the Netherlands: VROM-raad.

[55] Uyterlinde, M., and Van der Velden, J. (2017). Kwetsbare wijken in beeld. teh Hague, the Netherlands: Platform31.

[56] Gubbels, J., Van Langen, A., Maassen, N., and Meelissen, M. (2019). Resultaten PISA-2018 in vogelvlucht. Enschede, the Netherlands: Universiteit van Twente.

[57] Inspectie van het Onderwijs. (2017). De Staat van het Onderwijs 2015-2016: Onderwijsverslag [The State of the Education: Annual report]. Utrecht, The Netherlands: Inspectie van het onderwijs.

[58] Inspectie van het Onderwijs. (2021). De Staat van het Onderwijs [The State of the Education]. Utrecht, tje Netheerlands: Inspectie van het Onderwijs.

[59] Onderwijsraad. (2021). Later selecteren, beter differentiëren [Later selection, better diffrentiation]. The Hague, the Netherlands: Onderwijsraad.

[60] Brunstein, J. C., and Heckhausen, H. (2018). Achievement Motivation. In J. Heckhausen, and H. Heckhausen, Motivation and Action (third edition) (pp. 221-304). New York: Springer.

[61] Van Echtelt, P., Sadiraj, K., Hoff, S., Muns, S., Karpinska, K., Das, D., and Versantvoort, M. (2019). Eindevaluatie van de Participatiewet. the Hague, the Netherlands: Sociaal en Cultureel Planbureau.

[62] Louwerse, I., Van Rijssen, J., Huysmans, M., Van der Beek, A., and Abema, H. (2020). Predicting future changes in the work ability of individuals receiving a work disability benefit: Weighted analysis of longitudinal data (Article). Tijdschrift voor Bedrijfs - en Verzekeringsgeneeskunde, 28(3), 42Ilse Louwerse*, Jolanda van Rijssen, Maaike Huysmans, Allard van der Beek, Han Anema.

[63] Weel, I. (2021, May 18). Mensen die in de WIA terechtkomen, komen daar bijna niet meer uit. Trouw.

[64] Friele, R. D., et al. (2018). Eerste evaluatie Jeugdwet: Na de transitie nu de transformatie. the Hague, the Netherlands: ZonMW.

[65] Inspectiie Gezondheidszorg en Jeugd. (2021, March). Onvoldoende tijdige en juiste hulp voor hongeren met ernstige psychgische pboelem [Insufficient timely and appropriate assistance for young people with seirous psycis problems]. Utrecht, the Netherlands: IGJ.

[66] Marangos, A. (2018). Maatschappelijke ondersteuning: [Social support: choices of clients and local policy]. The Hague, the Netherlands: Sociaal en Cultureel Planbureau.

[67] Inspectie Gezondheidszorg en Jeugd. (2019). Wmo toezicht door de gemeenten in 2018. Utrecht, the Netherlands: IGJ.

[68] Vereniging Nederlandse Gemeenten. (2020). "een wereld te winnen": Over zorgfraude (Wmo en Jeugdwet). The Hague, the Netherlands: VNG.

[69] Tweede Kamer. (2020). Ongekend Onrecht: Verslag - Parlementaire ondervragingscommissie Kinderopvangtoeslag.

[70] Kleinnijenhuis, J. (2021, May 26). Kabinet neemt alsnog priveschulden over. Trouw.

[71] RTL Nieuws. (2020, May 10). Belastingdienst geeft toe: toch sprake van etnisch profileren. https://www.rtlnieuws.nl/nieuws/artikel/5117616/belastingdienst-toeslagen-profileren-nationaliteit.

[72] NU.nl. (2021, March 10). Belastingdienst stuurt excuses aan 240.000 mensen op zwarte lijst. https://www.nu.nl/economie/6124258/belastingdienst-stuurt-excuses-aan-240000-mensen-op-zwarte-lijst.html.

Chapter 8

Housing Law in Poland—From the Cooperative Model to Flat Ownership

Aleksandra Sikorska-Lewandowska

Abstract

In Poland, the housing system is currently based on cooperative apartments and the ownership of premises. This is due to historical conditions, because in the post-war period it was decided to foster cooperative housing; while the development of private property was impeded. After 1989, there were rapid economic and social changes, which also affected the real estate sector. It became possible to buy cooperative flats for ownership. The construction of new apartments was rapidly started and a development market was established. The owners of tenement houses made efforts to return the taken property, many of them regained the buildings, although their technical condition was very bad. There is a shortage of apartments in Poland, both available for purchase and for rent. In this chapter, I intend to present the current ownership status of apartments in Poland and the process of changes that took place, but did not solve the problems.

Keywords: cooperative housing, ownership of premises, flat, polish law, housing law, development system

1. Introduction

The rapid development of cities in Poland began with the beginnings of industry, at the end of the 19th century, when the country was under partitions. The inter-war period, i.e. the years from 1918 to 1939, was the time of the formation of the real estate market in Poland and of intense changes in the housing market. The shortage of flats continued throughout that time, and many of the rental flats available on the market were of a very low standard [1]. In the literature on the subject, attention was drawn to the very poor housing conditions of a large part of the society [2]. The influx of people from rural areas to cities in search of work increased the shortage of flats. The emerging cooperative movement became one of the pillars of the development of the housing sector. On October 29, 1920, the act on cooperatives was passed (Journal of Laws of 1920, No. 111, item 733), which created the legal basis for the establishment of housing cooperatives. Pursuant to that regulation, cooperative construction was initiated as part of housing and ownership-construction cooperatives. During the existence of the Second Polish Republic, housing cooperatives built about 100,000 flats. Importantly, the activities of the cooperatives were aimed at the needs of poorer people, who had been exploited by the owners of tenement houses. The estates were to meet high social standards, include playgrounds, schools, recreation places, cultural centers, cooperative shops and bars, or canteens. There were also cooperatives providing housing for intellectuals (officials, teachers) [3]. An alternative idea for the development of housing in Poland was the adoption of the concept of the legal separation of premises. On October 24, 1934, the Ordinance of the President of the Republic of Poland on the ownership of premises was issued (Journal of Laws of 1934, No. 94, item 848). It made it possible to separate premises within buildings and sell them freely, resulting in the creation of housing communities. However, the lack of wide access to loans meant that until the end of the interwar period, separate ownership of premises remained a sporadic phenomenon [4].

After the Second World War, the reconstruction and development of the housing stock were the basic challenges for the liberated country. The destroyed buildings were rebuilt and new ones were built. In Poland, called the Polish People's Republic (PRL), a vision of a socialist state was implemented, in which private property was treated as a relic of the pre-war system. For this reason, the owners of tenement houses that had survived the war or were rebuilt after the war, were deprived of the possibility of renting them freely. The institution of rental control was introduced as part of the public housing economy [5]. It consisted in the fact that the state authorities decided who was to occupy premises in private tenement houses, without the will of their owners. The tenant received an administrative decision, under which he was allocated an flat or a part of the flat, e.g. one room only. In this way, two or even three families were often accommodated in large flats. This system of administrative lease regulation was maintained in Poland for a very long time, and even in the 1990s, many premises were occupied by council tenants.

The people's authority fought against private property and legal regulations introduced control of real property transactions. In the nineteen sixties, control was focused on the development of cooperatives, recognizing that this was the right way to increase the number of flats on the market. Housing cooperatives were established in each locality and only they were permitted to build multi-family buildings with the help of state-owned construction companies. Private construction companies did not exist at that time. The free market did not exist and construction processes were centrally controlled within the model of the command-and-control economy. Anyone who wanted to get an flat had to join a housing cooperative and wait, sometimes for many years, for the flat to be allocated. The final cost of building flats was borne by their tenants - members of housing cooperatives, who had to pay for the so-called housing contribution (key money). This shifted the burden of financing the construction of new buildings from the state to members of housing cooperatives [6]. In the nineteen seventies, more than 123,400 cooperative flats were completed [7], which was a record number. The cooperative housing estates built in different towns looked very similar: the same technology and uniform designs were used. The socialist states at that time adopted the large slab (a method of constructing buildings, consisting in assembling them on the construction site from large-size prefabricated concrete or reinforced concrete elements) as the basic form of housing construction [8]. During that period, numerous prefabricated buildings were erected in Poland and the number of flats increased quite significantly, although there were still too few of them in relation to the needs. Individual construction was marginalized, only repetitive designs with specific parameters were allowed. The area of the house could not exceed 110 m2.

Company flats were a characteristic element of the housing reality of the Polish People's Republic. They were built close to workplaces, and the assignments were usually given to employees with long-term employment. They were part of the property of a given workplace. They were, depending on the size of the enterprise, single blocks of flats or even entire estates built up with residential buildings and infrastructure. A company flat was usually the only alternative to a cooperative flat.

2. Changes after 1989 in Poland and their impact on the housing market

After 1989, Poland underwent a political, economic, and social transformation. The command-and-control economy model was replaced with the free market economy model. Legal solutions were introduced that allowed the running of a business by private persons without restrictions. The possibility of creating commercial companies, and freedom of trade and services were restored. The process of privatization of enterprises which until then had been uniformly owned by the state was initiated. The banking sector was commercialized and the centralization of the economy was abandoned [9]. Restrictions on land ownership were lifted, and it became possible to acquire freely real estate and recover plots of land and buildings taken over by the state in the past. The rules of lease control were abandoned, but the privileges of tenants who, in the previous period, obtained the right to premises on the basis of housing decisions were retained, which significantly limited the owners of tenement houses in exercising their ownership rights.

The restoration of local government, in which local authorities were the basic units, was a very important element of the changes. Municipalities were equipped with land, including that built up with residential buildings. As a result of this process, known as communalization, municipalities became the owners of numerous residential buildings, taken over from workplaces or from the state, together with tenants. The buildings were very often in a very poor technical condition as a result of many years of neglect.

The transformation of the 1990s also had its negative side - bankruptcy of many enterprises, loss of jobs by part of society, and the rise of record breaking unemployment and hyperinflation. Many buildings built by state-owned enterprises were sold to other entities with tenants, or transferred in various forms, to local authorities, or housing cooperatives. There was still a shortage of flats on the market in Poland, both for sale and for rent.

The legal system in Poland was completely rebuilt, and over the years, all important legal acts have either been amended or enacted anew. In the field of civil law, private property has regained its prominent place. Modern construction law regulations and provisions regulating the rules of spatial development and real estate management have been developed. Real estate began to be perceived, not only in terms of utility for meeting the housing needs of the family, but also in terms of investment. On June 24, 1994, the Act on the ownership of premises was passed (Journal of Laws No. 85, item 388), which regulated, inter alia, the institution of a housing community, the construction of the right to separate ownership of premises, and the rights and obligations of owners of premises. Pursuant to this act, it became possible to separate premises in all multi-family buildings, both those newly built and those that existed at the time of entry into force of this act. A market of flats purchased as investment has emerged, and not only for satisfying one's own housing needs. This act initiated the growth of the development market in Poland, which was quickly gaining in importance. In this way, the monopoly of housing cooperatives for the construction of multi-family buildings was abolished.

3. Changes concerning cooperative housing

Until the 1990s, housing cooperatives played a dominant role in meeting housing needs in Poland. They functioned on the basis of the Act of September 16, 1982, Cooperative Law (Journal of Laws No. 30, item 210). It was not until December 15, 2000 that the act on housing cooperatives was passed (Journal of Laws 2001, No. 4, item 27), the aim of which was to create separate legal rules for the functioning of housing cooperatives, providing cooperatives with greater control of the cooperative's management.

The main goal of the operation of housing cooperatives is to satisfy the housing needs of members of the cooperative and their families. Setting up a cooperative is simple - the minimum requirement is ten founders. A housing cooperative is a legal person, and within it there is a general meeting of members of the cooperative as the highest body, as well as the supervisory board and management board that conducts the day-to-day affairs of the cooperative. The members of the cooperative acting jointly form the general meeting of cooperative members, which must be convened at least once a year. The equality of its members is the principle of the cooperative.

Currently, Polish regulations distinguish two types of rights to cooperative premises that may be granted to members of a cooperative: cooperative tenant rights to a dwelling and cooperative ownership right to a dwelling [10]. Previous regulations also concerned the right to a single-family house in a housing cooperative. The statutes of each cooperative indicate what types of rights the cooperative offers to its members. Each of these rights arises under an agreement concluded by the cooperative with its member. Membership in a housing cooperative is currently granted by law to everyone who is entitled to use the cooperative rights to the premises.

A co-operative tenant right to a dwelling is a right similar to tenancy. A contract specifies the amount of the housing contribution to be made by a member of the cooperative. Only one person may be entitled to it, with the exception of spouses. The cooperative remains the owner of the flat, and the tenant is entitled to use the premises and is obliged to pay the fees. This right is not transferable and not enforceable. In the event of the tenant's death, this right expires, in some cases it is possible for a relative (spouse, children) to enter into the right.

The cooperative ownership right to the premises is included in the category of limited property rights. This right is similar in content to ownership, although the housing cooperative remains the owner of the flat. It may belong to several people. This right is transferable and hereditary, and can also be enforced. A land and mortgage register may be established for such premises, and it may be the subject of collateral in the event of taking out a mortgage. Housing cooperatives are required to keep a register of premises for which a land and mortgage register has been established. Most of the cooperative flats are still operating in the legal form of the cooperative ownership right to the premises, despite the fact that since 1997 its creation has no longer been possible.

The cooperative ownership right to a dwelling was a surrogate for the ownership of a dwelling at a time when the provisions of Polish law did not provide for the legal structure of separate ownership of premises in multi-family buildings. Currently, the legislator considers the cooperative ownership right to the premises to be a redundant form in view of the most powerful form of holding the premises, which is the ownership of the premises. At the same time, the legislator does not interfere with existing rights, and does not force cooperative members to transform them into ownership, but provides such a possibility. Pursuant to the provisions of the Act on Housing Cooperatives, at the request of the entitled person, the housing cooperative is obliged to separate the premises and sell the right to ownership. A member of the cooperative is obliged to pay all fees related to the maintenance of such flat, as well as to reimburse the costs of building the flat, if they have not been covered yet. In addition, the housing cooperative is not allowed to charge any fees for the conversion of rights to the premises. In this way, a housing community may be established in a building previously managed by a housing cooperative, and a departure from the cooperative management regime to the rules of the Act on the

Ownership of Premises may occur. The trend is noticeable of the loss of management over some buildings by housing cooperatives. This is owing to the separation of the ownership of the premises in them and the decision to entrust management to an entity other than a housing cooperative. Thus, in the resources of housing cooperatives there may be separate premises, which have been created either as a result of the transformation of cooperative rights to premises, or as a result of the construction activity of the housing cooperative after 1997, i.e. when the possibility of creating cooperative ownership rights was abolished.

The number of cooperative flats made available for use in Poland is gradually decreasing. In 2000 there were 24,400 of them, and in 2010 only 5,025 [11]. According to the data as of December 31, 2018, there were 2,030,000 cooperative flats in total (out of a total of 14,615,000 flats in Poland) [12], which means that the number of cooperative flats is still very large. Large housing cooperatives operate in many Polish cities, managing housing estates built in the 1970s and 1980s. Currently, they focus on maintaining the good condition of buildings that require renovation, so their activities are limited to management only. In addition to large cooperatives, there are also small ones that have one or several buildings in their resources. Housing cooperatives in Poland may still carry out construction investments, and the premises may be handed over for use either on the basis of a cooperative tenant's right to a dwelling, or sold as a separate ownership of the premises. Currently, however, the activity of housing cooperatives in the area of erecting residential buildings is small.

4. A new trend - ownership of premises

The beginnings of the free market in the area of housing in Poland were characterized by dynamic processes consisting in the starting of activities in the area of housing construction by entities other than housing cooperatives. The construction of buildings for the purpose of creating company flats was stopped, as most of the workplaces were undergoing a deep restructuring and were getting rid of their property, or went bankrupt. In response to the needs of the housing market, a category of entrepreneurs which we call developers was created dealing with the construction of buildings and the sale of flats. They carried out the entire investment and construction process, partially from their own funds in conjunction with investment loans, and partially from the funds paid in by flat buyers, also from mortgage loans. The number of flats put into use in the developer system grew very rapidly. In 1995 it was 2,800 flats, in 1998–9,000 flats, and in 2004 - as many as 24,300 flats [11].

Many people in the late 1990s and early 2000s decided to buy a flat in the development system, either to meet the housing needs of their own family, or for investment purposes. At that time, the demand for new flats was huge, and it was considered that those built by developers were more attractive. This was owing to the fact that multi-family buildings were erected by developers using modern technologies, according to individual designs in the shaping of functional and spatial layouts of flats. As a result, they differed significantly from the uniform, repeatable shapes of blocks of flats erected earlier by housing co-operatives.

However, the lack of appropriate legal regulations, in particular in the area of consumer protection, sometimes resulted in negative phenomena, which entailed the loss of funds paid by those interested in buying flats. There were cases of bankruptcy of developers and other cases of failure to perform contracts owing to the fault of the developer, and the financial loss was suffered by people waiting for flats, which was a serious problem in the early period of growth of the development market in Poland. The imposition of contract templates by developers was an important problem, which resulted in a worse position of the consumer [13]. In this case, the free market led to behavior prejudicial to the weaker party to the contract, which was the buyer of the premises, and there was no legal regulation to protect the contractor.

The problem of the protection of flat buyers has been discerned in the literature [14] and in the jurisprudence of courts. On September 16, 2011, the act on the protection of the rights of buyers of a flat or a single-family house (Journal of Laws No. 232, item 1377), in Poland commonly known as the Developer Act, was passed. This regulation appeared relatively late, as appropriate legal provisions were already in force in other European Union countries. This act applies to the primary housing market. Its main purpose is to protect the rights of buyers of flats or single-family houses as part of legal relations with entrepreneurs professionally involved in erecting buildings. Before concluding a contract with a potential buyer of premises, the developer is obliged to provide the buyer with full information on its legal status, a history of completed investments, and details of the planned investment. The regulations impose an obligation on the developer to prepare a prospectus in accordance with the template specified by the legislator. The standard of performance of the flat and its price must be agreed at the stage of concluding the developer contract. This means that the buyer of the premises, even before the construction of the building, receives information about how the building was constructed, its interior and knows the price for which the apartment is to be purchased. Another example of securing the rights of the buyer of premises is the statutory definition of the minimum content of the development contract, which must be concluded in the form of a notarial deed. The claim to build the premises and transfer its ownership to the buyer is entered in the land and mortgage register. Another important factor is the obligation for developers to set up special bank accounts dedicated to a specific investment, the funds from which cannot be allocated to other projects. Owing to the introduction of this Act, the buyers of premises gained effective legal protection in their relations with developers. Currently, the developer construction market in Poland continues to grow, with 130,900 flats delivered for use in 2019 [11].

In April 2021, a new developer act was adopted - the act on the protection of the rights of buyers of a flat or single-family house and on the Development Guarantee Fund, which is to replace the 2011 Act. The purpose of introducing the new act is to increase the protection of buyers of premises, improve the security of legal transactions, and increase the level of acceptance of the regulations by entrepreneurs. The range of the act was extended to the purchase of garages and commercial premises. The new act provides for the establishment of a Development Guarantee Fund, to which developers will pay contributions. Guarantees of payment from this Fund will cover all contracts concluded by the developer with the buyer of premises. Raising the standards of protection of the flat buyer will make the purchase transactions of flats or premises for other purposes on the primary market safer for buyers.

Housing communities are established by law in multi-family buildings where the premises are separated for ownership. Polish law distinguishes between small and large housing communities. According to the current regulations, a small community has up to 3 separate premises, so there are few of them and they do not play a significant role at present. Housing communities in Poland do not have legal personality, but have been endowed with legal capacity, so they are entities separate from their members [15]. Their basic role is to manage the common property, i.e. to maintain the building and the area around the building in a proper condition, and conclude appropriate contracts with service providers. The owners of the premises are obliged to bear the costs of maintaining the common property by making advance payments to the bank account of the housing community. A share in the communal areas is a right related to the ownership of the premises, which means that the sale of these rights is only possible jointly.

The management regime in large housing communities is based on the distinction between essential activities, which are decided jointly by all owners of premises by a majority of votes, and ordinary activities, where the management decides [16]. The owners of the premises or people from outside the group of owners may be elected to the community board. The community board conducts current affairs and represents the housing community. In a housing community, a manager may be appointed under a contract, then the community board is not elected. Each owner has the right to share control of the activities of the board, which includes inspection of the community records and the right to know all relevant information. Voting in the housing community is proportional to the shares in the communal areas, which means that the person who has a larger flat has more votes. The same rules apply to the costs of the communal area - in proportion to the size of the share.

The functioning of housing communities in Poland is based on simple and clear rules. It works perfectly in the case of small and medium-sized communities, as it provides real influence on the part of each owner of the premises on the decisions made by all owners. Sometimes, however, housing communities have several hundred flats which, with such a large number of owners, may cause practical problems related to the management of the real property.

Flats in housing communities are very popular, especially in new buildings, in newly built housing estates. Buyers of flats very often use mortgage loans, thanks to which they can afford to buy a flat. In 2018, there were already more flats in housing communities in Poland than in housing cooperatives, the statistical data indicate the number of 2,967,000 flats [12].

5. The current status - the path to sustainable housing

After thirty years of development of the housing sector in Poland in free market conditions, it should be stated that it is very diverse and is constantly changing. The housing situation was undoubtedly culturally determined, but most of all it resulted from the adopted social policy and economic development strategy. The housing shortage turned out to be an inherent feature of Polish reality, characteristic of both the extreme inequalities of the interwar period, the egalitarianism of the Communist system, and the new economic situation [17]. The period of cooperative construction has irretrievably passed, but a huge number of flats still remain in the cooperative stock. In addition to modern, comfortable housing estates in many cities, there are districts with neglected buildings that require enormous expenditure on insulation or renovation. Housing estates erected by housing co-operatives in the 1960s and 1970s require a special renovation effort. Actions are required in the technological and architectural sphere, as it is necessary to improve the functional quality of flats, and sometimes to add balconies or lift shafts. It was only in 1997 that Polish law introduced an obligation to take into account the needs of disabled people when designing and constructing buildings. However, most multi-family buildings built before that year have not been adapted to the needs of disabled people, and their adaptation is often impossible because of narrow staircases or corridors inside the premises, or a very small size of the rooms.

For years, various types of instruments have been introduced to improve the housing conditions and the condition of buildings erected in the pre-war period, or in the times of the Polish People's Republic. Such instruments include bank loans granted on preferential terms to housing communities or housing cooperatives for renovation purposes, and renovation bonuses granted to tenement house owners who carry out their renovation. The problem concerns not only the flats and buildings in which the flats are located, but also the surroundings. The priority is to restore the balance in towns and cities so that they provide good conditions, not only for housing, but also for spending leisure time. Social expectations are growing not only with regard to the standard of the dwelling itself, but also the standard of the living environment.

Poland's accession to the European Union, which took place on May 1, 2004, caused changes also in the housing market. Numerous legal acts had to be adapted to the standards in force in the member states. Thanks to various types of aid programmes, it has become possible to restore valuable historic buildings, and the beneficiaries of the aid include municipalities and religious associations.

The condition of the historic parts of cities is currently very different, apart from the restored tenement houses, there are also some that require thorough renovation. Therefore, on October 9, 2015, the Revitalization Act (Journal of Laws 2021, item 485) was adopted, the purpose of which is to provide a legal framework for the processes of renewal of cities and smaller towns. These processes are carried out by the relevant municipalities, on the basis of communal revitalization programmes, and they consist in removing dilapidated areas from their critical condition through integrated activities concentrated territorially for the local communities, space and economy. The participants of the revitalization processes are residents of the revitalization areas as well as owners, perpetual usufructors of real estate and real estate managers located in that area, including housing cooperatives, housing communities, as well as the local and public authorities. As part of revitalization programmes, damaged city areas, such as postindustrial sites, are restored.

Gradually, measures are also taken to improve the air quality in cities. For years, subsidy programmes have been in place for individual building owners to replace coal stoves with other heating systems. From 2020, there is an obligation in Polish law to connect newly built buildings to the municipal heating network in order to avoid equipping new or renovated buildings with systems emitting exhaust fumes.

In many places, on the initiative of local authorities, numerous measures are taken to improve the quality of life, increase the amount of green areas, and build or restore areas for recreation. In the development programmes of numerous municipalities, the improvement of living conditions in cities is an important goal for the implementation of which various activities are undertaken. Parks, squares, and riverside areas, which for years have remained neglected and unused by the inhabitants, are being revitalized. Much emphasis is also placed on expanding telecommunications networks to ensure universal access to the Internet.

The development market is growing, but the demand for new flats is not weakening. According to the data from 2018, 84% of Poles live in their own flats, and only 16% in rented premises [11]. This proves that the ownership of a flat is of great importance as a right that ensures financial stability and the security of the family's existence. As it seems, for people born before 1980, this may be the result of experiences from before the transformation period, when house ownership was unavailable and housing cooperatives or work establishments remained almost the sole suppliers of housing. The Polish housing market still remains a market in a period of growth, which will probably take place until the housing needs of the society are met.

6. Summary

The history of housing law in Poland has resulted in numerous changes in order to search for solutions that would ensure an increase in the number of flats.

DOI: http://dx.doi.org/10.5772/intechopen.98644

Concepts changed as the socio-economic system changed. After the Second World War, the cooperative model was dominant, assuming obligatory membership in housing cooperatives. Until the 1990s, housing cooperatives played a dominant role in meeting housing needs in Poland. After 1989, Poland underwent a political, economic, and social transformation. The command-and-control economy model was replaced with the free market economy model. In response to the needs of the housing market, a category of entrepreneurs which we call developers was created dealing with the construction of buildings and the sale of flats. Then, legal regulations relating to the activities of developers were developed. Housing communities are established by law in multi-family buildings where the premises are separated for ownership. The period of cooperative construction has irretrievably passed, but a huge number of flats still remain in the cooperative stock. After thirty years of development of the housing sector in Poland in free market conditions, it should be stated that it is very diverse and is constantly changing. The housing situation was undoubtedly culturally determined, but most of all it resulted from the adopted social policy and economic development strategy. The housing market in Poland is still in a phase of dynamic growth and ways are being sought to increase the number of housing units available for purchase or rent.

Author details

Aleksandra Sikorska-Lewandowska
Department of Commercial and Maritime Law, Nicolaus Copernicus University, Torun, Poland

*Address all correspondence to: asl@umk.pl

References

[1] Andrzejewski A., Polityka mieszkaniowa, Warszawa 1987, pp. 119-129.

[2] Daszyńska-Golińska Z., Mieszkanie jako problem polityki społecznej, Ruch Prawniczy, Ekonomiczny i Społeczny 1928, No 3, pp. 279-284.

[3] Koryś P., Wolny rynek, państwo i ekonomia społeczna w Polsce w 20-leciu międzywojennym, http://www.tradycjegospodarcze.pl/tekst/47

[4] Radwański Z., Pozycja społeczna, treść i charakter prawny odrębnej własności lokali, in: Studia Cywilistyczne 1968, vol. XI

[5] Bednarek M., Prawo do mieszkania w konstytucji i ustawodawstwie, Warszawa 2007, p. 101

[6] Kisielowska H., (ed.) Prawo mieszkaniowe. Regulacje i praktyka, Warszawa 2015, p. 20

[7] Polska 1918-2018, Historia w liczbach, Warszawa 2018, p. 138. https://stat.gov.pl/files/gfx/portalinformacyjny/pl/defaultaktualnosci/5501/34/1/1/polska_19182018.pdf

[8] Zaniewska H., Thiel M., Środowisko mieszkaniowe w zrównoważonym rozwoju miast i osiedli – aspekt społeczny i urbanistyczny, Technical Transactions. Architecture, Czasopismo Techniczne. Architektura 2007, vol. 3, pp. 312-322.

[9] Kaliński J., Transformacja gospodarki polskiej w latach 1989-2004, Warszawa 2009.

[10] Mączyński A., Dawne i nowe instytucje polskiego prawa mieszkaniowego, Kwartalnik Prawa Prywatnego 2002, No 1, pp. 90.

[11] Rocznik statystyczny Rzeczypospolitej Polskiej 2002, GUS, Warszawa 2002, pp. 228.

[12] Rocznik statystyczny Rzeczpo spolitej Polskiej 2020, GUS, pp. 330 https://stat.gov.pl/obszary-tematyczne/roczniki-statystyczne/roczniki-statystyczne/rocznik-statystyczny-rzeczypospolitej-polskiej-2020,2,20.html

[13] Raport z kontroli wzorców umownych stosowanych w umowach deweloperskich, Warszawa 2007, pp. 23-68.

[14] Gliniecki B., Umowa deweloperska. Konstrukcja prawna i zabezpieczenie wzajemnych roszczeń stron, Warszawa 2012, pp. 297.

[15] Sikorska-Lewandowska A., Legal Status of Housing Communities in Poland, International Real Estate Review 2020, vol. 23, pp. 763-775.

[16] Sikorska-Lewandowska A., The housing community in Polish law: methods of management of communal areas, Studia Iuridica Thoruniensia 2017, vol. XX, pp. 299-311, http://dx.doi.org/10.12775/SIT.2017.015

[17] Wojtkun G., Standards of contemporary housing, Przestrzeń i forma 2012, No 17, pp. 301-322.

Chapter 9

Sustainability Features of Jeddah Traditional Housing

Maged Attia

Abstract

Sustainability is clearly represented in Jeddah old city where traditional builders have developed unique practices that promote environmental, social and economic qualities. The urban form and buildings proved efficiency towards the conservative society, the pattern of life and the prevailing hot-humid climate. The open space system, characterized by narrow walkways and wider intersections, provided shaded and ventilated places for walking and communicating. Houses were configured according to social traditions that imply the separation between private and public life. Walls were constructed of locally coral stone blocks known by its thermal insulation ability, then finished with white color which acts on reflecting sunlight and reduces the heat absorption. Large wooden latticed windows admit daylight but control heating and ventilation, while providing privacy for the family. Moreover, rainwater was collected and stored in basement reservoirs. However, attempts to record these manifestations of sustainability have always been subjective. Therefore, based on long experience of academic work in relation to historic Jeddah, the present chapter intends to reintroduce previous knowledge but supported by evidence whenever possible hoping that it can help formulating guidelines for effective and sustainable alternatives. This is of great benefit to current professionals.

Keywords: Sustainability, traditional housing, local architecture, Jeddah, Saudi Arabia

1. Introduction

Saudi Arabia occupies a location between Asia and Africa between latitude 18° south and 28° north inhabiting most of the Arabian Peninsula. The western side of the Arabian Peninsula, alongside the Red Sea, is occupied by "*Tehamah*" Plain with a width that ranges between 6 and 60 km. Jeddah city is located at 21° 30′ latitude (**Figure 1**) as the harbor for the holly city of Makkah and as a trade hub with Yemen and Egypt. It was bounded by a wall with a number of gateways that varied along time until the removing of this wall in 1947 [2]. Being a major seaport on the Red Sea, Jeddah has always been a cosmopolitan city whose culture and architecture has been influenced by others, mostly Egypt, Turkey, India, Indonesia and Yemen.

Climate in Jeddah is hot and humid almost throughout the year. The average maximum temperature is around 39°C. Minimum rates of temperature recorded at night decreases by 10°C from the maximum ones recorded at daytime [3].

A short visit to the historic core of Jeddah "*Al-balad*" can directly convey a comfortable feeling in both the architectural and urban context. This brings back to

Figure 1.
The location of Jeddah city in the Western Province [1].

mind the question of how these buildings could manage to respond to crucial concerns like the harsh climate, available building materials, construction technology as well as customs and traditions. Most of these items have become an area of research from different viewpoints, and sustainability is no exception.

Sustainability scope is not restricted to environmental practice but comprises issues related to social and economic interests. Environmental sustainability is concerned with sensitivity while interacting with the developmental location and its components. Social sustainability is concerned with maintaining identity and local socio-cultural traditions while enhancing the quality of life. From an economic viewpoint, sustainability emphasizes the optimum utilization of available resources, reducing operation and maintenance cost and commercial vitality [4].

Without knowing its meaning, sustainability was deeply inherited in the traditional built environment of Jeddah. Many passive strategies for improving thermal comfort on both the architectural and urban level were adopted utilizing local resources and technologies. Outdoor and indoor spaces are of reasonable size, flexible for more than one use and incrementally extend according to need and ability. Local culture, which is heavily influenced by Islam, was directly reflected on the city urban tissue, the house architecture down to the smallest details.

The orientation of masjids towards Makkah for prayer "*qiblah*" affected shaping the surrounding paths and buildings. The morals of privacy, hospitality and humility are essential in religious guidelines which urge the veiling "*hijab*" of women from strangers' eyes, separation between men and women, rights of families in communities, neighbors' rights and not hurting him, not looking upon others' houses and turning a blind eye (turning the gaze away) [5]. In a response, paying attention to guests' entrance and reception areas, isolating family movement, screening windows and openings as well as balustrade, simplicity, and abstraction are major design considerations in the historical house. These are agued below.

2. Urban pattern configuration and sustainability

Like old Islamic Arab cities, the urban pattern of historic Jeddah, is characterized by its organic open space system (**Figure 2**). This urban pattern has emerged with

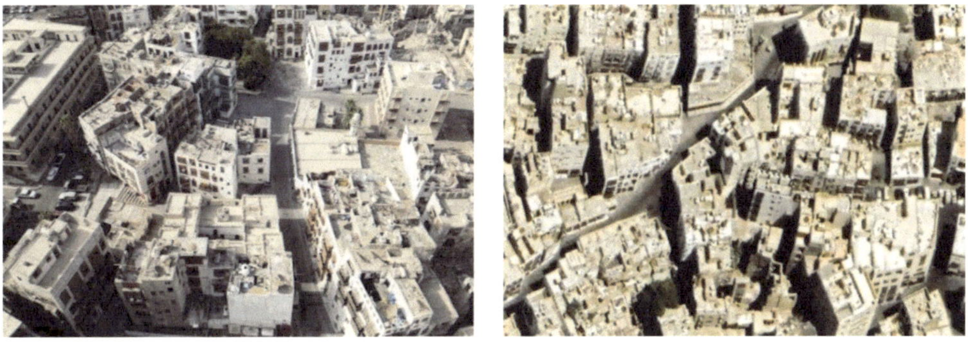

Figure 2.
Building forms in historic Jeddah.

Figure 3.
Map of historic Jeddah showing its organic open space system [7].

intended control over the microclimate of the area. Responding to the hot-humid climate, open space system adopted a network of narrow paths "*harat*" leading into intimate open spaces "*barhat*" [6]. While, buildings are intense, converge and adjacent (**Figure 3**). Paths vary from main or major (12–20 m), primary or local (4–10 m), and secondary (2–4 m). With a height of two to five storeys, building heights range between 8 and 20 m. Width to height aspect ratios are thus in the range of 1/2 and 1/5 or narrower in many cases (**Figure 4**).

This compact configuration of the urban tissue could adequately protect the buildings and the open spaces from the harsh weather by offering abundant areas of shade and shadows. Yet, complete shade was intentionally avoided. For outdoor spaces to be healthy, paths are exposed to sunlight for a short time, while the open spaces are receiving it for longer times along the day. The best-known benefit of sunlight is its ability to enhance providing the body with vitamin D. A half-hour in the sun can initiate enough amount for a whole day. But, most public health messages have focused on the hazards of too much sun exposure where it can contribute to sunburn or skin diseases [8].

However, open spaces will receive more intense solar radiation causing their contents to be heater than the shaded paths. Then, the hot and less dense air creates a dynamic thermal system motivated by the cooler and more dense air. Cool air masses accordingly flow from the narrow paths to replace the hot air with different velocities that help to alleviate the air temperature and the impact of humidity [9]. Narrowing paths cross-section contributes accelerating the air currents. Simulation with ANSYS R19.2 software accords with the phenomenon (**Figure 5**). Compared with modern streets, air temperature of Jeddah traditional open spaces is found to be cooler by more than 3°C, especially during peak heat hours [10]. Moreover, buildings are mostly configured with spaces in between to facilitate the movement of air around them.

Path orientation plays another role in influencing local climate. Relatively speaking, the urban tissue is configured taking the shape of stripes. Some of them are perpendicular to the coast allowing the sea breeze to smoothly penetrate through buildings, while the others are aligned with the north–south direction perpendicular to the path of sun which keeps their paths in shade most of the day and eases channeling airflow smoothly into the city's fabric.

Figure 4.
Paths with different aspect ratios.

The effect of path orientation and its aspect ratio on reducing the impact of the local climate was investigated by ENVI-met, a computer simulations software (**Figure 6**). The results show that streets with higher aspect ratio (1/2.5) have better outdoor thermal comfort conditions than the ones with lower aspect ratio (1/0.5). Due to more urban shading that the buildings create, temperature can be decreased to 22°C in PET index. The preferable order of street orientations was found to be

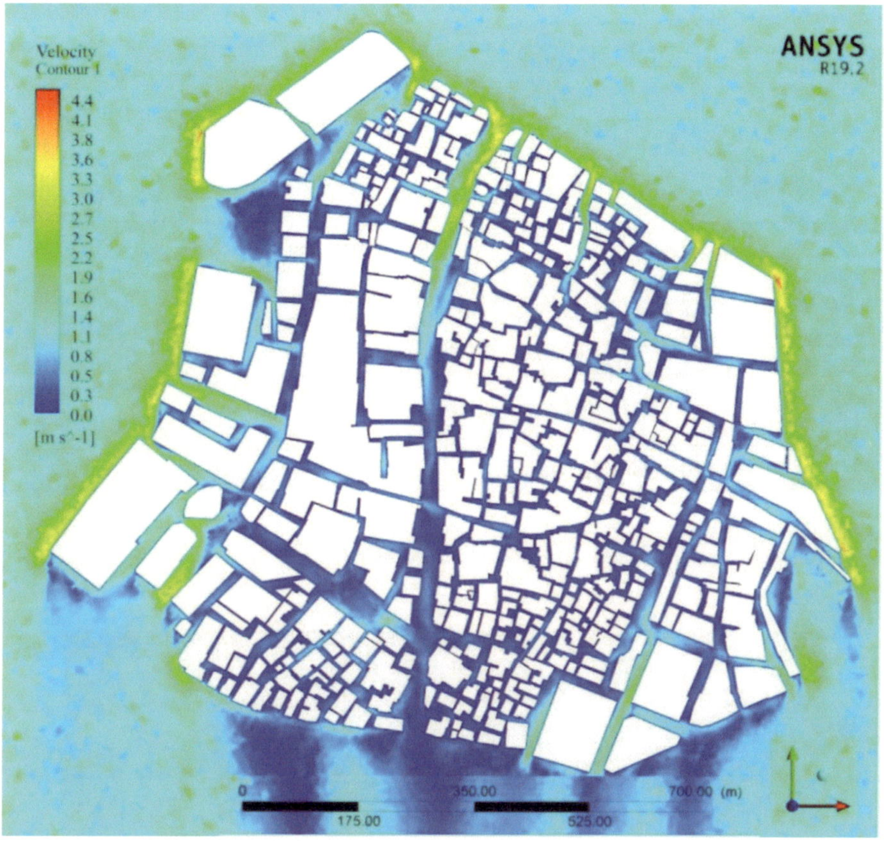

Figure 5.
Wind velocity in the open spaces of historic Jeddah.

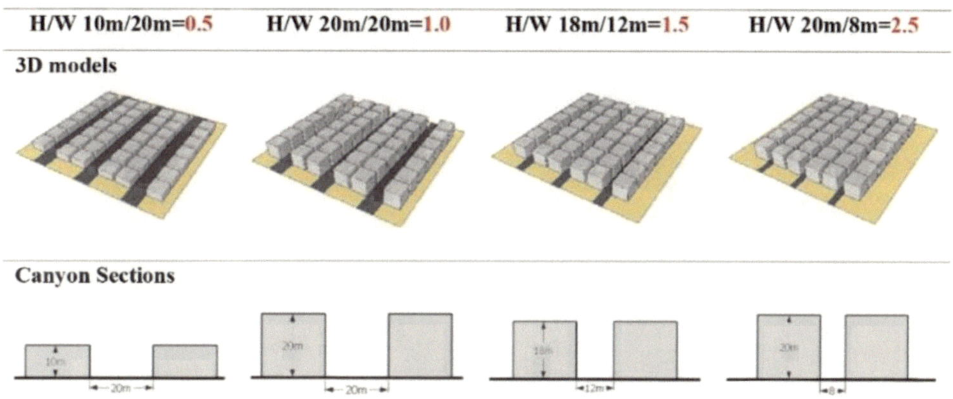

Figure 6.
Examined paths H/W aspect ratio.

N-S, NW-SE, E-W and NE–SW. Also, increasing the aspect ratio reduces, to varying degrees, the air temperature (up to 1.7°C), mean radiant temperature (up to 33° C), the wind velocity while increasing the relative humidity (up to 5%) [11]. However, decrease in wind velocity does deny its existence. To great extent, results are in line with the conditions in historic Jeddah.

The urban pattern of historical Jeddah did not handle with the environmental framework only. With the same efficiency, the urban pattern maintained social and

Figure 7.
Vitality of main paths and open spaces maintained by walkability.

Figure 8.
The axial map of historic Jeddah produced by integration-Rn.

economic sides of sustainability. The hierarchical and compact open spaces system provided different levels of spaces ranging from public, semi-private to private. Each level of the spaces promoted different form of social relationship among residents. Small open spaces and secondary paths were used as playing areas for children to have fun under the supervision of mothers who manage to assemble and chat around. In larger scale spaces, men settlers of adjacent houses sit in groups on raised floor. Also, in these spaces sons can safely play overseen by fathers. Moreover, allocating shops integrated with the residential use on main paths and open spaces, as well as communal services like masjids, enabled inhabitants to move easily in between. This approach, known as mixed uses, encourages walkability, ensures safety and natural surveillance and supports commercial vitality (**Figure 7**).

A Space Syntax study attempted to understand the effect of spatial configuration on movement patterns (**Figure 8**). Outcomes of the study illustrate that the total pedestrian density and the Intensity of moving people are concentrated on the major paths (red-orange in the figure), followed by the local paths (yellow-green lines), and then by the secondary paths (blue lines). This explains the appropriateness of each segment for the intended function i.e., privacy and safety for women and children, controlled openness for men and vitality for commercial movement [12].

3. House design and sustainability

Like the urban pattern, sustainability with its integrated concept has been considered in the historic house of Jeddah too. It was shaped to meet local traditions, local climate and economic abilities. The well-established social traditions dictated the separation of private and public life which was reflected on allocating spaces in the house. There is a spacious zone at the entrance "*salamlek*", which is donated to men settlers where strangers are received. While, the other section of the house "*haramlek*" is dedicated to women and their life to provide maximum protection from the eyes of strangers. This however produced a dual system of movement to allow using private and public spaces severally [13].

On the ground floor, the house has its bigger rooms, such as reception room "*dehleez*" and the guest room "*majlis*", in the front section (**Figure 9**). These rooms are generally higher than others and the floors are kept wet to maintain them cool in hot days. Houses on main paths contain shops, storage rooms and warehouses, each has separate entrances from the path and not connected to the rest of the house.

Other facilities, such as the staircase "*daraj*", the kitchen, storage areas, toilets as well as sleeping rooms share the rear part of the house. Number of rooms in the house varies according to the ability of the residents. However, the house usually has two entry points. The main entrance is placed on street preceded by a number of steps and mostly used for men. On one of the other sides, another entrance for women and family members is located. The family entrance leads to the staircase which does not have any visual contact into any room of the house to maintain privacy.

Reception room acts as a transitional zone that links the entrance door and the inner areas of the house. Men gusts may use first floor in special occasions, but upper floors are only confined to family life. Upper floors contain the women guest room which is the largest and most revered space in the house. Its location on the main facade offers the preferable overlook and ventilation provided by large wooden latticed windows "*rawsheen*".

Adjacent to this room, there is a small room known as the "*soffah*" which is used as a living room. A dual-use room, known as the rear "*moakher*", is placed at the

Figure 9.
Plans and main elevation of selected historic house.

Figure 10.
Terraces and balustrades details.

back end of the house. It is a medium- sized room which is used as a sitting room for family women by day, and it turns into a bedroom at night. Hence, the house is inhabited by many extended families; each family has its separate suite.

Whenever the house rises, the floor area of each higher storey retracts creating terraces on the rooftops of annulled rooms. However, terraces provide outdoor spaces for women to perform domestic activities like drying clothes. These terraces are visually protected by heigh balustrades with details that provide privacy but allow air to move through (**Figure 10**). Likewise, the building roof provides other outdoor spaces where inhabitants can use for sitting or sleeping in summer nights. As far as possible, houses are arranged to ensure that an overview of other houses' terraces is prevented according to Islamic sense of propriety.

Almost every house has two to five storeys in a cubical form that does not contain any protrusions, except for the "*rawasheen*". The building form is not straightly configured, but with broken masses so that adequate shade and shadows might be caught.

The direct light hitting the Nawar house and its context could be simulated at different times of the day and year. Results indicate that the direct sunlight during the morning and afternoon hours is largely blocked by the geometry of the house itself, and the neighboring buildings which largely shadow the facades in the lower area of the building (**Figure 11**) [14].

In Jeddah climate, a large amount of air is required to move through the house in order to alleviate the impact of high humidity. Stimulating air to flow requires having a positive pressure side and ensuring that the air outflows from the negative pressure sides carrying humidity and the unpleasant fumes. Therefore, most of the house had two to three facades and arranged to have at least one facade facing the preferred airflow, while others are exposed as possible.

Spaces were arranged so that walls do not obstruct the continuity of air flow by providing openings between the rooms. Openings range between *rawshan* or windows on the main facade, small openings at a heigh level on the opposite walls or integrated on the top of doors. The staircase also helps ventilating the house allowing air currents to vertically flow through running from the ground floor to the roof, which is known as the chimney effect. The staircase shaft also encourages the air circulation between floors.

In the same previous study of Nawar house, the air flow through the house was visualized with the Wind Tunnel Pro simulation software. The visualization of the air flow proves that the wind penetrates through the large *rawshan* in the front facade and continues to drive the air cylinder rotating in the room due to vertical temperature differences. This situation leads to mixing the air layers and intensifies the heat and moisture exchange with the wall and floor surfaces. The small high openings in the wall opposite the *rawshan* facade increase the airflow velocity even more and is literally sucked into outdoor (**Figure 12**).

Beside the role played by the staircase to enhance air movement through the house, some houses were provided with air shafts, a treatment that developed from the traditional wind catcher "*malqaf*" (**Figure 13**). Assessing the impact of air shaft on the relative humidity, air movement, air temperature and the CO2 level -an important indicator of air quality- have been a main interest for some studies.

Two scenarios were monitored in Nassif house "*bait Nassif*". In the first case, the window facing the air shaft was closed; the same window was left opened in the

Figure 11.
Nawar house physical model in the light laboratory.

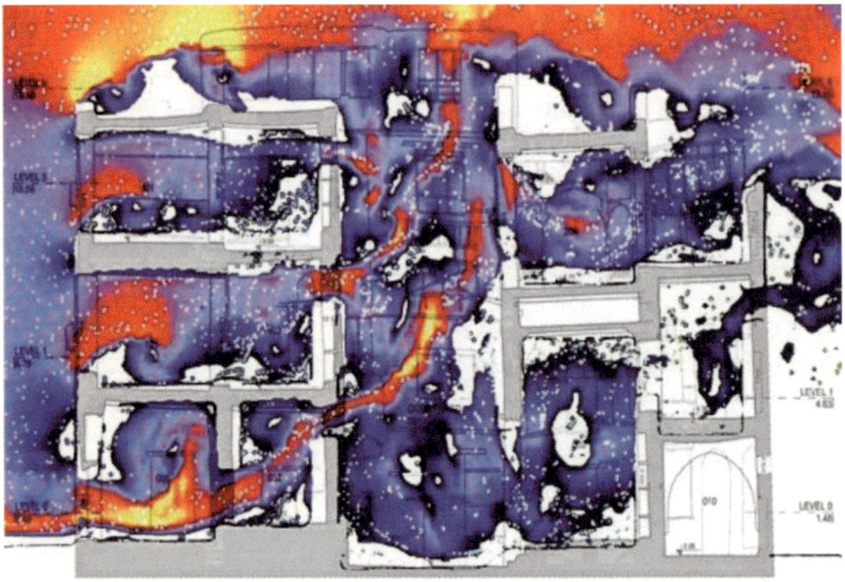

Figure 12.
A vertical section showing the air flow through the Nawar house produced by wind tunnel pro simulation software.

Figure 13.
Air shafts in Nasif house (left: Exterior view) and Nawar house (wright: Interior view).

second. Results proved the efficiency of thermal performance for the studied space in case of the opened window. Air velocity of 1.1 m/s was achieved in the first case, while it dropped to 0.0 m/s in the second. The average of CO2 which was measured at 395 ppm in the first case opposite to 560 ppm in the second. Relative humidity reduced to 59% while it was 65% with the closed window. Air temperature also dropped from 33°C to 32°C when the window was opened. Comparable results were monitored for Noor Wali house [15].

4. Building materials and construction technology

Historic buildings of Jeddah are mostly built with the same materials, technology and principles. Structure relies on a wall bearing system that transfers loads across its section. The walls are built of coral blocks "*Al-kashur*" or "*Al-Mangabi*" stones, then covered by plaster "*nourah*" for insulation and esthetic purposes. *Al-mangabi* stones are cut from coral fossils or from the sea, *Al-arba*^{c}*een* lake or *Al-Mangabi* lagoon, just north of Jeddah. Then, it is shaped in blocks form and used in constructing the walls. *Nourah* is a type of calcareous stone treated by fire and then dismantled, drained and mixed with water. It is used alone in the case of external polishing, while mixed with stone fragments for repairs. The silt, precipitated after rain, is utilized as mortar to cement blocks to each other. Notably, these are natural, renewable and recyclable materials.

Al-Mangabi stones proved efficiency as an insulation and resistance median. Empirical studies indicate that *Al-Mangabi* stone wall of 70 cm thickness gives high thermal mass to the building, which reduces and delays heat gain inside. A difference of 2.7°C (from 35°C to 32.3°C) between outer and inner surfaces of external wall is measured [16]. Also, the thermal properties of the stone were examined in comparison to contemporary building materials. The results confirmed that the change in the temperature of the internal surface of the stone was almost like double-brick wall section with insulation (**Figure 14**). Hence, it was recommended for use in contemporary buildings; the cost of insulation and the energy required for air conditioning can be lowered [17]. Results could be more enhanced with the white color of plaster which reflects the direct solar radiation and reduces the heat gained in the building's mass.

Wood is the other major construction material which was used in structure and decoration. Structurally, Mangrove is widely used in roofing for the length of its poles which reach about 2.8 m and accordingly dictate room widths. While, Teak "*sajj*" and sandalwood "*sandal*" were used to make windows, *rawasheen*, doors and different forms of lattice work. These woods were imported from India, Java or East Africa. Local woods available in the region, such as palm trees and leaves, acacia and juniper were also used. Sustainability of wood has been heavily documented. Though, being imported is a shortage.

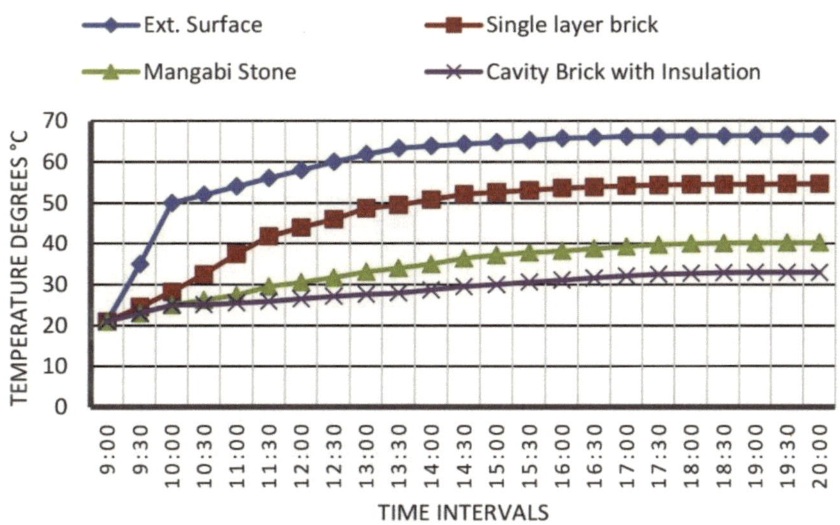

Figure 14.
Thermal performance of Al-Mangabi *stone compared with some contemporary materials.*

However, the construction process starts with a shallow foundation trench to the depth of one or two courses of *Al-Mangabi* blocks. At every six rows (120 cm), a bonding course of round wood beams "*takaleel*" is inserted and locked with short pieces at the ends to ensure homogenous load distribution. They support the building to resist the settlement stresses (**Figure 15**). Also, they enable maintaining the wall by supporting the wooden beams from the external and internal side and changing damaged stones underneath (**Figure 16**).

At ground level, the walls thickness is about 80 cm. The higher the building go, the thickness would be reduced by about 15 cm with each floor. This reduction is taken from the internal side of the wall to create a ledge with sufficient space to place the ceiling beams side by side. Spaces assigned for *rawasheen*, windows and doors would be left according to width. Wooden lintels or arches as well as wooden fixtures of *rawasheen* are then utilized in position. The internal walls are provided with niches "*taqat*" which are used as cupboards. Both openings area and niches served to alleviate the weight of the structure, not ignoring reduction of the walls thickness and areas taken out in balustrades. In many cases, voids covered by *rawasheen* are found to reach around 50% of the wall surfaces [18] (**Figure 17**).

For the roofing of each floor, wooden beams are placed resting on ledges of two parallel walls and firmly impeded inside the wall. Several layers are then implemented above the wooden beams including palm leaves or light weight wood planks, sackcloth, wet soil, crushed gravel, lime and finally clay mortar, which are natural materials are used too (**Figure 18**).

Figure 15.
External views of Al-Mangabi *blocks and wood beams.*

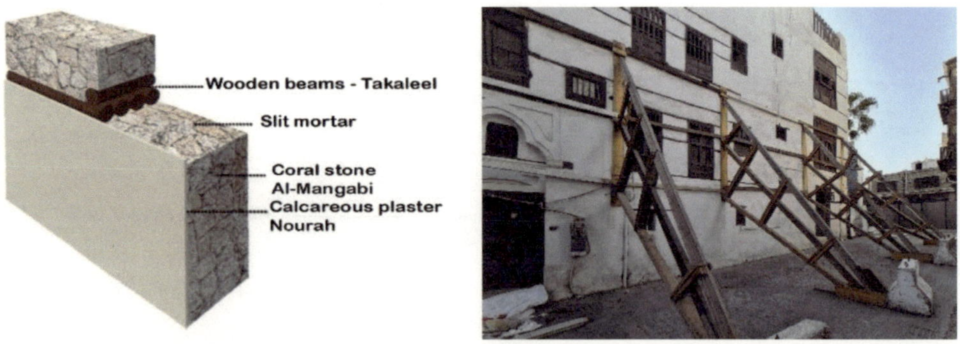

Figure 16.
Supporting traditional wall for maintenance.

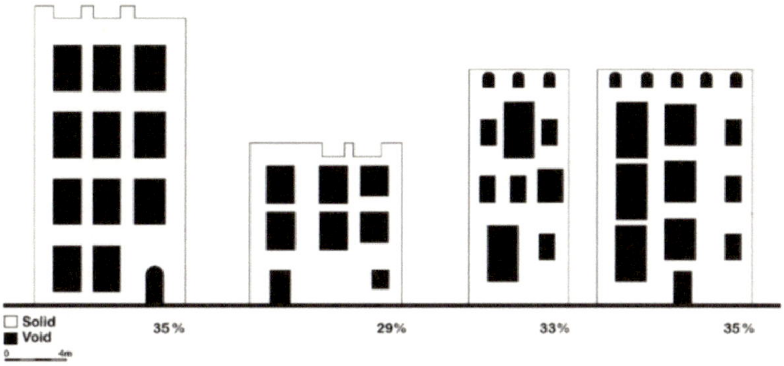

Figure 17.
Solid and void percentage of selected facades.

Figure 18.
Roofing system in the historic house.

From structure view, staircase, is the backbone of the building. Being rectangular stone walls with central column *"fahl el-daraj"* running vertically across the building height, it is considered as a main pillar. Whereas walls of staircase and external walls mutually support each other through the wooden beams.

5. Rawshan

One of the most noticeable elements in historical Jeddah architecture is the *rawshan* which covers large areas of the building's facade. As mentioned earlier, it is a large wooden structure with a recognizable latticed component. Some buildings are characterized by vertical *rawshan* that extends to 3–4 storeys (**Figure 19**), while others contain horizontal ones. *Rawasheen* are normally found projected about 60 cm and are fixed with the aid of wooden cantilever.

Rawshan has many functional and aesthetical roles in the building. Beside protection from the harsh climate and solar radiation, it provides a resting place for two persons lying full-length with the ability to watch outside without being noticed. This is consistent with the privacy values. The lower and upper stripes of the

Figure 19.
View of historical rawshan, *Noor Wali house.*

Rawshan are fixed wooden panels, while the middle contains shutters that can be lifted outwards and upwards to control the air to pass through creating gentle and continuous cross ventilation. The air passing through is employed to cool water contained in pottery grouped in identified place "*shurbah*". Around the top of the *rawshan* there is an ornate projected unit called a *rafraf*; Its projection promotes casting shade on the higher section. Another latticed component, known as the "*ghula*", is often hanging over the lower half of *rawshan* or windows to cover the interior spaces while the shutters are open providing additional privacy [19].

Rawashan provides natural gentle daylight with interesting patterns projected on internal walls and floors. Being finished with moderate soft colors that have a reasonable level of reflectance, *Rawshan* reduces glare in outdoor spaces caused by the building's white walls. Colors reflection determines the amount of light that will be directed inside. Its projection helps to shade the building's facade and the narrow surrounding paths.

Rowshan is also a flexible space that can be added to adjacent rooms. When necessary, it is flexible enough to perform the function of any type of rooms in the house. Too, activities taking place in a room can simply extend in *rawshan* as additional space (**Figure 20**). However, each *rawshan* is unique enough so that it is almost difficult to find two identical pieces in the whole area. There are endless varieties of sizes, shapes, treatments and organizations.

The indoor air temperature of two rooms with *rawshan* were examined from 5 to 31 August 2018. Rooms are typical, but one with opened *rawshan* and the other was closed. *Rawshan* could essentially regulate the indoor temperature during the high fluctuations. Median temperature in the rooms ranged between 32.3–38.4°C when the outdoor temperatures ranged between 30.9–48.7°C (**Figure 21**). Closed *rawshan* caused to delay heat transmission about three hours a day. But, when the *rawshan* was opened, the time regressed to one hour. The open *rawshan* allowed more airflow which mostly alleviated the rise of temperature. Similarly, night ventilation decreased the indoor air temperature assisted to decrease and postpone reaching peak time of temperatures in the space. Notably, these results are with the contribution of building's total thermal mass [20].

Results confirm that the relative humidity declines as air the temperature escalates and vice versa. Due to airflow into the room with opened *rawshan*, it lowered more heat acquisition and permitted relative humidity more than the room with

Figure 20.
Different functions of rawshan *in the inside space.*

closed *rawshan*. Air velocity was measured at 0–8.1 m/s in the outdoor, 0 to 6.9 m/s in room with opened *rawshan*, and 0 to 1.1 m/s in room with closed *rawshan*. The highest frequency of air velocity values measured were found to be (2 m/s - 18.27%)

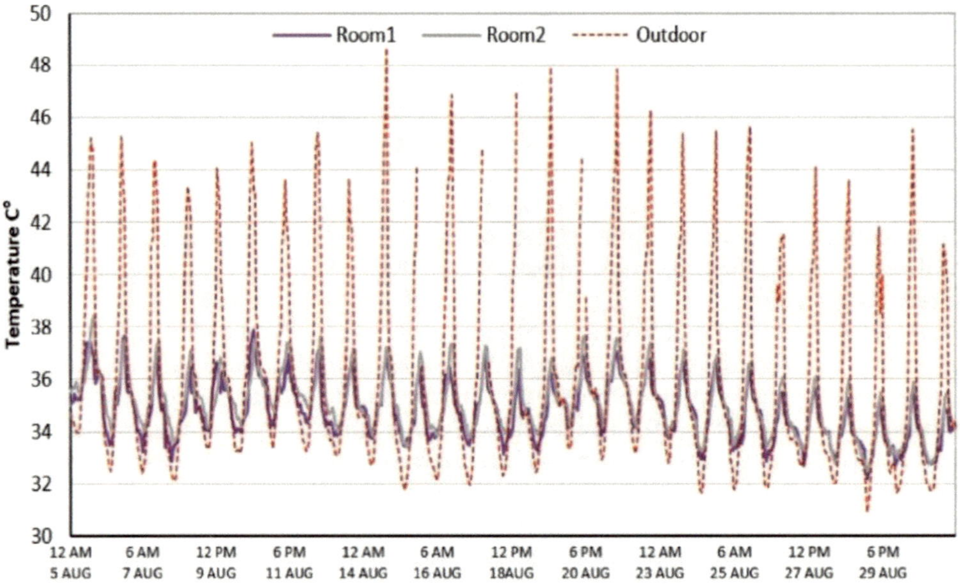

Figure 21.
A comparison between ndoor and outdoor air temperature during the investigated period.

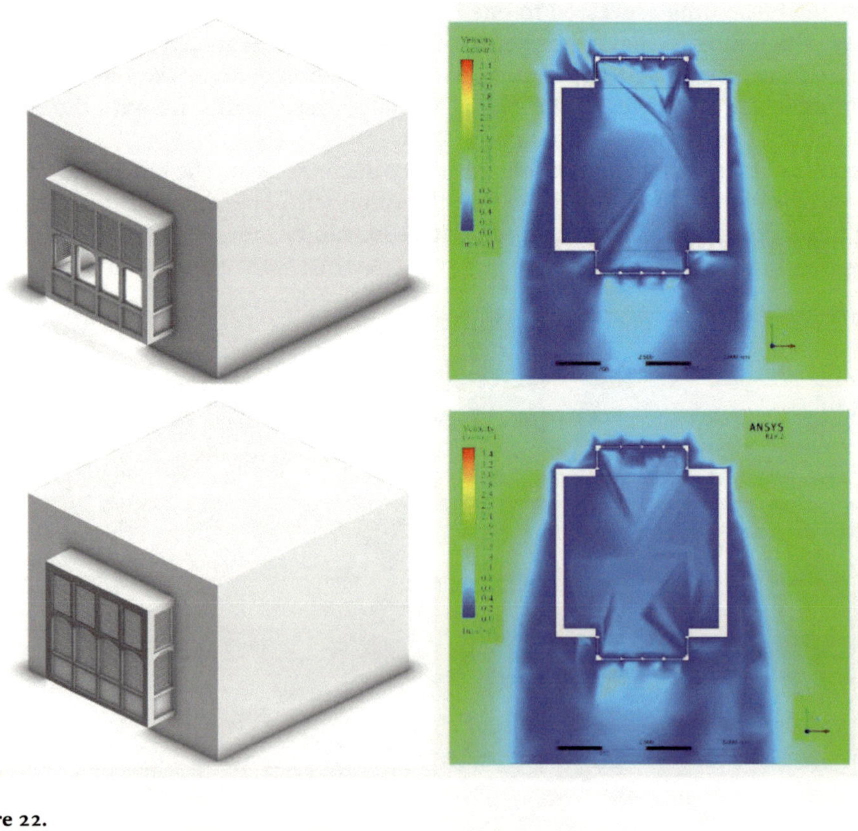

Figure 22.
Air movement through rawshan, *opened (above) and closed (down).*

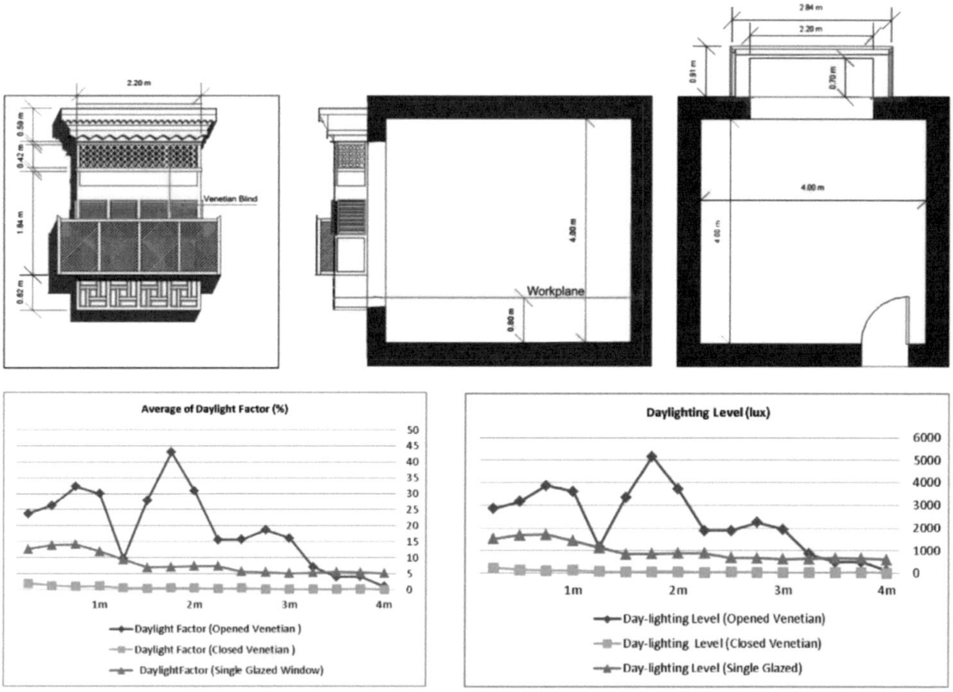

Figure 23.
Daylight factor and daylight level of examined single glaze, closed venetians and rawshan.

in the room with open *rawshan*, (0.5 m/s - 28.9%) in the outdoor and (0 m/s - 90.44%) in the room with closed *rawshsan*. Simulation by ANSYS R19.2 software proved comparable results (**Figure 22**).

The effects of *rawshan* on the performance of daylighting was examined with a computer model (Autodesk ECOTECT) to demonstrate its ability in reducing the glow of solar lighting in interior spaces. It was compared with closed venetians and single glazed windows of the same dimensions. *Rawshan* scenario proved efficiency on both daylight factor and daylight level (**Figure 23**) [21].

Rawshan is thus a passive mean that efficiently reduces buildings dependency on air-conditioning and artificial lighting and accordingly reduces energy consumption.

6. Gateways and ornamentation

Details used to reflect the social and economic strata of the residents. Beside richness of *rawshan*, this was also presented in gateways and ornamentation. Better houses usually have gateways with the finest elaborated woodwork. The principal external doors have two leaves. They are decorated with engraved panels; the right fold contains a smaller one "*khokhah*". The gateway is topped by pointed or semi-circular arches decorated with different motifs (**Figure 24**). The gateway has two purposes; first, it identifies the social status of the owner, and second, it reaches to the *dehleez* leading to the reception hall while controlling field view to protect privacy.

External plaster, added to protect *Al-Mangabi* stone, was developed by adding decorative carving especially on the lower level of the building facades framing the main gateways and windows. Plaster was employed to the stone and instantly

Figure 24.
Main gateways with different levels of details.

Figure 25.
Level of details that reflect the economic strata.

carved while it is still wet. The plaster was profoundly engraved with ornamental geometric or floral designs which were cut so that the upper surfaces sloped downwards to ease water to run off the plaster. The technique was more laborious and more permanent. Again, richness of decoration (**Figure 25**) satisfies a basic human need of self-expression, being noticed, demonstrating the social and economic excellency which is a key ingredient of sustainability.

7. Underground water cistern

Although Jeddah does not get too much rain, but it is subject to serious seasonal precipitation. In the historical houses, rainwater was harvested and collected in

Figure 26.
Rooftop and underground cistern of Nasif house.

underground massive cisterns *"saharij"* with vaulted ceilings for domestic use along the following months. This system works only on gravity without pumps, only pipes from exposed roofs to a tank below. The rate of rain may be as little as 0.5 cm in some seasons, yet it was very necessary. However, family members used to clean up the roof at times when rain is expected in communal gathering known as favor *"faz^c a"*. The rainwater will be directed to the drainpipes that carry it to the lower tank (**Figure 26**). This water harvesting system does not then require pipes throughout the city. These underground cisterns were also found in non-residential buildings like the masjid.

Nowadays, the only reliable source of water for Jeddah is desalination which provides 972,400 m3/day for 3.4 million residents. Desalination processes cost the Saudi authority about 1.87 US$/m3, at a total daily cost of 1,818,388 US$. Only 2% of desalination cost is paid by the residents at a rate of 0.04 US$/m3, while the rest is heavily subsidized by the government [22].

A recent study proves that rainwater harvesting is still feasible, and it can significantly decrease desalination cost if applied in each house. Relying on energy, and corresponding environmental impact, can thus be limited. The study illustrates that rainwater harvesting from rooftops can help reducing the effect of urban flooding and accordingly the amount of water emitted in the sea threating the marine life. It can also reduce the possibility of raising water table which threatens the durability of buildings [23]. It is argued however that the water cisterns on the ground floor may have helped to cool the lower part of the houses [24].

8. Conclusion

The current work raises the efficient sustainability of historical houses that can alleviate relying on fossil fuel and corresponding negative impacts. Keeping in mind the inclusiveness of the concept of sustainability, it can enhance the performance of Jedda's contemporary housing by considering the following:

Environmental issues: Shading in outdoor spaces and paths is a considerable objective; it can be provided by compact development which accordingly alleviate the impact of hot climate. Building configuration can also protect its envelop from overheating. Shading in open spaces and on buildings can be enhanced by projected masses. Likewise, enhancing air movement through the urban tissue is an effective way for creating comfortable outdoor spaces. Also, the natural ventilation for inner spaces of the houses can reduce air temperature and humidity. Passive tactics for promoting shade and natural ventilation, like "*rawshan*" and air shafts, can be examined, simulated and efficiently developed. Too, paths width to height aspect ratio and orientation is a vital tool to control both shading and natural ventilation in outdoor spaces. Research identifies the direction of N-S, NW-SE, E-W as the most appropriate, in association with aspect ratio of 1/2.5.

With the abundance of contemporary building materials, some traditional materials, like "*Al-Mangabi*" stone, continue to prove their efficiency in meeting the harsh environmental conditions. Making use of the properties of such materials, especially the thermal resistance, can inevitably open the scope for more sustainable products and applications.

Rainwater harvesting from rooftops is a feasible approach that can alleviate relying on desalination and reduce the effect of urban flooding; both have serious environmental impacts.

Social issues: Local culture, customs and traditions and religious teachings are inherited in Arab communities and mostly reflected in historic houses. Privacy suggested hierarchical open space system with tiny areas for women and children, private entrance and separate section for women in the house, and special treatments for windows, openings as well as balustrade to protect the family from strangers' eyes. Hospitality advocated fancy gates and spacious reception room for guests. Humility promoted simplicity and abstraction.

Economic issues: Feasibility and economic vitality are a main driver for society. In the historic house, spaces were rationally tailored, flexibly used and incrementally extended to meet family needs. On the urban level, mixed use, and corresponding active trade movement, brought vitality and safety for outdoor spaces. Not contrasting with humility, ornamentation demonstrated the distinction of economic classes.

The discussion herein does not mean to replicate the historical houses but to benefit from sustainability values inherited in. Historical houses were formulated affected by several environmental, social and economic forces. It is important to understand which of these forces are still acting and which are vanishing. In contemporary housing, the reflection of continues forces can be developed in the light of present knowledge.

Acknowledgements

I would like to acknowledge with thanks the role of Arch. Mohamed Hafazalla for processing wind analysis with the aid of ANSYS R19.2 software.

Author details

Maged Attia
Faculty of Architecture and Planning, King Abdulaziz University, Jeddah, Saudi Arabia

*Address all correspondence to: mattia@kau.edu.sa

References

[1] Ministry of Municipal and Rural Affairs. Jeddah City Profile. Riyadh: Ministry of Municipal and Rural Affairs-UN-Habitat; 2019. 133 p.

[2] King G. The Traditional Architecture of Saudi Arabia. London-New York: I. B. Tauris Publishers; 1998. 32-51p.

[3] Alawad A. Using the architectural style of heritage buildings as a tool to avoid health risks: An analytical study of Rowshan in traditional houses in the city of Jedda. Procedia Environmental Science. 2017;37:604-613. DOI: 10.1016/j.proenv.2017.03.047

[4] Attia M. LEED as a tool for enhancing affordable housing sustainability in Saudi Arabia: The case of Al-Ghala project. Smart and Sustainable Built Environment, 2013; 2 (3):224-250. DOI: 10.1108/SASBE-02-2013-0009

[5] Bokhari A, Hammad M, Beggas D. Impact of Islamic values and concepts in architecture: A case study of Islamic communities. In: Proceedings of Sustainable Development and Planning XI; 9–11 September 2020; Online; 2020. p.383–396. DOI: 10.2495/SDP200311

[6] Gruber P, Eissa M. Old town of Jeddah: Results of the building survey of the Al-Nawar house. Architectura Band. 2014; 44:1-16.

[7] Saudi Commission for Tourism and Antiquities Historic Jeddah, The Gate to Makkah: Nomination Document for the Inscription on the World Heritage List. January 2013; p. vi.

[8] Mead M. Benefits of Sunlight: A bright spot for human health. Environmental Health Perspectives. 2008; 116(4):A160-A167. DOI: 10.1289/ehp.116-a160

[9] Magbrabi A. Airflow characteristics of modulated louvered windows with reference to the rowshan of Jeddah, Saudi Arabia [Thesis]. Sheffield: University of Sheffield; 2000.

[10] Al-Lyaly S. The traditional house of Jeddah: a study of the interaction between climate, form and living patterns [Thesis]. Edinburgh: University of Edinburgh; 1990.

[11] Zamzam E. The Influence of canyon orientation and aspect ratio on outdoor thermal comfort in hot humid climate: case study city of Jeddah [Thesis]. Jeddah: Faculty of Architecture and Planning, King Abdulaziz University; 2021.

[12] Rashid M, and Bindajam A. Space, movement, and heritage planning of the historic cities in Islamic societies: Learning from the Old City of Jeddah, Saudi Arabia. Urban Design International. 2015;20:107-129. DOI: 10.1057/udi.2014.6

[13] Kamal M. The morphology of traditional architecture of Jeddah: Climatic design and environmental sustainability. Global Built Environment Review GBER. 2014;9(1):4–26.

[14] Jäger-Klein C, Radinger G, Stumpf W, Styhler-Aydın G. Untersuchungen zur sonneneinstrahlung und natürlichen lüftung an einem typischen historischen wohnhaus in der altstadt von Dschiddah. Bauphysik. 2015; 37(6): 303-362. DOI: 10.1002/bapi.201510041

[15] Mohamed M, Mohamed M. Investigating the environmental performance of the wind catcher in Jeddah. In: Proceedings of Islamic Heritage Architecture and Art II; 17-19 April 2018; Malta: WIT Press; 2018. p.15-26. DOI:10.2495/IHA180021

[16] Ali Z, Alamoudi A, Alajmi B, Khayat E, Alshraim S. Traditional sustainability: environmental design in

the traditional buildings of the Middle East. In: Proceedings of the 30th International PLEA Conference; 16-18 December 2014; Ahmedabad: CEPT University. p.1-8

[17] Bagader M, Mohamed M. A comparison study for the thermal and physical properties between "al-mangabi" and the available building materials for the external walls in Jeddah. International Journal of Engineering Research and Technology. 2020;13(6):1319-1329. DOI:10.37624/IJERT/13.6.2020.1319-1329

[18] Sabbagh M. Sustainable qualities in traditional architectural elements, a study of the rowshan [Thesis]. Montréal: School of Architecture, McGill University; 2007.

[19] Alitany A, Redondo E, Adas A. A new methodology for a detailed 3D modeling and documentation of the complex architectural heritage elements: a feasibility and case study of wooden projected window "the Roshan" in the historical city of Jeddah. Expresión Gráfica Arquitectónica .2014;19(24): 176-187. DOI: 10.4995/ega.2014.2316

[20] Bagasi A, Calautit J, Karban A. Evaluation of the integration of the traditional architectural element Mashrabiya into the ventilation strategy for buildings in hot climates. Energies. 2021;14:530-561. DOI: 10.3390/en14030530

[21] Al-hashimi A, Semidor C. Virtual study of the day-lighting performance of rawshan in residential buildings of Jeddah. In: Proceedings of CIE Centenary Conference (Towards a New Century of Light); 15-16 April 2013; Paris. France: CIE x038:2013. p. 689-696

[22] Al-Ekhbariah TV Channel. Desalination in Saudi Arabia. 2012, Sep 11 [Video File]. Retrieved from https://www.youtube.com/watch?v=92i8-UQsI30

[23] Bogis A, Bork D, Miller P. Are green infrastructure strategies suitable in arid climates? a design feasibility study from Jeddah city, Saudi Arabia International Journal of Architecture and Planning. 2021;1(1):9-18. DOI: 10.51483/IJARP.1.1.2021.9-18

[24] Al-Ban A. Architecture and cultural identity in the traditional homes of Jeddah [Thesis]. Denver: University of Colorado; 2016.

Chapter 10

Proposal of Rural Housing and Habitat Improvement of the Town El Encanto, in Tapachula, Chiapas, Mexico

Lorenzo Franco Escamirosa Montalvo, Carlos Uriel del Carpio Penagos, María de Lourdes Ocampo García, Ángel René Estrada Arévalo, Arturo López González and Roberto Arroyo Matus

Abstract

In the state of Chiapas, Mexico, there are numerous rural communities located in isolated territories and away from important population centers. Families, in essence, have indigenous roots and low economic incomes and, because of this condition, their homes are precarious, unsafe and unhealthy, with many limitations to access basic water and sanitation services, as well as basic health services, recreation, education, communication, etc. This study analyzed the housing and habitat problem of the El Encanto community, located on the coast of Tapachula, Chiapas. As a result, structurally safe, economical and healthy housing proposals were developed, with water and sanitation services, functional spaces with sociocultural characteristics, typology and appropriate to environmental conditions, with the use of ecotechnologies and building materials of the place for the conservation of the environment and biodiversity, in addition, proposals for community equipment for habitat improvement were developed.

Keywords: rural housing, security, habitat, sanitation, sustainability

1. Introduction

In Mexico, the state of Chiapas is characterized by its vast diversity of ecosystems, the result of its geographical location, soil types, climate, etc., too, is recognized as one of the most culturally rich entities in the country, cradle of original villages such as tsotsiles, tseltales, zoques, lacandones, mames, tojolabales, among others, which have resulted in an extraordinary variety of villages that are located throughout the territory of Chiapas, configured by mountains, highlands, depressions, plains and coast. In this regard, there are more than 20 thousand localities, of which 99% have fewer than 2,500 inhabitants and of those 85% have less than 250 and 74% less than 100 inhabitants [1]. In this sense, the dispersion of the population of the entity and the existing orography, are factors that limit the access of localities to the basic services of water and sanitation, health, education, culture, recreation, communication, among other aspects.

The study of the house and the habitat that is presented, corresponds to the town of El Encanto of the municipality of Tapachula in the state of Chiapas, Mexico. Population data indicate that the town has 1,726 inhabitants, of which 51.8% are men and 48.2% women, there are 446 dwellings that, according to their conditions, lack of property and deprivation of the inhabitants to education, the indicators of marginalization and social lag place the locality with the degree of marginalization "Very High" and the degree of social lag "Medium" [2], also, the recent estimates of the Human Development Index, which evaluates the health conditions, education and income of the population, places Chiapas in the last place throughout the country [3, 4].

El Encanto is a town that does not have basic water and domestic sanitation services, public lighting is deficient, its streets have no folder and lacks recreational spaces for family coexistence. Today, it is observed that the families of El Encanto are peasants, fishermen and low-income employees, living at constant risk to health, as a result of the pollution of the water they consume and the environment in general, due to the poor disposal of liquid and solid waste generated. Most of the houses are unsafe and unhealthy, built by the inhabitants with inadequate materials and technical processes, with high vulnerability in relation to the site where they are located, corresponding to the coast of Chiapas with the Pacific Ocean and in the region of greatest teluric activity existing throughout the country, so they are exposed to strong winds, hurricanes and seismic action. On the other hand, El Encanto is located in an area of estuaries and lagoon system with high ecological diversity, especially with extensive mangrove ecosystems, which are being affected by the domestic and economic activities of the inhabitants of the town and other nearby human settlements, since there are processes of environmental pollution and degradation of the original conditions in existing ecosystems, due to the inadequate management of the waste generated.

The analyses of the town of El Encanto were carried out with the application of the Model of evaluation of the conditions of rural housing and the environment [5], which, at first, focuses on the analysis of the current situation of the site, with information obtained through surveys, registration cards, etc., in order to identify and interpret the problems of housing and its immediate environment. The elements analyzed are the socio-economic situation of low-income families, the characteristics and conditions of the housing, as well as the water and sanitation services, the management of the waste generated, the health risks, the structural security of the house, the typological characteristics of the community, the materials of the place suitable for the construction of new houses Etc. In a second moment, the transformation process begins, which in this case, consists of the elaboration of prototypes of safe, economic and healthy housing, for the benefit of low-income families and in accordance with the environmental, socioeconomic and cultural conditions of the inhabitants. The typology, the application of ecotechnologies and the use of local materials with low environmental impact were considered in the prototypes. Proposals for habitat improvement in general were also developed in accordance with the physical and socio-cultural characteristics of the site. In general, the intervention proposals are aimed at improving the quality of life and well-being of people, as well as preserving the ecological design of the area. With this, it is intended to contribute to the development of the inhabitants of El Encanto and to the conservation of the environmental conditions of that region.

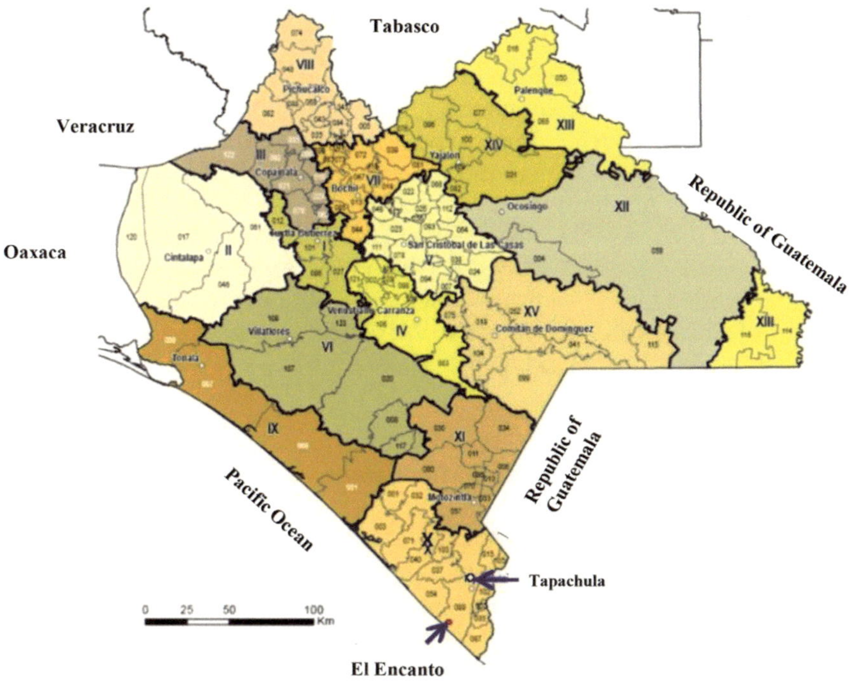

Figure 1.
Location of the "El Encanto", Tapachula, Chiapas. Source: NISG [1].

2. Geographical location of the town El Encanto

El Encanto is part of the municipality of Tapachula de Córdova and Ordoñez, Chiapas, Mexico. It is located 392 km from the city of Tuxtla Gutiérrez, capital of Chiapas, by the road section Tuxtla Gutiérrez-Arriaga and then the stretch Arriaga, Tonalá, Pijijiapán until you reach the city of Tapachula. El Encanto is located 26.5 km from Tapachula by road to Puerto Madero and 2.9 km before reaching the town of Puerto Madero (see **Figure 1**).

3. Background

In January 2019, through one of the members of the team, Dr. Angel René Estrada, communication was established with Ms. Elena Matías Molina, representative of the El Encanto community, with the aim of expressing the interest of the team of academics and students of the Faculty of Architecture, Autonomous Universidad of Chiapas, to carry out research work and develop intervention proposals aimed at improving housing, sanitation, recreational spaces, for the benefit of the inhabitants. On that visit, the first approaches were made to the town, where the lack of basic water and sanitation services, the lack of public recreational and family coexistence spaces, streets without folder, and with little lighting were observed. Most of the homes are inhabited by low-income families and were built by them with the resources available and traditionally with organic materials such as cane, wood, palm, etc., others with conventional materials with cement-sand blocks or solid brick produced on site by the inhabitants, ceilings with galvanized sheet; also, precarious housings built with waste material, such as cardboard, sheets, etc. were identified. In general, it was observed that families live in unsafe and unhealthy conditions, there is contamination of bodies of water due to poor disposal of wastewater, presence of excrement of domestic animals (dogs, hens, etc.) and poor disposition of the excreta of the inhabitants.

With the acceptance of the community, the team of collaborators of the Academic Body Urban Development of the Faculty of Architecture of the Autonomous University of Chiapas and the Academic Body Natural Risks and Geotechnology of the Academic Engineering Unit, Autonomous Universidad of Guerrero, created the research project: *Proposal to improve the habitat of the colony El Encanto of city Puerto Madero, Tapachula, Chiapas.* In August 2019, the project was approved and funded by the Institute of Science, Technology and Innovation of the Government of Chiapas and the Faculty of Architecture, which was carried out in the period: September 2, 2019 to January 31, 2020 [6].

4. Analysis of homes and their environment (habitat)

In the analysis of the conditions of the households of families, with low economic income, the Model of Assessment of the Conditions of Rural Housing and the Environment was applied [5], which includes surveys, registration cards and photographic report. The model proposes that initially the analysis of the current situation (zero time- "t_0") should be carried out, *which consists in obtaining the information on site and determining* the conditions of the dwellings and their housing environment. The analyses include the socio-economic aspects of low-income families, the characteristics, conditions and basic services of housing, the management of the waste generated and the conditions of the natural environment, also, at this stage, the health risks, structural safety of housing, the typological characteristics of the community, the flora and fauna, the materials of the place with the possibility of being used in the construction of new homes are recognized.

On **Figure 2**, you can see that the town is located on the side of the Tapachula-Puerto Madero road. The study area is the town of El Encanto and the analysis unit is the houses. The analysis of the conditions of the houses and their surroundings was carried out from a random and representative sample of the homes of the community. In this regard, data were taken from the Population and Housing Count 2010, which, in the case of El Encanto, has a population of 1,726 inhabitants, 446 dwellings and an average of 3,86 inhabitants per dwelling [1].

The conditions and statistical arrangements taken for the sample size were as follows: a) The homes considered in the sample are those inhabited by low-income families, who receive less than \$102.68 national currency per day (5 dollars) or between \$102.68 and \$205.36 national currency per day (between 5 and 10 dollars), and b) The sample size was obtained with the following mathematical model [8].

$$n = \frac{n_0}{1 + \frac{n_0}{N}} \qquad n_0 = P(1\text{-}P)\left[\frac{z\left(1\text{-}\frac{\alpha}{2}\right)}{e}\right]^2 \tag{1}$$

Where:

z = Degree of reliability (95%, cumulative normal distribution value is 1.96).

N = Population size (446 dwellings).

Q = Probability of success (98%, as a family is considered to generate organic and inorganic waste and makes use of water to meet their biological needs).

q = Probability of failure (2%, since: q + P x 100%).

e = Experimental error allowed (may vary between 0.03 and 0.07; in this case it was taken 0.05).

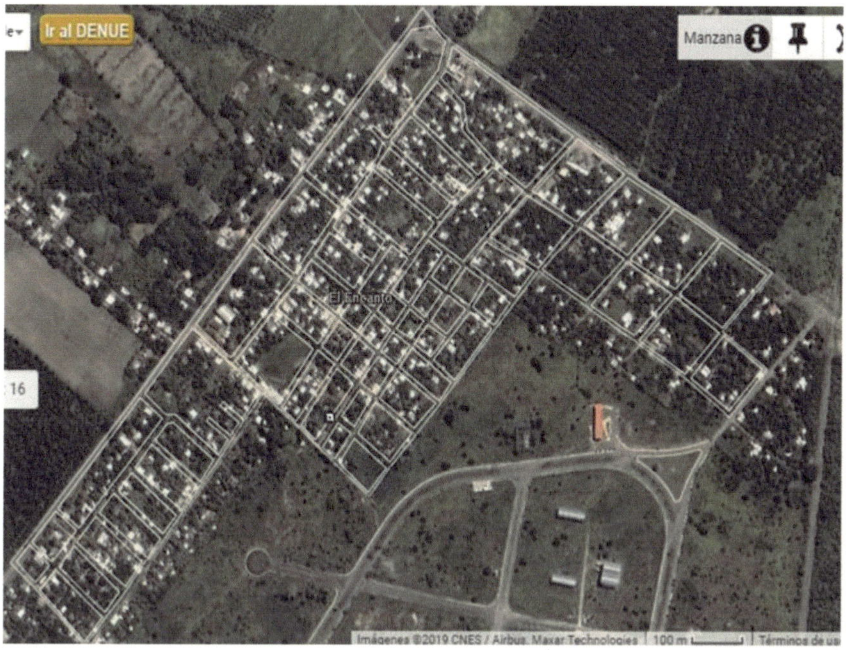

Figure 2.
El Encanto, Tapachula, Chiapas. Source: NISG [1, 7].

PR = Percent rejection (considered 10%).

n = Calculated sample size (28 dwellings, plus 10% rejection, therefore, the minimum sample is 31 dwellings. In this case 62 surveys were applied, which is equivalent to double calculated).

Fieldwork was carried out on 11 and 12 October 2019, as set out in the research project. At this stage, the participation of architecture students was held, to support in the activities of obtaining and analyzing information, and with this, the teaching is strengthened, based on the information obtained in the field, in a real context that will be useful in the application of exercises established in the bachelor's degree, in the subjects: Workshop of Construction Materials Zero Impact and Sustainable

Figure 3.
Student training.

Architecture. Subsequently, students also participate in the development of alternative housing proposals and public spaces. Before starting fieldwork, students were trained for the proper interpretation and recording of the required information, in accordance with site analysis tools, too, they were instructed to behave and address people with respect (see **Figures 3–6**).

Figure 4.
Group of students and teachers.

Figure 5.
On-site work team *with locals.*

Figure 6.
Students in interviews (surveys).

5. Results obtained

In the process of validation and systematization of the information obtained in the home surveys applied in El Encanto, 56 were validated that are well above 31 required, based on the model used, therefore the reliability of the information obtained was successful. The analysis of the surveys allowed to identify and interpret the situation of the houses and their immediate housing environment. Below are the results of the information obtained on site:

5.1 Socioeconomic records

Of the total homes surveyed, 76.8% responded that the head of the family is originally from Chiapas, 10.7% come from another state and 7.1% are foreigners. Regarding the time of residence, 37.5% mentioned that they have always lived in El Encanto, 41.10% have more than 10 years, 12.5% between 5 and 10 years and only 7.2% less than 5 years. This indicates that 78.6% of the inhabitants have 10 or more years of retaining residence in El Encanto, so the population is consolidated on the site.

The number of inhabitants per dwelling, 3.6% register a person, 16.1% 2 people, 26.8% 3 people, 16.1% 4 people, 14.3% 5 people, 8.9% 6 people, 3.6% 8 people, 1.8% 9 and 1.8% 15 people; 12.5% are enabled by 2 families, 1.8% by 3 and 1.8% by 4. The religion they profess, 39.3% are Catholic, 10.7% Jehovah's Witnesses, 37.6% other Christian religion and 12.5% are atheists. The marital relationship, 39.3% are married, 32.1% in free union, 10.7% divorced, 10.7% widowed and 3.6% single.

There are 55.38% women and 44.62% men. The ages of the inhabitants: 28.72% are under the age of 15, 23.08% from 15 to 29 years old, 26.67% from 30 to 49 years old, 11.28% from 50 to 59 years old and 16.26% register 60 years or older. According to the ages, there is a population decrease of 20 to 29 years, representing 12.82%, possibly due to the change of residence to other places for work reasons. Of the total number of over-14 s, 20.87% are illiterate and, of the inhabitants with studies, more than 30% have primary, 35% secondary, 18.5% high school and 2.9% have professional studies.

With regard to work activities, aged 14 years or older, 80% of people work and, of this, 30% receive less than a minimum wage (MW), 60% between 1 to 2 MW and 10% more than one MW. On average they have income of 1.43 MW, which, in monetary terms, in 2019 was equivalent to $102.68 national currency per day [9], which is equivalent to approximately 5 dollars. On the other hand, most work in activities in the primary sector: fishermen, barkers, day laborers, etc., with the following working conditions; 32.91% temporary, 22.78% per contract, 39.24% definitive and 5.06% are working on their own. For the purposes of the project, the production and marketing of cooked clay partitions by some inhabitants is highlighted. In relation to income with the monthly expenses made by families, they minimally manage to meet their basic food needs and with extreme limitations address the other aspects such as clothing, housing, health, transportation, etc. Access to consumer goods, linked to people's lifestyles, it is an important aspect that encourages and protects health habits; however, the acquisition of consumer goods is conditioned on the economic capacity of families, in this sense, the results show that 62.5% of homes have refrigerators, 58.9% have blenders, 33.9% have washing machine, 53.6% television and 37.5% radio transistor.

Another important aspect was the legal possession of the property where they live, since the possession of the land offers certainty, security and reduces the tension of the inhabitants with respect to the housing heritage. In this sense, 89.3% own homes, 5.4% lend and 1.8% rent.

5.2 Housing conditions (physical-spatial)

Data record that, housing is inhabited 4.06 people, average. This figure suggests that housing proposals will consider at least 2 bedrooms with the right space to accommodate 2 people per bedroom. Also, the adequacy fora future increase in the number of inhabitants will have to be considered, as 30.4% of the dwellings are inhabited by more than 5 people, which suggests a possible future spatial growth of housing. In relation to the surface, the land is flat with soil type at the site: 71.4% the houses are found on clay, 23.2% on sand and 5.4% on rocky material. The housing floor, 32.1% in floor of soil, shows precari condition of habitat, 46.4% has concrete, 17.9% polished concrete and 1.8% have mosaic coating (see **Figures 7** and **8**).

In the various types of roofs of the homes analyzed, more than 50% use galvanized sheet, 35.7% use local palm and 3.6% sheet of cardboard considered as precariousness. It is important to note the use of galvanized sheet, due to the low cost of acquisition and maintenance, however, do not have any additional elements that decrease the implications of this material, related to the high temperatures generated in the heat season, which make uninhabitable the spaces built with this type of roof (see **Figures 9–12**).

Figure 7.
Walls and roofs in housing.

Figure 8.
Wooden walls and palm-sheet cover.

Figure 9.
Concrete block walls.

Figure 10.
Galvanized sheet roof.

Figure 11.
Housing; foil walls and palm roof.

Figure 13 shows the existing conditions inside the dwellings. Spatial fragility, overcrowding, disorder, precariousness, etc., which disturbs people's well-being are observed. Also, it was observed that only 48.3% of the dwellings have some structural confinement, essentially the construction with bricks or blocks of cement-sand, however, only 26.8% of the dwellings have adequate confinement vertical and horizontal, in the rest of the houses the structural confinement is insufficient and inadequate, which represents a situation of risk and vulnerability structural.

Figure 12.
Housing; cane walls and palm roof.

Figure 13.
Housing interiors; stay, bedroom, kitchen-eat.

Regarding the conditions of the structure, 48.2% of the houses have problems: 21.4% register cracks in walls, 7.2% in doors and windows. The causes are diverse: incorrect construction procedures, lack of structural reinforcement, etc., in addition, there is damage to the wooden structure used in the ceilings, 35.7% of the roof have deterioration and structural damage: 19.6% with rotting woods and 10.7% with fractured woods.

5.3 Immediate environment conditions

El Encanto is located in a region of high ecological diversity, in this sense, 89.3% of the houses have fruit trees planted in the surrounding areas with their property; However, only 60.7% of families are careful to maintain vegetation and control organic waste generated in the yard (see **Figures 14** and **15**).

The waste generated, liquids, solid or gaseous, product of daily domestic activities, only 55.4% of the inhabitants have some control, however, 51.8% present health risks to the health at the poor disposal of the water used in body grooming,

Figure 14.
Abundant vegetation on the property.

Figure 15.
Cleaning the property and burning waste.

washing of waste and clothing, in addition, it was identified that 3.6% burns organic solid waste: branches and leaves of trees of the property.

Of the total dwellings, 69.60% register having animals for consumption or trade: 57.1% have poultry and 12.5% hey have goats. That aggravates the situation of mis management of organic waste, basically the droppings of animals generated, which directly harm the health of people. Others have pets; 60.7% have dogs and 17.9% cats (see **Figures 16** and **17**).

Figure 16.
Liquid waste accumulated on the site.

Figure 17.
Bird corral for self-consumption and sale.

The town has no sanitary sewerage network, 85.8% have makeshift sheds with toilet furniture and use water for the eviction of excreta, which is transported to and deposited in a poorly constructed septic tank that contaminates the body of underground water, 17.9% use common latrines, 1.8% defecate outdoors and, the rest, did not specify (see **Figures 18** and **19**).

The materials used in the construction of toilet sheds and latrines are very varied, 26.8% have concrete signature, 3.6% polished concrete, 10.7% use bocks, 5.4% waste material and12.5% nothing. The walls and ceilings are precarious as seen in the figures. In terms of quality of service, all do not meet minimum sanitary conditions: 92.8% have no lid or seat, odors are perceived, discharges of excreta pollute groundwater level (depth of 3 to 4 m) due to filtration, latrine shed conditions or toilets are improvised, poor quality and unsafe (see **Figure 18**).

For body grooming, most have improvised and fragile sheds, as shown in **Figures 16** and **19**. The waste water, deposited in the yard, accumulates and that causes an unhealthy environment with bad odor, attracts the harmful fauna that puts at risk the health of the inhabitants.

Water is obtained by the inhabitants, to meet basic needs, through artesian wells, some have square protection, but the generality is circular (see **Figure 20**). The material used for protection is bricks of cooked clay settled with lime mortar or cement-sand. The water, the inhabitants store in plastic container; however, there are houses, the least, that have tanks built with bricks masonry, as shown in **Figures 21** and 22.

Figure 18.
Toilet and septic tank.

Figure 19.
Bathroom area.

Figure 20.
Types of artisan wells for water extraction.

In relation to cooking food, all homes use handcrafted stoves, which consume significant amounts of firewood. The characteristics of the stoves are very diverse, as shown in **Figures 23** and **24**, observing energy inefficiency, high consumption of firewood, exposure and contact of the inhabitants to smoke emissions (CO, CO_2, others), which affect the health, well-being and quality of life of people.

5.4 Urban structure (urban services)

Although El Encanto has full electricity coverage, only 76.8% of the homes surveyed are connected to the electricity grid. Only 26.8% of homes have street lighting. Previously, it is established that the town has no water or sanitary drainage, however, at the time of field work, the construction of the well, treatment, storage and distribution of drinking water, covering the total area of the town, had been completed. Today (2021), the inhabitants do not yet receive drinking water,

Figure 21.
Water storage in plastic container.

Figure 22.
Storage tank features.

due to an administrative and operating problem by the municipality of Tapachula, Chiapas.

5.5 Health services

There is no basic health unit in the locality to care for illnesses or receive first aid from doctors, a consequence of any ailment or injury caused by a work or domestic accident. In this regard, the information obtained in the surveys indicates that, 73.2% of the inhabitants of the dwellings have health service, through the "People's Insurance", which this year (2020) became the Institute of Health for Welfare. On

Figure 23.
Food processing stove.

Figure 24.
Type of stove precariously built.

the other hand, 8.9% have IMSS, 5.4% ISSSTE and 10.7% without the right to a public service. Common diseases suffered by the inhabitants: 51.8% flu, 14.3% diarrhea, 10.7% cough, 5.4% dengue, 3.6% parasite and 3.6% hypertension.

Health institutions where residents receive medical care for illness, despite the fact that most have public health services, 41.1% of users say they go with private doctors and others; 28.6% to the SSA, 7.1% to IMSS, 1.8% to ISSSTE. The sites to which they travel for medical care, 60.7% head to Puerto Madero, to 2.9 km, and 35.7% to the city of Tapachula at 26.5 km.

5.6 Road and transport

The streets of the town are reticular with anchovies of 7.0 m and bounded by some fences and trees planted by the inhabitants who define its properties. The streets have no cladding, except for a small section in the main access of El Encanto. This problem causes the soil to soften in the rainy season and prevent people and vehicles from circulating (see **Figures 25** and **26**).

5.7 Solid waste collection

The solid waste collection service is done through a collection vehicle of the town hall of Tapachula, which makes a route of 26.5 km from Tapachula plus 6 km where the municipal dump is located, near the town of Viva Mexico, which means that routes of 65 km are made. The service is only performed on Wednesdays of each week, due to the low amount of waste and high operating costs, in addition, the service is limited to the streets of the main and central access, so, the inhabitants move their waste to the collection points. For this reason, it causes the inhabitants to continue the practices of burning garbage.

Figure 25.
Street without coating.

Figure 26.
Fences of property borders.

5.8 Community equipment or public spaces

During the field work, the facilities of the only public spaces that exist in the town were also inspected, which correspond to the Preprimaria and Primary schools (see **Figures 27** and **28**), which is located one side of the Tapachula - Puerto Madero road and in the main access street to the town.

In pre-primary and primary schools, it is noted that all classroom buildings where classes are taught require urgent basic maintenance: cleaning and painting in general, installation of new luminaires and electrical installations, construction of toilets for children, construction of new classrooms, kitchen construction for the production of school breakfasts, in addition, there is an opinion on civil protection, which prohibit the use of one of the classrooms of the pre-primary school, due to the severe damage that the structure has, a consequence of the sism that occurred on September 7, 2019, and since then has not been repaired, so the old kitchen for the preparation of children's school breakfasts was adapted as a classroom.

Figure 27.
El Encanto pre-primary school.

Figure 28.
El Encanto primary school.

5.9 Problems identified by the inhabitants

The inhabitants surveyed are asked what are the problems of the services that the locality has? 44.6% manifest the lack of security, 39% the accumulation of garbage (solid waste), because the collection takes place only on Wednesdays and 7.1% that there are problems in transport, since the medium used is the tricycle adapted with a motorcycle and when transiting on the road is unsafe.

When asking the inhabitants what are the problems of public spaces? 55.4% believe that the absence of a health unit to receive emergency medical services, 28.6% believe that it is missing in high school, since children move to Puerto Madero, 2.9 km from El Encanto, by a path located next to the road, representing a constant risk and 7.1% say that a local market is missing.

In relation to the question of what are the problems that have been done in the service infrastructure? 25% of the inhabitants pointed out that the problem is paving, since currently the streets, with the exception of a small section of the main access, have no cladding, are only plotted at ground level, and, in rainy times, the soil softens and are impassable. With an almost similar percentage, 23.2% say that the main problem is the absence of the drinking water supply network, which is already solved with home intakes located on the premises, however, to date they do not yet receive drinking water service. On the other hand, 17.9% of the inhabitants pointed out that the absence of health drainage is a health problem to be addressed,612.5% say that street lighting is the main problem to be addressed, as it is related to the safety of people, and 10.7% who do not have electricity means a problem for them.

Finally, he wondered what other problems were considered important in the town? 55.4% have employment problems, and 5.4% of them have economic problems, 16.1% are manifested in land or property tenure and 1.8% report food problems.

5.10 Construction materials identified at the site

The materials identified in the locality and in the region for use in the construction of both the homes and the proposed community spaces, are as follows: sand to make cement-sand mortar to glue solid bricks of cooked clay and other masonry structures, river stone used in the concrete of cyclopean processing in foundation,

Figure 29.
Cooked clay bricks.

Figure 30.
Type of handcr afted oven used on site.

concrete with crushed river stone, coconut palm stem wood used in roof structures, cane (bamboo) used in walls and roof structures, etc.

The bricks are made by locals with local clay, cooked in handcrafted ovens (see **Figures 29** and **30**). A sample of solid clay bricks parts was analyzed in the Materials Laboratory of the Faculty of Architecture and obtained an average compressure resistance of 9.31 MPa (93.10 kg/cm^2), which complies with the Complementary Technical Standards [10], however, the furnaces will be improved for greater heat efficiency, lower consumption of firewood and adequate control of emissions during production.

6. Proposals for a solution: alternative housing and community equipment

The realization of the diagnosis of the habitat of El Encanto and the results obtained were fundamental to the elaboration of the proposals of alternative housing and community equipment. The characteristics of the houses, their identity values, materials, habitability, form, function, structure and construction systems, structural quality, sanitation in general and environmental components, were essential for the design of the prototypes. From the above, the traditional elements, the functional spaces of the house are defined: bedrooms, stay, kitchen, sanitary service, etc., in addition to the construction of the elements with the use of local materials and others with low environmental impact, the application of ecotechnologies according to the traditions of the inhabitants, the type and environmental conditions. The housing proposals were created with the idea that the inhabitants have the possibility of self-construction, with the technical assistance of a facilitator (an academic, technician or student of engineering or architecture). Also, to make them structurally safe, economical and healthy and for the benefit of low-income families. Similarly, the characteristics of the community equipment proposals were prepared in accordance with the socio-economic conditions and environmental of the site. In this sense and based on the diagnosis, proposals for community equipment for health, art and culture services, recreational, were designed, in addition to the improvement of trails of El Encanto, essentially in the sections of interconnection with public spaces.

6.1 Considerations for the development of alternative housing and its surroundings

In the process of drawing up the proposals, the different aspects and conditions identified in the homes and environment are contemplated. The criteria and guidelines adopted for the design and construction proposals for community housing and equipment are then related.

A. Alternative housing proposal (physical space):

a. Respect for traditional identity, culture and activities.

b. Use of local materials, low environmental impact and characterizing the site, harmony with the context.

c. Creation of open spaces for estufas, corridors and hammocks.

d. Balance of door and window with solid walls, also unify heights.

e. Housing access hierarchy.

f. Facades of houses with a focus on neo-vernacular architecture, with garden in front that offers transparency and shade for the circulation of passers-by.

g. Use of colors that match the context and make the locality an identity.

h. The principles of housing hygiene and access to an adequate housing environment, according to the World Health Organization [11] and the Sustainable Development Goals of the United Nations Development Programme [3, 4] will be addressed.

In the area of housing (construction):

a. Maximum construction area 36 m^2, includes: stay and 2 bedrooms.

b. Two-water cover with galvanized sheet with thermal insulator, preferably organic.

c. Use ecotechnologies for eco-friendly stove construction with greater heat efficiency, reduced firewood consumption and gas emission control.

d. Electric network with lower power consumption luminaires.

e. Construction of a 60 cm high platform for the housing floor, with the aim of reducing the risks and vulnerability caused by heavy rains or hurricanes.[1]

f. Consideration of the housing growth area (from 2 to 3 bedrooms).

In the area adjacent to housing, the yard (construction):

---

[1] During the heavy rains that occurred on October 4, 2005, by Hurricane Stan, which caused river overflows and bridge collapses off the coast of Chiapas, Mexico, the inhabitants of El Encanto comment that the water was drained in the form of a 50 cm sheet over the surface and swept everything, until it reached the Pacific Ocean, in Puerto Madero, Chiapas.

a. Body grooming área.

b. Area of washing clothes and kitchen utensils.

c. Water storage tank.

d. Dry latrine of one or two chambers at the boundaries of the property.

e. Consideration of the area for the rearing of animals (hens) and control of animal excreta.

f. Control of wastewater generated by domestic activities: washing clothes, body grooming, etc., a starting of wetland construction.

Materials used in construction:

a. Use of site materials. Inorganic materials: stone, sand, clay, etc. Organic materials: coconut palm stem, cane, coconut shell, etc.

b. Use of solid bricks of cooked clay produced by inhabitants of El Encanto, with oven modified for mayor heat efficiency and lower consumption of firewood.

B. Proposal for Community equipment and streets improvement.

a. The socio-economic characteristics of the site, existing environmental conditions, use of site materials, application of ecotechnologies, etc. overflows shall be considered.

b. It is proposed to set up a health unit for urgent medical care.

c. For recreation and family coexistence, it is proposed to develop a unity of art and culture, a multi-purpose sports court in the open space for the practice of: fast football, volleyball, basketball and inclusion of children's games.

d. Hierarchy of streets for a better distribution of the town, it is proposed to create a public streets circuit with coating, which interconnects the community equipment proposed, improves the circulation in rainy season of both pedestrians and used vehicles.

e. Urban furniture is required: Chairs, trash cans, lighting, etc., as well as the implementation of an efficient garbage collection system.

f. In existing, pre-primary and primary educational facilities, a maintenance and construction programme for health services and school canteens should be implemented.

6.2 Proposal-making process

The housing proposals and public spaces were developed with the participation of students of the Faculty of Architecture, with the advice and direction of the academics of the working group (see **Figures 31** and **32**).

Figure 31.
Academic advice to students.

Figure 32.
Review of housing proposals.

6.3 Alternative housing 1

5 alternative homes of 36 m^2 of construction are proposed, which includes 2 bedrooms (3x3m each), living areas, dining room and kitchen (see **Figures 33–38**). Foundation: stone and concrete masonry of 15 MPa (150 kg/cm^2) and section: width 50 cm, depth 50 cm and floor height 60 cm. Walls: solid clay brick masonry, section: 7x14x28cm, made *in situ*, reinforced with vertical and horizontal concrete elements of 15 Mpa (150 kg/cm^2), reinforced with steel. Vertical elements are placed at the intersections of walls and on the sides of doors or windows, with section: 15x15cm; the horizontal elements or dalas, with section: 15x25cm, are placed at the height of 2.00 m of doors and windows and on the crest of 3.40 m height. All reinforced concrete elements shall have: 4 rods of 3/8″, además de ¼” stirrups @15 cm. Ceiling: coconut palm stem structure that supports thermal insulation made with organic layer of otate rod, soil mixture with coconut shell and Covered with galvanized sheet. In the patio will be located: bathroom, water storage tank, laundry room of kitchen utensils and hands, in addition, the dry latrine

- **Alternative housing 1**

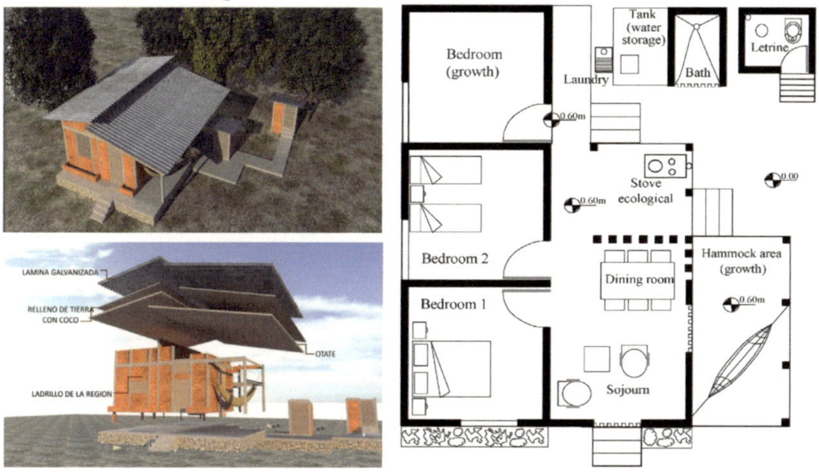

Figure 33.
Alternative housing 1, architectural plant.

Figure 34.
Facade alternative housing 2.

Figure 35.
Facade alternative housing 3.

Figure 36.
Facade alternative housing 4.

Figure 37.
Facade alternative housing 5.

(double chamber), with height of 90 cm from the floor for maintenance. In addition, future growth is considered: a fourth bedroom and the rest area (hamacas).

6.4 Community public spaces

In the streets proposal for the interconnection of public spaces, is proposed a height of 0.60 m above the ground. The coating with river stone joined with mortar and walkers with blocks.

The public spaces service proposed are: Health Unit, Center for Art and Culture, Multi-Purpose Sports Court and Children's Games (see **Figures 39–44**). **Figure 40** shows the location of real estate and the improvement of the road of interconnection of real estate.

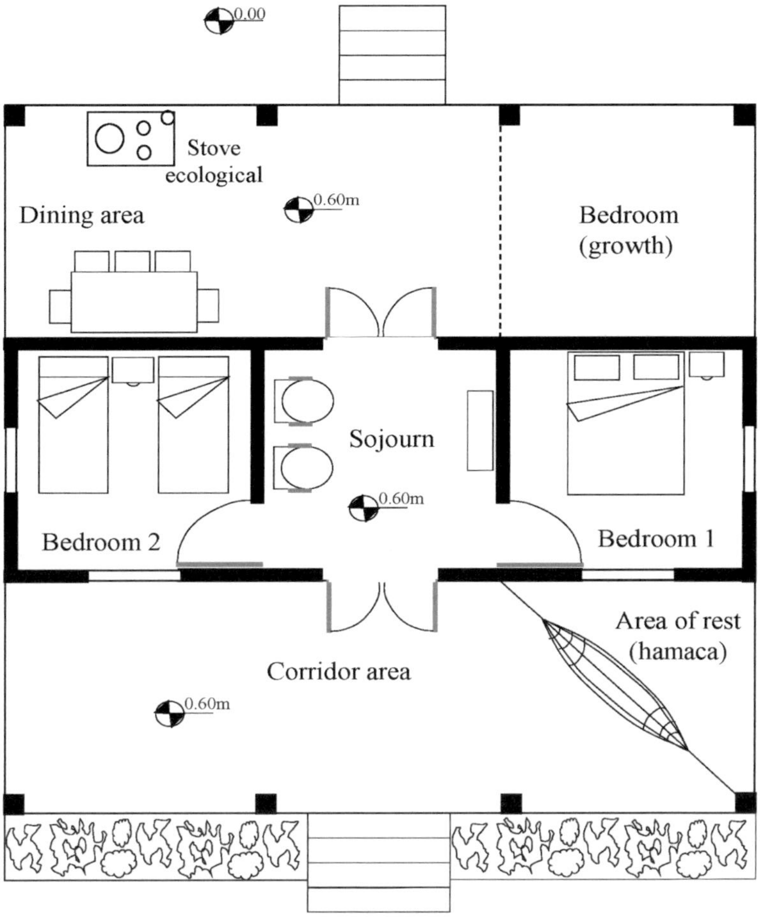

Figure 38.
Alternative housing 4, architectural plant.

Figure 39.
Improving the streets.

Figure 40.
Location of community public spaces.

Figure 41.
Health unit.

6.5 Process of conciling proposals with the community

On 5 December 2019, a working meeting was organized in El Encanto to present housing proposals and community equipment. Ms. Elena Matías Medina, the town's representative, and locals participated. The characteristics of the proposals are detailed with graphic information posters and models of the houses prepared by students and academics of architecture: Health Unit, Center of Art and Culture, Multi-purpose sports court and Children's Games. The objective of the meeting was to know the opinion of the inhabitants and to specify the characteristics of the roofs, sanitation, etc. (see **Figures 45** and **46**).

Figure 42.
Center for art and culture.

Figure 43.
Multi-purpose sports court.

Figure 44.
Children's games.

Figure 45.
Proposals submitted to the inhabitants.

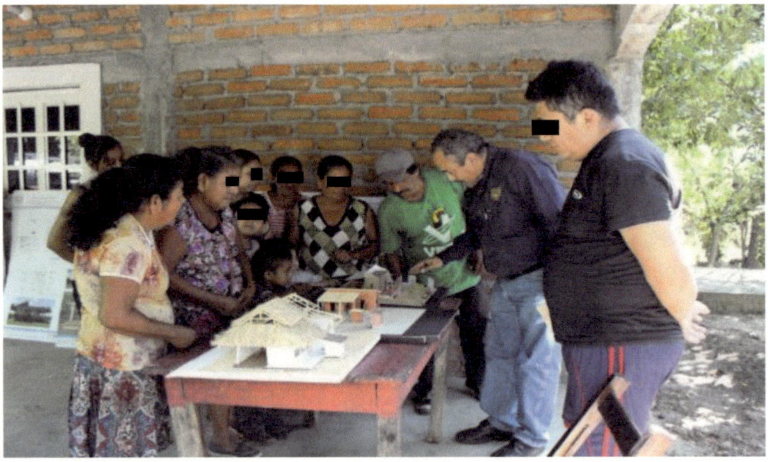

Figure 46.
Analysis of houses models.

7. Conclusions

The houses and community equipment presented for the village of El Encanto, Tapachula, Chiapas, were elaborated based on the environmental conditions and socioeconomic aspects of the inhabitants. It respected the typology, the domestic customs, the use of building materials of the place, tales such as: sand, river stone, coconut palm stem wood, cane, clay for the elaboration on the site of cooked bricks. In addition, the use of ecotechnologies, ecological stoves, water and sanitation services, covered with galvanized sheet and organic thermal insulation, etc. was implemented. The site's analyses and results, obtained by the team, are essential to the design of the proposals: structurally safe, economical and healthy.

The project, in its various stages, constituted a pedagogical exercise framed in a vision of architecture as an academic discipline, committed to improving the living conditions of the low-income population and environmental conservation. In this sense, the design of the house and habitat is the product of the interaction between academics and the inhabitants of the town. The design of the house and habitat is the product of the interaction between the academics and the inhabitants of the town. In this case, the main conditions have been the economic and social precariousness of the inhabitants of El Encanto and the physical environment, such as the zero slope of the terrain, the situation of vulnerability to the events of nature such as rains, cyclones and tsunamis; are challenges for architecture professionals. In this work, the teaching of architecture is included with practical exercises in real situations, aimed at improving the quality of life and well-being of poor families.

Undoubtedly, social participation at all stages of the project brings not only empirical knowledge, but also helps to ensure the social appropriation of the products contributed by the project, such as an valuing the knowledge and changes generated by them. Faced with a situation such as the current one, the use of local materials and the experience of the inhabitants in the production of cooked clay bricks, constitutes a low environmental impact solution for the construction of new housing.

With regard to urban facilities, since they have to comply with official standards, in the case of schools or health centres, the experience of the inhabitants in participating collectively in the design of their homes provide a good starting point to consolidate a community organization that allows them to successfully manage before the relevant authorities, the resources for such equipment.

Acknowledgements

The research work was carried out thanks to funding from the Institute of Science, Technology and Innovation of the State of Chiapas and the Faculty of Architecture of the Autonomous University of Chiapas, Mexico. Likewise, we thank the inhabitants of the town of El Encanto, Tapachula, Chiapas, for the facilities and information provided, especially Mrs. Elena Matías Molina, representative of the community, as well as Eleazar Matías, Tito Rivera, Manolo Pascacio, Idolina Santiago, Edi, Eva and many more, who kindly shared their testimonies and knowledge. Finally, our recognition of the enthusiastic participation of the students of the Autonomous University of Chiapas: 7th Semester "A" and "B" of the school year of architecture August-December 2019, to the students of social service, Alondra Castillo Gómez and Saraín Domínguez López, to Roberto Román Cárdenas of the Master in Architecture and Urbanism, and to the professors Eddy González and Nguyen Molina for their collaboration.

Author details

Lorenzo Franco Escamirosa Montalvo[1*], Carlos Uriel del Carpio Penagos[1], María de Lourdes Ocampo García[1], Ángel René Estrada Arévalo[1], Arturo López González[1] and Roberto Arroyo Matus[2]

1 Autonomous University of Chiapas, Mexico

2 Autonomous University of Guerrero, Mexico

*Address all correspondence to: franco@unach.mx

References

[1] National Institute of Statistics and Geography (NISG, 2010 and 2015). *Mexico in numbers*. Chiapas - Getting to know Chiapas (online): https://www.inegi.org.mx/app/areasgeograficas/?ag=07; http://internet.contenidos.inegi.org.mx/contenidos/Productos/prod_serv/contenidos/espanol/bvinegi/productos/estudios/conociendo/702825217815.pdf

[2] National Population Council (NPC, 2012), Marginalization indexes by locality 2010, Mexico. (online): http://www.conapo.gob.mx/es/CONAPO/Indice_de_Marginacion_por_Localidad_2010

[3] United Nations Development Programme (UNDP, 2015), *Human Development Index for Federal Entities, Mexico 2015* (online): http://www.latinamerica.undp.org/content/rblac/es/home/library/human_development/indice-de-desarrollo-humano-para-las-entidades-federativas–mexi.html

[4] United Nations Development Programme (UNDP, 2015), Sustainable Development Goals (online): https://www.undp.org/content/undp/es/home/sustainable-development-goals.html

[5] Escamirosa Montalvo, Lorenzo Franco (2015). *Rural housing and healthy environment. Ocuilapa case of Juarez, Chiapas*. Mexico, Ed.: Autonomous university of Chiapas, Council of Science and Technology of the State of Chiapas and Porrúa.

[6] Escamirosa Montalvo, L., C. Del Carpio, M. Ocampo, R. Estrada, *et al*, (2020). "Technical Report: Proposal to Improve the Habitat of the Colony El Encanto of the City of Puerto Madero, Tapachula, Chiapas", Institute of Sciences, Technology and Innovation of the State of Chiapas, Mexico.

[7] National Institute of Statistics and Geography (NISG, 2015). *Catalogue of keys of federative entities, municipalities and localities*. October 2015 (online): http://geoweb.inegi.org.mx/mgn2k/catalogo.jsp

[8] Larson, Harold J. (1981), *Introduction to Probability Theory and Statistical Inference*, Ed. Limusa

[9] Government of Mexico (2018), *Official Journal of the Federation* (OJF), published in press release No. 009/2018, of December 26, 2018.

[10] Complementary Technical Standards (CTS, 15 December 2017). CTS for the Design and Construction of Masonry Structures, *Gaceta Oficila of Mexico City*, Government of Mexico City.

[11] World Health Organization (WHO, 1990). *Principles of Housing Hygiene*, Geneva, Switzerland.

Chapter 11

Housing Challenges in Nigeria

Uyi Ezeanah

Abstract

One of the policy goals of sustainable development is the delivery of adequate housing for Countries in the global south. This is because in many countries of the global south housing is delivered more through private mechanisms, consequently this poses challenges to adequate housing delivery for most countries of the global south and the Nigeria populace. Some of the challenges faced in providing decent and adequate housing for the people include problems of unskilled workmen, unachievable westernized building standards, housing finance, high level of urbanization, poor policy programmes, and contravention of building standards in addition to poor building materials. This chapter explores housing challenges in Nigeria. In this chapter problems of housing in Nigeria will be explored; the trajectory of the history of national housing delivery and housing policies will be presented. Also, various challenges bedeviling adequate housing delivery in Nigeria is explored and explains how the quality and quantity of housing delivered in Nigeria has affected by these challenges.

Keywords: Challenges, public housing, housing finance, housing programmes

1. Introduction

Housing is second most essential basic needs of humankind outside food. Its impact on the health and welfare and output of man is profound. Therefore housing plays a vital role in a person's standard of living and place in the society [1]. Nigeria has an estimated population of about 200 million people and this implies that urgent attention should be placed on the country housing sector if the housing needs of the inhabitant are to be met. One key issue affecting housing delivery in Nigeria is that the level of housing shortage has not been adequately presented. This is as a result of inadequate and inappropriate statistics and data by the managers of housing in Nigeria [2].

However, there are still attempts to estimate the magnitude of the housing shortage in Nigeria. In 1991, the National Housing Policy specified in detail that to achieve the goal of providing 8 million housing units by the year 2000, 700,000 housing units would have to be built each year; it concluded that this number is necessary to compensate for the housing shortage in the country [3–5]. According to Okonjo-Iweala [5], around 100,000 housing units are built each year, and an average of 80% of Nigerians live in informal housing, which is plagued by problems related to poor quality and inadequate infrastructure that have hitherto been discussed [6].

It is estimated that, in terms of funding, 12 trillion naira (about 45.3 billion pounds) is needed to solve the existing housing scarcity problem [2]. It is projected that in Nigeria's 20-year time frame, an average of 56 trillion naira (approximately

211 billion pounds) is needed each year to meet the demand for housing [2, 7, 8]. Although the exact reasons for the housing shortage vary across the country, the main problem in Nigeria is the low income of residents. This is problematic since privately constructed houses are expected to comply with official planning laws and other costs incurred during the construction of the house [7, 9, 10]. High occupancy rate ranging from 5 to 8 persons per room which is above the official recommended range of 2-4 persons per room in many Nigeria cities is one factor that shows there is housing shortage [11, 12]. One reason for this is that most of the housing stock is traditional in form, and the sharing of one room by more than two persons along with the sharing of individual housing units by several households is a central feature of housing in in many cities in Nigeria.

Since then, huge resources including effort, time, materials and money, have been devoted to planning the Nigerian environment at the national and sub-national levels. Nonetheless, the various challenges that have been, and are being addressed have hardly diminished [13]. In fact, the problems of housing shortages, like those of unemployment, physical deterioration, poverty, inadequacies and inequality in service delivery systems have escalated. The incidence and growth of these problems seem to outpace the capacity of government to take them. Nigerians are faced with the fact that their cities are in trouble and that there is an urgent need to do something that will ameliorate the emerging problems.

2. Finance and housing delivery

Housing finance is one of the most important factors affecting housing delivery and urban policies. Several authors believe that establishing a suitable financing system can improve housing in any given society [5, 7, 10, 14]. Some governments have updated their financial institutions to encourage private sector financing organizations to participate in the provision of housing delivery financing [14]. As Makinde [7] and Mukhtar et al. [14] stated, in Nigeria, housing delivery is affected by the lack of an adequate housing financing system, which undermines housing production.

In Nigeria, housing financing is divided into formal and informal [7, 14, 15]. Formal institutions include financial institutions such as commercial banks, Federal Mortgage Bank of Nigeria (FMBN), and Specialized Development Banks. In contrast, informal sources of housing finance are typified by individuals saving money for extensive periods of time in order to build a house. Most of the houses (80%) constructed in Nigeria were provided through long-term personal savings [7, 10, 14]. Other informal funding sources include: voluntary housing drives and different credit union organizations and individual money lenders [5, 7, 10]. Although informal sources often help provide the necessary funds for housing delivery, because such practices are not planned or documented, it is impossible to statistically measure their impact on housing delivery. Their scope of operation differs [16].

Institutions embedded in formal financial mechanisms make it difficult for many people who want to build houses to obtain this form of financing. In 1991, the Nigerian National Housing Policy (NHP) established a dual system of housing finance. In the 1991 housing policy, the first housing finance system was the Primary Mortgage Bank (PMB), and the second system was the Federal Mortgage Bank of Nigeria (FMBN) created in 1977 to offer financing for the provision of standard housing to the people. One of the main functions of FMBN is to guide and manage the PMBs. Nevertheless, this function has been reallocated to the Central Bank of Nigeria [7]. The two goals of FMBN are to strengthen the formation and expansion of the PMB Bank nationwide and provide funding for housing. It was established to oversee the National Housing Fund (NHF), which ensures that public and private employees make mandatory contributions to the fund [7, 17]. In 1992, the NHF was established to provide long-term mortgages at an affordable interest rate to people. However, the achievement level were not successful because only 12,000 people benefited from a total of 3.8 million qualified taxpayers who applied for loans [5]. In order to apply for an NHF loan, certain requirements must be met, including an annual income of more than 3,000 naira and a donation of 2.5% of the employee's basic monthly income. These compulsory contributions and the long wait for obtaining these funds often limit the ability of workers to obtain such funding sources [5]. In addition, the interest paid on such loans must not exceed 9%, and the repayment period is approximately 30 years. In addition, the funds granted to any individual applicant must not surpass 5 million naira [10]. At present, this is not enough to build high-quality houses, and it is also unaffordable for most of the Nigerian population. The capacity to provide housing in Nigeria through access to formal financial mechanisms has been affected by several challenges. These include housing finance problems, for example; people are constrained by various institutional bottlenecks, bureaucratic procedures to obtain land ownership hamper the payment of NHF funds, high inflation and interest rates in the country, and the inability of people and developers to obtain long-term funds loans [10, 14].

2.1 The National Housing Policy in Nigeria

A number of programmes and policies have been articulated and presented in a bid to meet the housing challenges facing the built milieu in Nigeria. In this light, the first national policy on housing in Nigeria was launched in 1991 with the aim of providing housing for all Nigerians by the year 2000. The policy encountered major obstacles in the implementation process, and failed to achieve the expected effect on the social environment to provide decent housing for all in the target year of 2000. This is due to the deficiencies of the Primary Mortgage Institution (PMI) and the lack of access to land and the affordability of mortgage loans etc. as evidenced in the housing delivery programmes embarked upon by the government of Nigeria from the pre-independence era till the year 2000 which was a massive failure [18].

Given the significance of housing in the national economy and because of the inability of previous policies and programmes to efficiently resolve the logjam of housing problems in the country, there was need to practically review the 1991 National Housing Policy. "The draft policy was subjected to critical comments and inputs and the New National Housing Policy was published in the year 2006" (8 p 64). Therefore, the new National Housing Policy objectives amongst others are:

- Develop and sustain the political will of the government for the provision of housing for Nigerians.
- Provide adequate incentives and an enabling environment for greater private sector (formal and informal) participation in the provision of housing.
- Strengthen all existing public institutions involved in the housing delivery at the federal level.
- Encourage and promote active participation of other tiers of government in housing delivery.
- Make land for housing development easily accessible and affordable.

The 2006 housing policy did not live up to expectation because the imperfections of the 1978 land-use act on land administration process nationwide, this could not provide a panacea for the cost of housing construction amongst others as land was very expensive to buy. Again, a bid was made to revise the 2002 national housing policy in September 2011, with important evidences affecting the housing sector. The reason for revisiting the previous policy was to enable a revitalisation of the housing sector to enable it to serve as a panacea for effective socio-economic development.(National Housing Policy (NHP) [19], therefore, in 2012, a new national policy on housing was enacted to meet the challenges of housing delivery through more public private-partnership, encourage active participation of all the three tiers of government in the provision of housing, to reduce the cost of producing houses by promoting the use of locally made building materials and also to improve the quantity and quality of rural housing, amongst others [19].

As assessed in this chapter, the impact of the national housing policy and the government's resolve to provide housing for the Nigerian society traverse the numerous housing programmes implemented by the government. Here, various development programs have been launched under the NHP to provide housing that meets various income levels, from the lowest to the highest income levels. However, from the previous debates [7, 8, 20], it can be seen that there has been little success in achieving these policy goals. This is because of the diverse challenges faced, reasons for failure include; production costs, government changes, political influence, corruption, and shortage of skilled workers.

2.2 History of public housing programmes in Nigeria

This section examines the trajectory of different housing initiatives in Nigeria and that despite these laudable programmes public housing delivery is beset with a number of challenges with has compromised the quality and quantity of housing delivered in Nigeria.

The international urban housing situations of the bulk of urban dwellers have continued to wane in both measurable and qualitative footings. (United Nations Human Settlement Programme [21] this trend has been bothersome and has continued to stimulate and contest ideas around different housing provision approaches over the last four decades, for periods discussed below. The involvement of the public sector in housing in Nigeria has been more of policy formulation than housing delivery. Below is step by step account on efforts by the government to provide housing in Nigeria.

The pre-independence era marked the first effort towards housing delivery in Nigeria. This marked the beginning of the Government Residential areas known as GRAs, where houses were built to provide adequate comfort for the residents as the "housing forms and spatial patterns of the GRAs reflected the English nostalgia for the garden city" ([22], p. 3). However, following the pre-independence era is the Post-independence Housing Era, 1960-1972. In this era, there was no marked difference in the provision of housing during the pre-independence era which was characterized by colonial government. In the Post-independence Housing Era, 1960-1972 [23] housing units in the government reservation areas (GRA) formally occupied by the colonialist became the abode of the new Nigerian administrative and political elites without any form of objection by these elites [24]. In the post –Independence era, the National Development plans were proposed which are discussed below:

The First National Development Plan (1962-1968) was launched to cater for the provision of 61,000 housing units, of which only a total number of 500 housing units were constructed by the Federal Government, which ended as a result of the civil war. Consequently, the second national development plan was initiated in

(1971-1974) to cater for 59,000 low-cost housing again 7,080 housing units were built. Reasons offered for the failure are insufficient planning and formation, insufficient funding, errors in execution, under-pricing and costing etc. [22].

In order to provide housing to cater for the need of the Nigerian populace the Third National Development Plan (1975-1980) was launched. Here 202,000 low-cost housing units principally for the low income groups were to be provided, in all 28,000 housing units were delivered [25]. This then gave rise to the Fourth National Development Plan (1981-1985) which was embarked upon by the Civilian government. Only 23.6% of the initially intended 160,000 housing units were built [26] as a result of corruption, politicization and uncooperative attitude of state executives [27]. Following the post –Independence era is the Post Second Republic till present, it was characterized by the following periods. Military Governments (1986-1999) and Civilian Governments (1999-to date). The military government embarked on housing delivery efforts of delivering 121,000 housing units across the federation, 5,500 units were delivered, which was brought to a halt by an inauguration of a democratically elected government [28].

Following the election of a democratically elected administration in May 1999, the Federal Ministry of Works and Housing (FMWH) and several state governments indicated their willingness to participate in housing programs [29]. As a result, in 1999, the National Housing and Urban Development Policy was established to provide an additional 10,271 housing units through a public-private partnership and 4,440 housing units were provided under a public-private partnership [28]. Following the inauguration of a new civilian government on May 29, 2011, a variety of housing interventions/programs were implemented in the federal capital area, with a focus on public-private partnerships (PPPs), with the goal of providing (**Table 1**):

- 208 housing units under the prototype plan.
- 20,009 housing units through Federal Housing Authority (FHA)
- To provide through mortgages provided by Federal Mortgage Bank of Nigeria (FMBN), a total of 43, 934 housing units
- To provide through estate development loans provided by FMBN a total of 7510 housing units. (FMWH, 2014).

The Nigerian Government again set out to remedy the challenges of housing delivery within different states in the country by engaging with PPP (Public-Private Partnership) or through Federal Housing Corporations (FHA). Here, deliveries were made of some prototype housing programmes at the state and federal level respectively where 10 completed luxury town houses in Lagos State were provided by public-private partnership and 2&3 bedroom bungalows in Kaduna State were built by Federal Housing Authority (FHA) amongst others. In spite of the government efforts in this direction, to deliver on intended number of housing units, the problem of housing delivery remain insurmountable as unimpressive results have been recorded in the provision of housing in Nigeria, despite huge allocations of money to the housing sector in the National Development Plans as discussed earlier [30–32].

2.3 Challenges of different types of housing delivery in Nigeria

In Nigeria, the delivery of housing is provided by both formal and informal sectors, houses provided by both the public and private sector are regarded as formal

Period/establishment	Total number of housing unit	Achievement level
Pre-independence	• Houses were built for the expatriates but number of units is unknown	• Houses were built for the expatriates but number of units is unknown
Post-Independence		
First National Development Plan (1962-1968)	• Planned construction of 61,000 housing units	• The political instability and the resulting civil were contributing factors for delivering only 500 units.
Second National Development Plan (1970-1974)	• The National Council of housing was established in (1972) to advise government on housing matters. • 59,000 housing units were planned for direct construction • Low-cost housing units across the federation.	• Only 7,080 housing units were provided
Third National Development Plan (1975-1980)	• Federal Ministry of Housing, Urban Development and Environment was created. • Land use Decree was promulgated (1978) • Planned construction of 202,000 low-cost housing units nationwide.	• Only 28,000 planned housing units were provided.
Fourth National Development Plan (1981-1986)	• National Housing Programmed was launched for the first time in 1980 • 160,000 low-income housing units to be constructed were allocated for N1.9 billion naira. • The second stage of the housing programme under this plan was planned to construct 20,000 housing units across the country.	• 23.6% representing a total of about 47,234 housing units were constructed in the first phase • The second face was cut short by military coup.
Post Second Republic		
Military Government (1986-1999)	• 121,000 housing units were planned for construction on site and services housing programme between 1993 and 1995 • 1991 the National Housing Policy was launched.	• 5500 housing units were delivered.
Civilian Government (1999-date) Ministry Prototype Housing Scheme within the Federation	• 262 prototype housing units • 2,140 housing units planned to be constructed	• 238 housing units were completed • 1,756 housing units were completed
FHA signed up with some states to provide housing units.		
FMBN a. Estate Development Loan b. Primary Mortgage Institution (NHF Mortgage)	• 7,510 housing units to be constructed • 4,934 housing units to be constructed	• 7,510 housing units were provided • 4,934 housing units were constructed
PPP- Construction Initiative	• 3,284 housing units to be provided	• 2009 housing units so far have been constructed.

Source: Adapted from Ibem et al. [28].

Table 1.
Overview of housing provision in Nigeria (pre- Independence- to date).

sector housing while those built or delivered by individuals, co-operatives, families or through community development efforts that to not comply to official building standards are referred to as informal sector housing

In Nigeria, according to [25] the majority of housing provision is executed by the private sector. Here 90% houses delivered are by individuals (self-built) and this signifies a high proportion of housing units produced by individuals. However, housing delivered by organized formal private sector, as well as the state (real estate developers) are insignificant. In Nigeria, many households in the urban areas are characterized by a mix of mix of middle-income earners and low-income earners [7, 33]. The range of socio-economic classes present in cities echoes the diversity of housing types delivered within Nigeria cities. This means that people with middle income to the lower income earners live in rented, informal low quality houses, while high income earners occupy luxury owner-occupied housing [34]. This implies that notwithstanding all efforts made by the Nigeria populace at providing housing through private mechanisms, housing delivery in qualitative and quantitative terms remains a mirage and this is aggravated by the presence official standards that are alien to Nigerian culture [2, 7, 29, 35–37]. Housing development in some Nigeria states; Edo, Lagos, Delta, Bayelsa, and Imo is limited by local practices which further worsens the engagement of individuals in the process of housing delivery. This is because of certain demands made by the Community Development Association (CDA) in form of levies. These levies are all informal payments demanded by the (CDAs) from housing developers before they can commence building in addition to other formal payments made to the appropriate official agency to obtain building permits. Ezeanah [38] showed that large sums of money are collected from housebuilders before they are allowed to build houses, hence posing a great challenge to housing delivery within Edo and some other Nigeria states.

Moreover, the rental type of housing categorized as public and private is a type of house delivered in Nigeria and 80% of households in Nigeria live in the private rental houses [9, 29, 35, 36]. This type of housing is plagued with various issues such as issues of finance, poor building materials, and demands for high levies, bureaucratic bottlenecks, extensive importation of building materials and this accounts for housing shortages within the Nigeria shortages [7, 10, 37, 38]. Consequently, in Nigeria despite efforts made at delivering housing through both the formal and informal sector, the housing deficit in Nigeria is huge as there is still a housing shortfall of approximately 16 million.

While the private initiatives enabled more houses to be delivered, these private initiatives were limited by local practices that worsened the engagement of some individuals in Nigeria with the house building processes, thereby limiting the quantity of housing delivered within Nigeria. Therefore, housing delivered are usually not enough to meet with the demands of housing in Nigeria with a shortfall of approximately 16 million housing units.

3. Conclusions

In Nigeria, diverse housing initiatives and programmes embarked on by the federal government to deliver housing for the populace have been saddled with a lot of challenges and has failed to provide the intended number of houses for the people. Again efforts at providing housing for the people through the diverse development plans in Nigeria failed which regrettable demonstrates the uninspiring attitude of the Nigerian Government in providing housing for the people. The failure of the government in ensuring that formulated policies and programmes shown in the development plans from 1962 till to date are sustained and implemented is one key cause for its failure at delivering houses for the people. Furthermore, institutions embedded in formal financial mechanisms made it difficult for many people who want to build houses to obtain formal form of financing which again limited the peoples effort at constructing houses.

In conclusion, housing shortages in Nigeria for both the middle and low income earners is observed notwithstanding the various housing initiatives carried out by the government. Also, it is shown that most of the populace in Nigeria housed either through rental housing or self-built housing are faced with a lot of housing challenges such as; limited access to land, high cost of building materials, high cost of levies, bureaucratic bottlenecks, institutional problems, enforcement issues amongst others which constrain the delivery of housing in Nigeria.

Acknowledgements

My thanks goes to Dr. Maren Mallo Daniel whose consistent push and encouragement has helped me to conclude this Chapter. Finally, I wish to thank the anonymous peer reviewers at IntechOpen for their time, insightful comments and expertise which has helped to improve this Chapter.

Conflict of interest

The author declares no conflict of interest.

Author details

Uyi Ezeanah
University of Jos, Jos, Plateau State, Nigeria

*Address all correspondence to: ezeanahu@gmail.com; ezeanahu@unijos.edu.ng

References

[1] Jiboye, A. D. Evaluating Tenants' Satisfaction with Public Housing in Lagos, Nigeria Urbanistika*ir architektūra Town Planning and Architecture;* 2009 33(4), p. 239-247

[2] Alagbe, O. A., and Opoko, P. A. Housing Nigerian Urban Poor through Self-Build Housing Concept Using Compressed Stabilized Laterine Bricks. International Journal of Research in Social Sciences; 2013, 2(4), p. 13-18.

[3] Federal Republic of Nigeria National Housing Policy, February 1991.

[4] COHRE (2004) - Housing Rights in West Africa: Report of Four Fact-Finding Missions – A Draft Report, for Consultation, Discussion, and Networking Purposes. Centre on Housing Rights and Evictions (COHRE). In Alagbe, O.A, Opoko, A.P. (2013) Housing Nigerian Urban Poor Through Self-Build Housing Concept Using Compressed Stabilized Laterite Bricks. International Journal of Research in Social Sciences. 2 (4) pp. 13-18. www.ijsk.org/ijrss

[5] Okonjo-Iweala, N. Unleashing the Housing Sector in Nigeria and in Africa. In *6th Global Housing Finance Conference*; 2014.

[6] Federal Government of Nigeria *Report of the Vision 2020 National Technical Working Group on Housing*, Federal Government of Nigeria; 2009.

[7] Makinde, O.O. Housing Delivery System Need and Demand, Journal of Environmental Development and Sustainability; 2013, 16 (1), p. 49-69.

[8] Waziri, A.G. and Roosli, R. Housing Policies and Programmes in Nigeria: A Review of the Concept and Implementation. Journal of Business Management Dynamics; 2013, 3(2), p. 60-68.

[9] Ogu, V. I. Housing Enablement in a Developing World City: The Case Study Of Benin City, Nigeria. Habitat International; 1999, *23*(2), p. 231-248.

[10] Udechukwu, C.E. Obstacles to individual home ownership in Nigeria, International of Housing Markets and Analysis; 2008, 1(2), p. 182-194.

[11] Omuta, G. E. D. Minimum versus affordable environmental standards in third world cities. Cities; 1986, 3(1), p. 58-71.

[12] Fiadzo, E. Estimating the determinants of housing quality: The case of Ghana (No. 6). Joint Centre for Housing Studies, Graduate School of Design [and] John F. Kennedy School of Government, Harvard University; 2004.

[13] Fokolade, A. and Coblentz, H.S. Citizen Participation in Urban and Regional Planning in Nigeria. Community Development Journal; 1981, 2(3), p. 45-55

[14] Mukhtar, M.M, Amirudin, R. and Mohamad, I. Housing delivery problems in developing countries: a case study of Nigeria ", Journal of Facilities Management; 2016, 14(4), p. 315 – 329.

[15] Nubi, T.O. 'Procuring, Managing and Financing Urban Infrastructure: Towards an Integrated Approach Land Management and Property Tax Reform in Nigeria, in `Omirin *et al.,* (ed.) Department of Estate Management, University of Lagos, Akoka; 2002.

[16] Adedeji, Y. D. and Olotuah, A. O. An evaluation of Accessibility of Low-income Earners to Housing Finance in Nigeria. *European Scientific Journal,* ESJ; 2012, 8(12), p.80-95

[17] Sanusi, J.O. Mortgage Financing in Nigeria: Issues and Challenges. A Paper Presented at the 9th Edition Nigerian

Institution of Estate Surveyors and Valers Organised John Wood Ekpenyong Memorial Lecture on 29th January; 2003.

[18] Okewole, I.A and Aribigbola, A. Innovations and sustainability in Housing Policy Conception and Implementation in Nigeria; 2006, p. 414-420.

[19] National Housing Policy Federal Republic of Nigeria; 2012.

[20] Lekwot V.E, Vachaku B.N, Usman T.M, Ifeanyi E. An Assessment of Government Intervention in Urban Housing Development in Nigeria. International Scientific Research Journal; 2012, 1(2), p. 34-41.

[21] UN-HABITAT. The Challenge of Slums - Global Report on Human Settlements; 2003.

[22] Olayiwola, L. M., Olurotimi Adeleye, and L. Ogunshakin. "Public housing delivery in Nigeria: Problems and challenges." XXXIII IAHS. World congress on Housing Transforming Housing Environments through the Design September 27-30, Pretoria South Africa; 2005.

[23] Abiodun J.O. The Provision of Housing and Urban Environmental Problems in Nigerian. Urban and Regional Planning Problems in Nigeria. University of Ife Press Ltd; 1985. p. 174-191.

[24] Aribigbola, A. Conceptual Issues in Housing and Housing Provision in Nigeria. *Effective Housing in the 21st Century Nigeria*. The Environmental Forum FUTA; 2000, p. 1-8.

[25] Federal Government of Nigeria. Government White Paper on the Report of the Presidential Committee on Urban Development and Housing, Lagos, Government Printing Press; 2002.

[26] Aina, T. Housing and Health in Olaleye-Iponri, a Low-income Settlement in Lagos, Nigeria; 1990, p. 56-88.

[27] Olotuah, A.O. The challenge of housing in Nigeria. In *Effective Housing in the 21st Century Nigeria, Akure: The Environmental Forum, Federal University of Technology*; 2000, p. 16-21.

[28] Ibem, E. O., Anosike, M. N., and Azuh, D. E. Challenges in Public Housing Provision in the Post-Independence Era in Nigeria. Journal of Human Sciences; 2011, *8*(2), p. 421-443.

[29] Ogu, V. I., and Ogbuozobe, J. E. Housing Policy in Nigeria: Towards Enablement of Private Housing Development. Habitat International; 2001, 25(4), p. 473-492.

[30] Muoghalu, L. N. The urban poor and Accessibility to public Housing in Nigeria. *Poverty in Nigeria, A Multi-Dimensional Perspective. The Urban Poor in Nigeria: Evans Brothers Publishers Ltd, Ibadan*; 1987, p. 161-171.

[31] Atolagbe, A.M.O. Affordable Shelter for the Urban Low-Income Classes in Nigeria: Local Resources Strategies. In Amole Bayo (ed.). *The House in Nigeria, Conference Proceedings,* Obafemi Awolowo University (O.A.U.) Ile-Ife, Nigeria; 1997, p. 274-279.

[32] Jiboye, A. A Critique of Official Housing Policy in Nigeria" The House in Nigeria, Bayo Amole (Ed.), Proceedings of the National Symposium, Obafemi Awolowo University (O.A.U.) Ile – Ife 23 – 24 July. 1997, p. 284-288.

[33] Olatubara, C.O. and Fatoye, E.O. "Residential satisfaction in public Estates in Lagos State, Nigeria". Journal of Nigerian Institute of Town Planners; 2006, 1(19), p. 103-124.

[34] Ndubueze, O. J. Urban housing affordability and housing policy

dilemmas in Nigeria (Doctoral dissertation, University of Birmingham); 2009.

[35] Agbola, T. The housing construction process in Nigeria: Implications for urban growth and development. Cities; 1988, *5*(2), p. 184-192.

[36] Ikejiofor, U. The God that Failed: A Critique of Public Housing in Nigeria, 1975-1995. Habitat International; 1999, *23*(2), p. 177-188.

[37] Collier, P., and Venables, A. J. Housing and urbanization in Africa: Unleashing a Formal Market Process. Policy Research Working Paper 6871; 2014.

[38] Ezeanah, U. The Delivery of Quality Housing in Benin City: The Influence of Formal and Informal Institutions, an unpublished PhD Thesis in the Department of Urban Studies and Planning, The University of Sheffield, Sheffield, United Kingdom; 2018.

Chapter 12

Typological Analysis of Gated Communities Characteristics in Ibadan, Nigeria

Olusola Oladapo Makinde

Abstract

Given the fact that the fear of crime is growing in Ibadan, and the number of gated communities or enclosed neighbourhoods is growing daily, need for an in-depth study of this phenomenon is essential to make available understanding into the architect's thought procedures, design values and establish a typological dimension of the prevailing enclosed neighbourhood. The aim of the study is to examine the types and characteristics of gated communities in Ibadan, with a view to informing policy on neighbourhood design and gated community development. This study identifies and assesses the types and characteristics of gated communities in Ibadan; this was assessed using field survey through direct observation check and structured questionnaire methods. The result of the typological classification of gated communities through a variety of enclosure in Ibadan shows nine deferent types of gated communities, this include type A (Ornamental gating), type B (Walled subdivisions, type C (Faux-gated entries), type D (Barricaded streets), type E (partially gated roads), type F (Full gated roads), type G (Restricted entry, bounded area), type H (Restricted entry, guarded area), type I (Condominiums). The result of the defensive physical characteristics of the gated communities shows that Old Bodija Scheme has the strongest characteristics with an average index of 3.58 DPC while Agodi GRA appears weakest with an index of 2.63 DPC. The eight typologies, which were identified, gave understanding to the design philosophy of the architects by showing the elements which they manipulated in the design of gated communities. Finally, the paper examined the level of importance of defensive physical characteristics that include Territoriality, Surveillance, Milieu and Image in the development and design of gated communities.

Keywords: Gated Communities, Typology, Characteristics, Architecture, Design

1. Introduction

The study of the types and characteristics of Gated communities (GCs) in Ibadan is very significant in housing development. Several problems, as well as advantages regarding GCs have been acknowledged in the literature, these include: reducing crime temporarily or permanently; decreasing the fear of crime or make available psychological respite; which can lead to an improved sense of community; which enhanced a sense of ownership and obligation [1]. The undesirable aspects are: generating a false sense of privacy and safety; relocating crime; segregating communities; decreasing response times of emergency vehicles; causing tension and conflict between urban residents; enhancing the fear of crime; triggering social segregation; increasing urban separation and fragmentation; causing problems with regards to services and maintenance [2].

A GC is a housing community that has a security gate situated at the front entrance. Occasionally the whole neighbourhood is fenced inside a perimeter of gates. It is a housing development with controlled access and exit [3]. These physical developments, in combination with security guards, substitute the older social control devices, which are centred on social unity within the community concerned. In its current form, a GC is a form of a housing estate or residential community comprising strictly-controlled entries for walkers, bicycles, and cars, and regularly branded and characterised by a closed perimeter of fences and walls [4]. GCs typically consist of minor residential roads and comprise numerous shared facilities. For smaller communities, this might be only a garden or other common area. For larger communities, it might be likely for inhabitants to stay inside the community for most day-to-day doings. GCs are a kind of communal interest development but are different from deliberate communities [5].

Enclosed neighbourhoods denote existing neighbourhoods that ensure controlled access through booms or gates across main roads. Many are enclosed with fence or wall as well, with a limited number of controlled entrances and exits with security guards at each point in some circumstances [6]. In certain cases the streets within these neighbourhoods were before, or still is public assets and in several cases, the local council is still accountable for public amenities to the community within. The roads in these areas are private, and most of the time, the maintenance and management are carried out by a private management organisation [7]. Enclosed neighbourhoods have become an occurrence to be taken earnestly in Nigeria; not only those who desire to live in such development but also those who are concerned with urban management and planning, in addition to the social tendencies that determine human social relations and social dynamics within urban areas in Nigeria [8].

The important issues arising from the study will make known the true image of these gated housing designs in term of their spatial organisation of spaces and level of safety. It will help in making urgent and necessary changes in the current system in prevailing gated communities and also aid in evading these challenges and problems for gated estate developers and government on policies implementation. Combined with an understanding of the characteristics identified, a study of the types of GC in Ibadan can go a long way to assist policymaker and role players, especially local governments, to make more informed decisions. It will be important for future policy decision making in Ibadan to understand the types and characteristics of GCs have on the urban environment and the implications for urban management and maintenance. This study will pave the way for future research in terms of neighbourhood safety development and sustainability of the neighbourhood system. It is now up to policymaker to interpret this information that had been gathered to investigate the local situation.

2. Literature review

The concept of GCs is a fast-growing one especially in response to safety and security all over the country. Equally one can found the road and street closures in major cities of Nigeria. They came up in response to the fear of crime and insecurity within the non-GCs. One of the concerns is the government's inability to protect the property and the life of all its citizens especially in developing countries like Nigeria

[8]. This necessitated the formation of private alternatives to crime control and prevention. A relatively innovative trend as an architectural conception is that of the GC. The growth and development of GCs in Nigeria is more of a reaction to the high level of fear of crime and insecurity in the nation's city centres. It is common these days to see many important streets close up with gates and booms all in the bid for a secure and safe environment [9]. The need for enhanced property value, privacy, safety and security resulted in an enclosed area, including the need to feel safer. Security, social connectivity and safety are also linked to a greater sense of identity and community, the necessity to bring the entire community together to enhanced common ideals and values [5, 10].

The notion of GC is a complex issue that can be described by sets of cultural, architectural, economic, demographic, societal, political and psychological factors that are unpredictable. GCs are defined, according to Low, [11] as a residential neighbourhood with gates and walls enclosing the development, which excludes non-residents access to all inside amenities plus residences, activities and open space. As crime turn out to be more of a threat to that feeling of security, privacy and safety, inhabitants are, in line with Blakely and Snyder, [12] turning to high walls and gates enclosing residential areas, thereby privatising and segregating parts of the urban area. The growth of safety enclaves is a global phenomenon [13].

Landman, [14] described a GC as a physical area that is fenced or walled off from its environments, elimination or controlling entrance to these areas employing booms or gates. In numerous cases, the concept can refer to as a residential area with controlled access, so that common public spaces have their uses restricted and privatised (García de Alba, [15]). In the area were the crime rates is high, enclosed neighbourhoods are perceived by many as the only alternative for crime deterrence [7]. It takes place in many forms in a large number of countries. Within a short period, GCs have rapidly increased in Nigeria. This has been the case for several different types of GC. Even though the major evolution of enclosed neighbourhoods has happened in city areas, especially in Oshogbo, Kaduna, Kano, Lagos, Port Harcourt and Ibadan among others, the occurrence has not been restricted to these cities. Several applications for neighbourhood enclosures have also been received in municipalities [5]. In the context of high crime rates, many people consider security villages or enclosed neighbourhoods the only choice for safety living in metropolises. Nevertheless, not all people come to an understanding that this is the best or only solution to crime prevention in the country [9].

The major motivating and determining factors behind enclosure and gating, according to Low [11], are fear of crime, property values, and nice environments, the search for safety, privacy, security, prestige, lifestyle, control, exclusivity and community attachment as driving factors to GC developments. Residing in segregated housing spaces in the city is a multifaceted social process that is not only the consequence of the fear of crime also is a way to relocate from urban disorder, to establish seclusion of social homogeneity and to establish new forms of local government [12].

Grant and Mittelsteadt, [16] acknowledged eight factors that distinguish GCs from their neighbourhood to include safety features and barriers, functions of the enclosure, facilities and amenities included, tenure, type of residents, location, policy context and size. Several authors agreed that gated developments reveal a rising trend to private governance and private communities [17]. GCs are believed to have characteristics of social interaction, safety, good life, privacy, freedom from maintenance, exclusivity and homogeneity [18]. Residents choose to move into CGs for reasons such as increased privacy and safety, traffic and noise reduction, yard maintenance and aesthetics, prestige, exclusivity, control, and the increase in property values. By their implication, GCs isolated residents from the larger urban environment [14]. Planners encourage GCs because they lack mixed uses, diversity, connected street systems and public open space. Planners can make alternatives against the appeal of GCs development by providing local community meeting places and better public safety [19].

GCs, which are dispersed all around the world, vary from country to country, with reverence to their characteristics and in specific to different motives for development viz. safety, privacy, security, prestige and ethnicity [20]. These GCs are way out for everyday problems of ethnic conflicts and high crime rate. They intensely restructured the urban forms of many cities. GCs signify the hope of privacy, safety, offer an important position to marketing strategy for developers in a competitive environment, attractive to consumers looking for identity and a sense of community, increase property values and exclude none resident, from attractive amenities (Blakely and Snyder 1998; [21]).

According to Górczyńska [22] there exist various classifications of GCs that relate to their organisational, functional, and morphological parameters. The classification explained by Blakely and Snyder, (1998) presents three key types of GCs to inhabitants' motivations and characteristics: GCs is inhabited by those searching for safety and by people with the same lifestyles and by elites. From this study, the typologies of GCs vary slightly from those offered for Western countries. Brabec and Sýkora [23] have categorised three main categories of GCs: (1) guarded, (2) guarded and walled, and (3) walled communities that relate to the level of luxury of each type of community and the first type are the most luxurious. Another classification identified by Hegedűs, [24] was established on the supplementary functions of GCs (facilities for inhabitants such as swimming pools and playgrounds) and the existence of particular fences and gates. With these features, three types of GCs were identified: (1) GCs with incomplete functions (2) GCs with complete functions, and (3) lacking any physical separation but delivering amenities.

The prevailing typologies of gated neighbourhoods are mostly grounded on the morphological features of estates [25, 26] or the types of protection used [27]. In line with Glasze [25], there exist two main types of GCs: condominiums (usually luxurious buildings in the central districts) and gated estates composed of multi-family houses, single-family houses and terraced houses [25]. Hence, this type agrees to the elite type identified by Blakely and Snyder (1998). Regarding gated estates composed of numerous buildings, their diversity in terms of standard, size, and quality is tremendously large, and their inhabitants signify both middle and upper classes [22].

A more multifaceted typology of gated estates was suggested by Chabowski [28], who concentrated on eight features: (1) the period of construction, (2) the extent of closed space, (3) the scale of closure, (4) the number of storeys, (5) the period of closure, (6) characteristics of surrounding areas, (7) architectural types and (8) quality and price of dwellings. This typology matches up to the great diversity of GCs. Within those categories of functions and other factors the typology add considerations of the character of facilities and amenities, spatial patterns, level of affluent and the type of security features. Each of these physical characteristics relates to a single function. For example, Blakely and Snyder, [12] discuss the level of affluence as the main factor in prestige communities but give it lesser important in lifestyle and security zone communities. Aulia and Suryani, (2020) identified the typology of gated housing based on the development and growth in the city of Medan which were classified into three sets, specifically large-scale gated housing, medium-scale gated housing and small-scale gated housing. Given the limited study on typological analysis of GCs and little attention been paid to the defensive physical characteristics in the context of gated communities holistically in a developing country like Nigeria; hence, the need to exploit and look critically at these issues in Ibadan. This study attempted to fill the gap with a particular interest in a quantitative study which is relevant and important in housing.

3. Research method

There are 30 residential areas with Gated Communities (GCs) in Ibadan. Five (5) residential areas consisting of fifty-seven (57) GCs were purposively selected for demonstrating typical GCs characteristics (see **Table 1**). These include Agodi Government Reservation Area (AGRA), with ten (10) GCs; New Bodija Scheme (NBS) with fifteen (15) GCs; Old Bodija Scheme (OBS) with thirty (30) GCs; Kolapo Ishola Scheme (KIS) with one (1) GC and Alalubosa Government Reservation Area (ALGRA), with one (1) GC. Thirty GCs representing 52.6% of the 57 GCs were sampled from the five residential areas. There are a total of four thousand, nine hundred and twenty-two (4,922) residential buildings in the selected GCs. Four

S/ NO	Local Government Area	Government Reservation Area	Local Govt. Scheme	Property Dev. Corp Scheme	Total GCs Areas
1	IBADAN NORTH	Agodi GRA Mokola Low-Cost Housing Scheme Samonda Scheme (Old-Airport) Oke-Aremo Housing Scheme	Sabo Housing scheme Mokola Layout	Old Bodija Scheme New Bodija Scheme	8
2	IBADAN NORTHWEST	Jericho GRA Onireke Comm. & Links Reservation Onireke Housing Estate	None	None	3
3	IBADAN SOUTHWEST	Iyaganku GRA Alesinloye GRA Alalubosa GRA Ring Road HOP. GRA	Ring Road Layout Liberty Layout Oluyole Scheme Lagos Bye Pass Layout (Mixed Dev.)	Owode Housing Scheme now in Ido Local Government Area	10
4	IBADAN SOUTHEAST	None	Lagelu Residential Scheme, Felele Express	None	1
5	AKINYELE	None	Idi-Ose Layout	None	1
6	EGBEDA	Ife Road Scheme	None	Olubadan Scheme, AJODA	2
7	ONA ARA	Ogbere Housing Scheme	*Local Government Residential Layout	None	2
8	LAGELU	Kolapo Ishola (Old Dairy Farm) Scheme	Okebadan Scheme –Akobo/Alegongo	Akobo Scheme Iwo Road (Lalupon)	3
				Total	30

Table 1.
Locations of residential areas with gated communities (GCs) in Ibadan metropolitan area.

hundred and ninety-three (493) representing 10% of the residential buildings were selected and from which each household head was sampled using a systematic sampling technique. The data collected were analysed using descriptive and inferential statistics. **Table 1** shows the locations of residential areas with Gated Communities (GCs) in Ibadan metropolitan area and **Table 2** shows the target population for the study while **Table 3** shows the distribution of administered and returned questionnaires.

About 493 questionnaires were distributed out of which 396 (80.4%) were returned. **Table 3** shows the distribution of returned questionnaires across the GCs selected for the study.

S/NO	The Study Population (20% of the target population selected purposively	The number of Gated Communities (GC) Identified in the Study Population.	The number of Gated Communities (GC) selected for the study. According to proportion	Sampling Frame (No of houses)	Sampling Size (10% of the household head selected using systematic sampling)
1	Old Bodija Scheme	30	15	2,495	250
2	Agodi GRA	10	5	492	49
3	New Bodija scheme	15	8	800	80
4	Kolapo Ishola Scheme	1	1	300	30
5	Alalubosa GRA	1	1	835	84
	Total	57	30	4,922	493

Table 2.
Target population for the study.

S/no	The Study Population	The number of Gated Communities (GC) selected for the study.	Number of Administered Questionnaires	Percentage of Questionnaires Administered	Number of Questionnaires Returned	Percentage of Questionnaires Returned
1	Old Bodija Scheme	15	250	50.7	202	41.0
2	Agodi GRA	5	49	9.9	39	7.9
3	New Bodija scheme	8	80	16.2	64	13.0
4	Kolapo Ishola Scheme	1	30	6.1	24	4.9
5	Alalubosa GRA	1	84	17.0	67	13.6
	Total	30	493	100	396	80.4

Table 3.
Distribution of administered and returned questionnaires.

4. Data analysis and results findings

4.1 Physical characteristics of GCs in Ibadan

This section aims to describe the physical characteristics of GCs in the study areas. In other to achieve this, 10 randomly selected experts from academia and practising firms who are members of Nigeria Institute of Architects and who are conversant with the study areas were selected for the study. The selection requirements for the panel are that such personality must be a design expert and urban planner from academia in addition to professional practice based on their technical know-how and proficiency. These were selected from catalogue of Architect Registration Council of Nigeria (ARCON). These experts comprehensively and objectively measured the subjective physical characteristics of the neighbourhood using ratings from an expert panel from the variables that comprised: the four (4) key elements of physical characteristics which are territoriality, surveillance, milieu and image. These were used as parameters in describing the characteristics of the study areas. This means that the study measured these elements in the areas to be able to describe the characteristics of each area.

4.2. Defensive physical characteristics of the Neighbourhoods

The results of the observation of the neighbourhoods that constitutes the communities are discussed under the four (4) elements using the indicators earlier enumerated.

4.2.1. Territoriality of the neighbourhoods

Looking at the first indicator to measure territoriality of the study areas which is the use of landscape to create clear boundaries (see **Table 4**) it seems that KIS has the highest index at 2.8 while NBS has the least with 1.6. However, having the highest index at 2.8 out of a possible 5.0 seems to means that all the study area did not make much use of landscape as a good tool to create clear boundaries between the respective units. In terms of maintenance and cleanliness of the neighbourhood in the study area, KIS (again) seems to be the area that takes a keen interest in keeping highly maintained neighbourhood given its index of 4.4 while NBS neighbourhood appears not to take good care of its unit with an index of 2.4.

Also, there appears to be a wide gap in maintenance culture that include roads maintenance and building maintenance and landscape maintenance among others between the best-maintained areas and the least maintained. **Figure 1** above shows an example of overgrown grass in the open spaces of the Old Bodija area. Also, as shown in **Table 4** OBS and ALGRA seems to have the clearest definition of territories as evidence by its index of 3.0 each while Agodi GRA New and Bodija Scheme seems to have a relatively weak definition of its territories given an index of 2.2 the result also goes to show that it seems that all the areas do not have their territories clearly defined given a 3.0 index as the highest. In terms of elements used to define those territories, ALGRA & KIS seems to have an excellent return, this is manifested by an index of 5.0, while on the other end, Agodi GRA did not seem to care about the use element to define their territories. This can be seen in their index of 1.0. The wide gap between these areas underlines the ease with which areas seem to use simple elements to define their territories in their neighbourhoods, in the case of signs to define ownership; ALGRA seems to make the best effort at an index of 1.8 while all others areas have lower than this. However, the highest index of 1.8

Indicator	Neighbourhood									
	Old Bodija Scheme		Agodi GRA		New Bodija Scheme		Kolapo Ishola Scheme		Alalubosa GRA	
	Count	Index	Count	Index	Count	Index	Count	Index	Count	Index
Use of landscape to create clear boundaries	12	2.4	10	2.0	8	1.6	14	2.8	10	2.0
Maintenance and cleanliness of the premises	17	3.4	17	3.4	12	2.4	22	4.4	15	3.0
Clear definition of territories	15	3.0	11	2.2	11	2.2	12	2.4	15	3.0
Elements used to definc territories (such as walls, furniture and paving stones)	25	5.0	5	1.0	13	2.6	25	5.0	25	5.0
Definition of ownership	5	1.0	5	1.0	5	1.0	5	1.0	7	1.8
Security sings at the entrance	20	4.0	5	1.0	10	2.0	20	4.0	25	5.0
Elements to restrict access	20	4	16	3.2	8	1.6	17	3.4	20	4.0
Total		3.50		1.97		1.91		3.29		3.40

Table 4.
Territoriality of the Neighbourhood.

Figure 1.
Picture showing the low level of maintenance in the neighbourhood.

shows that perhaps not much effort is invested by the residents in putting signs that define ownership. **Figures 2–4**, below show evidence of ownership in ALGRA.

As for security signage at the entrance to the areas and neighbourhoods in the study areas, ALGRA seems to be in the forefront with an index of 5.0 while lowest is Agodi GRA with an index of 1.0. For Elements to restrict access ALGRA & KIS seems to have a better usage with an index of 4.0 and the lowest was New Bodija with an index of 1.6.

Figure 2.
Demonstration of ownership in ALGRA.

Figure 3.
Wall around the GC and the security gate in ALGRA.

Figure 4.
Street and entrance to ALGRA.

In summary, ALGRA has the highest Territoriality index with 3.4 this was as a result of Elements used to define territory and Security sings at the entrance while NBS has the lowest index of 1.91.

4.2.2 Surveillance of the neighbourhood

The study shows that looking at the first indicator of this element, which has to do with the placement of windows to overlook sidewalks; Agodi GRA had highest with an index of 3.4 while KIS and ALGRA do not seem to have their windows placed in such a way as to overlook sidewalks with an index of 2.6. The average values of this index also seem to confirm the fact that the ability to overlook sidewalks may not be a primary consideration during the design and construction of most of the units in the study areas. Considering the height of the fence to permit surveillance in the neighbourhoods, Agodi GRA and NBS has the highest index of 4.2 while KIS is the least at 2.6. **Figures 5** and **6** shows pictures of security gate and guardhouse, exits gate under lock and key. In the case of external light to eliminate blind spots, OBS seems to fare better than all the other neighbourhoods with an index of 3.6 while NBS does not seem to enjoy much of external lighting and thus the least index of 1.8 (see **Table 5**) But for vehicular traffic, OBS seems to enjoy a sizable amount over and above the others with an index of 3.8 while NBS seems to have the least with an index of 2.2. This may be explained by the fact that OBS is surrounded by an impressive network of important roads, signage, derelict gate, school, abandon building and street view and an enclosed street and view of exits gate and security guide house and avenues in the area as evidence in **Figures 7–15**.

In the case of surveillance potential of land use, **Table 5** shows that NBS seems to have the advantage with an index of 3.6 while OBS and ALGRA did not seem to do well at an index of 3.0. However, the proximity of the indices around an average 3.0 shows that the study areas are not far apart from each other and that they did not do too badly especially as the least is 3.0. Considering the level of commercial activities in or near a neighbourhood in the study areas, OBS and NBS seem to take the upper

Figure 5.
The exits gate under lock and key in Kolapo Ishola GC.

Figure 6.
Security gate and guardhouse in kolapo Ishola GRA.

Indicator	Neighbourhood									
	Old Bodija Scheme		Agodi GRA		New Bodija Scheme		Kolapo Ishola Scheme		Alalubosa GRA	
	Count	Index	Count	Index	Count	Index	Count	Index	Count	Index
Placement of windows	14	2.8	17	3.4	16	3.2	13	2.6	13	2.6
Height of fence to permit surveillance	17	3.4	21	4.2	21	4.2	13	2.6	20	4.0
External Light to eliminate blind spots	18	3.6	13	2.6	1.2	2.4	15	3.0	9	1.8
Type of vehicular traffic	15	3.0	17	3.4	19	3.8	16	3.2	19	3.8
Surveillance potential of land use	15	3.0	17	3.4	18	3.6	17	3.4	15	3.0
Level of commercial activities in or near the neighbourhood	25	5.0	15	3.0	25	5.0	14	2.8	20	4.0
Quality of surveillance	20	4.0	16	3.2	20	4.0	10	2.0	15	3.0
Total		3.54		3.31		3.74		2.80		3.17

Table 5.
Surveillance potential of the Neighbourhood.

Figure 7.
Picture showing Oba Olagbegi Neighbourhood in OBS.

Figure 8.
Picture showing derelict gate at Oshuntokun road, old Bodija.

Figure 9.
Picture showing gating and security guard at Awogboro Neighbourhood in OBS.

Figure 10.
Picture: A view of a school and an enclosed street in OBS.

Figure 11.
showing an example of signage informing of the closure time. Security levy payment directed by the neighbourhood executive committee and house number.

Figure 12.
View of the entrance gate and signage in Awosika neighbourhood.

Figure 13.
View of exits gate and security guide house in Awosika neighbourhood.

Figure 14.
View of streets in Awosika that shows an example of well maintains neighbourhood.

hand with an index of 5.0 each while KIS takes the rear with an index of 2.8. This shows that even though all the study areas enjoyed some level of commercial activities, OBS and NBS are exceptional. In terms of quality of surveillance, which is exemplified by the number of eyes on the street OBS and NBS again take the lead with an index of 4.0 while KIS again brings up the rear at 2.0. In summary, NBS seems to have the highest index of surveillance with a value of 3.74. The highest contributor to this being the type of Level of commercial activities in or near neighbourhood at a value of 5.0 while KIS seems to have the least index of surveillance at 2.80, the biggest contributor to this being the level of Surveillance potential of land use in or around the units with a value of 3.4 OBS, Agodi GRA and ALGRA seems to be strong on surveillance in their units.

Figure 15.
A road closure using a palisade gate at Adeyi neighbourhood, old Bodija.

Indicator	Neighbourhood									
	Old Bodija Scheme		Agodi GRA		New Bodija Scheme		Kolapo Ishola Scheme		Alalubosa GRA	
	Count	Index	Count	Index	Count	Index	Count	Index	Count	Index
Presence of 'safe' properties	20	4.0	15	3.0	20	4.0	14	2.8	6	1.2
Presence of undesirable properties	10	2.0	8	1.6	11	2.2	10	2.0	6	1.2
Total		3.0		2.3		3.1		2.4		1.2

Table 6.
Milieu of the Neighbourhood in the study areas.

Indicator	Neighbourhood									
	Old Bodija Scheme		Agodi GRA		New Bodija Scheme		Kolapo Ishola Scheme		Alalubosa GRA	
	Count	Index	Count	Index	Count	Index	Count	Index	Count	Index
Physical condition of neighbourhood	20	4.0	20	4.0	16	3.2	21	4.2	23	4.6
The physical condition of the adjoining neighbourhood	19	3.8	18	3.6	14	2.8	22	4.4	25	5.0
Total		3.9		3.8		3.0		4.3		4.8

Table 7.
Image of the Neighbourhoods.

4.2.3 Milieu of the Neighbourhood

Under the indicator labelled 'presence of properties believed to be safe' as shown in **Table 6**, OBS and NBS seemingly enjoyed a preponderance of such properties, this is reflected in its index of 4.0 each while the neighbourhood that seems to have the least of such properties is ALGRA with an index of 1.2. This is evidenced by the presence of such properties as divisional/area Policy command headquarters in OBS and NBS schools churches and mosques. As for the neighbourhood whose units have the highest number of undesirable properties NBS seems to have the highest incidence of such as can be seen in the index of 2.2 while ALGRA seems to have the least at 1.2. In summary, NBS outshone the other neighbourhoods in a milieu with an overall index of 3.1 while ALGRA has the least at an index at 1.2. The implication of this is that there is a need to incorporate the presence of safe properties toward planning neighbourhood environment and having in mind to eliminate undesirable properties in developed areas.

4.2.4 Image of the Neighbourhood

When one looks at the physical conditions of the neighbourhood as illustrated **Table 7**, ALGRA seems to take the top spot with an index of 4.8 while NBS has the least at 3.0 as shown in **Table 7**. While in the case of the physical condition of adjoining units ALGRA takes the lead with an excellent index of 5.0 and E again the least at 2.8. This is to be expected. In summary, logically neighbourhood A is top with an aggregate index of 4.8 and NBS is the lowest at 2.8.

4.3 Summary of the physical characteristics of neighbourhoods

To summarise the physical characteristics of the neighbourhoods in all the study areas as shown in (**Table 8**), one can see that OBS seems to have the highest aggregate value of physical characteristics at 3.49 NPI while NBS has the least at 2.94 NPI. This shows that OBS has better-organised neighbourhoods which were reflected with the strong Neighbourhoods profile exhibited. The results of the Neighbourhoods Profile Indices (NPI) based on the factors of Territoriality, Surveillance, Milieu and Image in the study areas shows that OBS had the highest with (3.49), while Agodi GRA had the least with (2.85) NPI in the study area.

4.4 Physical characteristics of the GCs

The result of the observation of the GCs was discussed under three (3) of the four (4) physical elements using applicable indicators. The elements used in this case are territoriality under five (5) indicators, surveillance using four (4) indicators and milieu in which two (2) indicator were also used.

Indicator	Neighbourhood				
	Old Bodija Scheme	Agodi GRA	New Bodija Scheme	Kolapo Ishola Scheme	Alalubosa GRA
Territoriality	3.50	1.97	1.91	3.29	3.40
Surveillance	3.54	3.31	3.74	2.80	3.17
Milieu	3.00	2.30	3.10	2.40	1.20
Image	3.90	3.80	3.00	4.30	4.80
Average	3.49	2.85	2.94	3.20	3.14

Table 8.
Summary for the Neighbourhoods profile indices (NPI) in the study areas.

4.4.1 Territoriality of the GCs

The result of this study (see **Table 9**) showed that, as regards the maintenance of the GCs in the study areas ALGRA appears to be the best maintained, this is indicated by an index of 4.8 while NBS, on the other hand, appears to be the least maintained with an index of 3.0. This result indicated that, but for pockets of areas in the neighbourhoods that are not well maintained, there appears to be a generally conscious effort at maintaining the GCs. **Figure 16** shows evidence of view of a well-maintained section and in ALGRA.

Also, when discussing the clear definition of territories, ALGRA (again) appears to take top spot given its index of 3.6 while NBS (again) appears to take the rear with a value of 2.2. However, when it comes to elements used to define those territories, Agodi GRA seems to take the lead with a value of 4.0 while NBS seem to experience a near-total absence of elements to define territories as predicted with an index of 1.0 (see **Table 9**). Considering security at the entrance to the GCs, ALGRA seems to do a lot better than other GCs with an excellent index of 5.0; this implies

Indicator	Gated Communities									
	Old Bodija Scheme		Agodi GRA		New Bodija Scheme		Kolapo Ishola Scheme		Alalubosa GRA	
	Count	Index	Count	Index	Count	Index	Count	Index	Count	Index
Maintenance and cleanliness of GCs	21	4.2	18	3.6	15	3.0	20	4.0	24	4.8
Clear definition of territories	16	3.2	14	2.8	11	2.2	17	3.4	18	3.6
Elements used to define territories	13	2.6	20	4.0	5	1.0	5	1.0	15	3.0
Security signage at the entrance of the neighbourhood	21	4.2	9	1.8	9	1.8	21	4.2	25	5.0
Elements to restrict access	21	4.2	15	3.0	17	3.4	22	4.4	20	4.0
Average		3.68		3.04		2.28		3.40		4.08

Table 9.
Territoriality in the GCs.

Figure 16.
Shows a view of a well-maintained section of ALGRA.

that virtually all the constituent part of this area hosted security signage at the point of entry to them. On the other and NBS (GCs) seem to do poorly in this regard as shown in the index of 1.0. Similarly, when it comes to elements to restrict access to the GCs, KIS seems to take the top spot with an index of 4.4 while Agodi GRA brings up the rear with an index of 3.0. In summary, it seems ALGRA has the highest index on territoriality having a value of 4.08 while NBS appears to have the least at 2.28.

4.4.2 Surveillance of the GCs

The result in **Table 10** shows the first indicator for consideration is the use of external light to eliminate blind spots and in this case, KIS seems to have the best advantage as can be seen in the index at 3.6 while NBS seems to have the least at an index of 2.0. This result may signal the effect of decayed infrastructure in the area which may have manifested in the inability to service the neighbourhoods, with sufficient external lighting. In respect of vehicular traffic, Agodi GRA, KIS & ALGRA seem to experience the highest level of traffic than others with an index of 3.4 while OBS and NBS appear to have the lowest at 3.0. When it comes to the level to the level of commercial activities in or near the neighbourhood, NBS seem to take the top spot with an index of 4.6 while ALGRA is the least at 1.6. As to the quality of surveillance, OBS seems to have the advantage over others with an index of 3.8 while NBS appears to come less with an index of 2.4. Therefore, looking at the combination of all the indicators for surveillance, it then appears to be that neighbourhood OBS is in high positions with an index of 3.45 while NBS comes up at the rear with an index of 2.85.

4.4.3 Milieu of the GCs

The result as shown in **Table 11** indicated in that OBS seem to have the highest presence of properties believed to be safe having an index of 3.6 while ALGRA having the lowest at an index of 2.2. This tallies with the result of that of the neighbourhood. As for the presence of undesirable properties, OBS appears to take the lead with an index of 3.6 while Agodi GRA comes last with an index of 1.0.

Indicator	Gated Communities									
	Old Bodija Scheme		Agodi GRA		New Bodija Scheme		Kolapo Ishola Scheme		Alalubosa GRA	
	Count	Index	Count	Index	Count	Index	Count	Index	Count	Index
External light to eliminate the blind spot	15	3.0	14	2.8	10	2.0	18	3.6	20	4.0
Type of vehicular traffic	15	3.0	17	3.4	15	3.0	17	3.4	17	3.4
Level of commercial activities	20	4.0	11	2.2	23	4.6	14	2.8	8	1.6
Quality of surveillance	19	3.8	15	3.0	12	2.4	15	3.0	18	3.6
Average		3.45		2.85		3.0		3.2		3.15

Table 10.
Surveillance in the GCs.

Indicator	Gated Communities									
	Old Bodija Scheme		Agodi GRA		New Bodija Scheme		Kolapo Ishola Scheme		Alalubosa GRA	
	Count	Index	Count	Index	Count	Index	Count	Index	Count	Index
Presence of 'safe' properties	18	3.6	15	3.0	17	3.6	15	3.0	11	2.2
Presence of 'undesirable' properties	18	3.6	5	1.0	15	3.0	11	2.2	6	1.2
Average		3.6		2.0		3.3		2.6		1.7

Table 11.
Milieu in the GCs.

Figures 17–20 above show evidence of entrance and exit gate to a closed neighbourhood, poor road, office and unoccupied building with overgrown grass in a neighbourhood in Agodi GRA. Therefore in terms of milieu, OBS seems to be the best having an aggregate index of 3.6 while ALGRA appears to be the least at an index of 1.6.

Figure 17.
View of the entrance gate and poor road in Agodi GRA CG Ibadan.

Figure 18.
View of the entrance gate to a closed neighbourhood in Agodi GRA.

Figure 19.
View of the entrance gate and gatehouse to a neighbourhood in Agodi GRA.

Figure 20.
View of one of the exit gate in Agodi GRA.

4.5 The features of gated communities in the study

The review of literature on types of GCs and experiences with enclaves lead to the suggestion of several variables and functions that differentiate kinds of gating in Ibadan; these include the function of the enclosure; security feature and barriers; amenities and facilities included; types of residents; tenure; location; size; policy context (see **Table 12**). These eight characteristics were expanded into a checklist for this study. Although the features of GCs vary, they all have gates to regulate entrance into the community. A considerable number of GCs maintain around-the-clock, on-site security, and many of these communities are walled in as well. If a GC is designed for retired residents, additional amenities were included which include: the clubhouses, recreational centres and the like. At the front entrance gate of a gated community, there is usually either a security guard, an intercom on which you punch in a private access code, or a card reader. Upon admittance, the gate will be open for you. The size of GCs varies dramatically, with small, compact communities at one extreme and large and comprehensive GC on the other extreme. The large communities include not only residential properties but also recreational and entertainment centres, dining, retail and other lifestyle opportunities.

After physical observation of the study areas, the following classification was arrived at as shown in **Table 13** and using the checklist of features defining GCs as shown in **Table 11**. Although walls and gates may look similar across the study areas

Functions of Enclosure	Physical	Economic	Social	Psychological or Symbolic
	Secure people and property Create an identity for the project	**Enhance property value. Protect club amenities**	**Give visual or spatial privacy. Control those insides**	**Display status and power. Control those outsides**
Safety Feature	*Nature of boundary* Wall Low fence, chain or bollard Faux guard Station Hedge or vegetation Swing harm gate *Nature of security* Guard at all times Auto opener entry	Fence-opaque Fence-barbed Mirror glass on the guardhouse Topographic features Lift-arm gate Patrolling guards Surveillance cameras	Physical Fence visually opened Speed bumps or chicanes Private properties Signs Water, Ravines, Forest Slide gate Devices in the roadbed Card entry Armed guards	Symbolic Fence-electric Pavement texture or colour No parking Signs Desert Swing gate Guards at designated time Code entry House alarms
Amenities and Facilities	Private roads Open spaces Institutional facilities	Meeting place Landscape maintenance Guards	Activities centres Quality design	Recreational facilities Commercial facilities
Types of Residents	Homogeneous by age	Homogeneous by class	Homogeneous by ethnicity, race and status	Shared activities (for example, golf)
Tenure	Principal residence Fee simple ownership	Secondary residence Condominium ownership	Seasonal residence Land lease	Public housing Rental
Location	Urban Infill	Suburban Greenfield	Exurban Resort destination	Rural Inner-city
Size	Cul-de-sac pod	Neighbourhood (ten to hundreds of units)	Village (hundreds of units, some commercial)	Town (thousands of units and mix uses
Policy Context	Restricting gating	Enables gating	Growing area	Stable or declining area

Source: Adopted by the Author from Grant and Mittelsteadt [16].

Table 12.
Checklist of features defining gated communities in the study areas.

DOI: http://dx.doi.org/10.5772/intechopen.97365

S/N	Type	Boundary	Road access	Notes
A	Ornamental gating	No mark boundary	Landmark gate at entry	Feature gates showing the subdivision name that is placed at the major entries to give identity to an area.
B	Walled subdivisions	Opaque fence or wall	Open	Full walled subdivisions that are common urban features. Cars and pedestrian may enter.
C	Faux-gated entries	Opaque wall or fence	Narrow entry, removable chains or bollard, guardhouse	Some subdivisions have physical features that look like guard houses or private entries to discourage uninvited vehicles from entering.
D	Barricaded streets	No marked boundary	Public streets closed by planters or concrete barriers	Many neighbourhoods barricade streets and creating cul-de-sac streets within the grid as a form of traffic control. Pedestrian access is open.
E	Partially gated roads	No marked boundary	Lifts or swing arm	Rural cottage subdivisions may feature gates that are only closed for part of the year. May have gates but no walls. Pedestrian access is open.
F	Full gated roads	Natural features such as water, ravines, forest and mountains	Lifts or swing arm	Prestige communities on islands, peninsular, or remote may limit access through combined natural and man-made features.
G	Restricted entry, bounded area	Fence or wall and/ or natural features that limit access	Gate with limited control access	Communities may completely restrict public access; video or telephone systems may also allow visitors to be vetted by residents.
H	Restricted entry, guarded area	Fence or wall and/ or natural features that limit access	Gate with limited control access; security guards, police or army	Communities may completely restrict public access; video or telephone systems may also allow visitors to be vetted by residents. GCs have guards at the gates or patrolling the premises. In some zones, guards may carry automatic weapons

Source: Adopted from Grant and Mittelsteadt [16] and Modified by the Author 2018.

Table 13.
Classifying gated communities through variety of enclosure in Ibadan.

they have a range of functions that include: physical, economic, social psychological or symbolic as shown in **Table 5** below. **Table 6** shows the features of GCs in the study areas which include: the functions of enclosure, safety feature, amenities and facilities, types of residents, tenure, location, and size and policy context. The gate provides the architecture of control as demonstrated from the physical characteristics of the study areas for both the insider and outsider; it reinforces the need for surveillance and importance for a social order where everybody knows his or her place. Walled and GCs are seen as synonymous but involved different levels of enclosures. This lead to the suggestion that a variety of enclosure in GCs in the study areas as outlined in **Table 13**, which shows the varieties the degree of enclosure proceeds from largely symbolic or psychological, to the full physical, as an architecture of control, became more explicit.

4.5.1 Types and frequencies of GCs

Table 14 as clearly shows that GC type H (Restricted entry, guarded area) has the highest frequency having a total of 11 GCs representing 19.3%, followed by type D (Barricaded streets) having 10 GCs representing 17.5%. 8 GCs representing 14.0% used type E (partially gated roads), while 7, 6, 5, 4 and 1 communities representing 12.3%, 10,5%, 8.8%, 7.0% and 1.8% respectively made used of type G (Restricted entry, bounded area), type A (Ornamental gating), type B (Walled subdivisions, type C (Faux-gated entries), and finally type I (Condominiums) which is the least in the logged. The implication of this is that people accept their neighbourhood to be restricted by the use of gate and wall and also to be guarded at all-time by the security agents for them to have feeling safety in their community.

A. **Ornamental Gating:** This is landmark gate features with the marketing name of the GC prominently display. Examples include the Legacy Estate Gate, the entrance gate to Carlton Gate Estate and Kolapo Ishola GRA and as shown in **Figures 21** and **22** respectively. **Figure 23** shows the street view of Carlton Gate Estate, Akobo (GRA) Ibadan.

B. **Walled subdivisions:** This is a type of neighbourhood that has a fence or wall that separated them from neighbouring suburbs. The wall runs alongside collector roads. Local streets into such developments remain open and are usually public. It makes use of private road build to narrower standards than public thoroughfares which make visitors hesitate to enter. The wall development is easily converted to fully enclosed settlements with the addition of gates or guards. **Figures 24** and **25**.

C. **Faux-gated entries:** these are GCs that are surrounded with Opaque wall or fence. The entrance into this area is narrow, and its make use of removable chains or bollard and guardhouse. Some subdivisions have physical features that look like guard houses or private entries to discourage uninvited vehicles from entering.

D. **Barricaded streets:** Generally appears in the inner city areas where existing streets are closed to reduced traffics. The pedestrian can still move through the developments as walls are seldom constructed to the barricades. In some cases, the barricades enclosed those within and seen to be controlling problems outside the neighbourhood. This type is as shown in **Figures 26–28** below.

The Study Areas	Types of GCs																		Total
	A		B		C		D		E		F		G		H		I		
	Freq	%	Freq	%	Freq	%	Freq	%	Freq	%	Freq	%	Freq	%	Freq	%	Freq	%	
Old Bodija Scheme	3	1.0	2	6.7	3	1.0	6	2.0	4	13.3	2	6.7	4	13.3	5	16.7	1	3.3	30
Agodi GRA	1	10	1	10	—	—	2	20	2	20	1	10	1	10	2	20	—	—	10
New Bodija Scheme	1	6.7	2	13.3	1	6.7	2	13.3	2	13.3	3	20	2	13.3	2	13.3	—	—	15
Kolapo Ishola Scheme	—	—	—	—	—	—	—	—	—	—	—	—	—	—	1	100	—	—	1
Alalubosa GRA	—	—	—	—	—	—	—	—	—	—	—	—	—	—	1	100	—	—	1
Total	6	10.5	5	8.8	4	7.0	10	17.5	8	14.0	6	10.5	7	12.3	11	19.3	1	1.8	57

Table 14.
Types and frequencies of GCs identified in the study areas.

Figure 21.
View of legacy estate gate within Kolapo Ishola GC.

Figure 22.
Entrance gate to Carlton gate estate.

E. **Partially gated roads:** This GC has no marked boundary, and is with lifts or swing arm and may feature gates that are only closed for part of the year. It may have gates but no walls surround the area; usually, the pedestrian access is open. Such example is as shown in **Figures 29–31**.

F. **Full gated roads:** this is a GC which surrounded with natural features such as water, ravines, forest and mountains and makes use of Lifts or swing arm to control access into the community. These communities may limit access through combined natural and man-made features that eliminate the need for a full wall.

G. **Restricted entry, bounded area (with full perimeter fencing)**: Fence or wall and natural features, strictly limits access to the areas. The entrance gate is with limited control access. The communities may completely restrict public access; video or telephone systems may also allow visitors to be vetted by residents. In such an area, physical boundaries replace the psychological boundaries and strictly segregate the place. Technology devices managed by the residents control access. This type is as shown in **Figure 32** below.

H. **Restricted entry, guarded area:** Fence or wall and/ or natural features that limit access to GC. The gate is with limited control access together with security guards, police or army. In these types of communities, the residents completely restrict public access. Video or telephone systems were used to

Figure 23.
View of Carlton gate estate, Akobo (GRA) Ibadan.

Figure 24.
Wall around Gaton gate estate.

allow visitors to be vetted by residents. GCs have guards at the gates and also security agent patrolling the premises. In some areas, guards may carry automatic weapons.

I. **Condominiums:** Vertical closed condominiums can vary from one apartment block enclosed by fences, with access control, either in the form of an intercom system or manned by private security guards, to a collection of high rises in a complex; the condominiums usually include only one or two high-rise buildings with a range of security mechanisms to improve safety. Horizontal closed condominiums usually take the form of low-density developments spread over large areas of land. The size of these developments varies depending on the need and area of land available. These are very similar to typical luxury estate

Figure 25.
Wall around legacy GC.

Figure 26.
View of Oyo-state, housing corporation; Basorun estate zone 7, GATE 3.

Figure 27.
Awogbore gate, Akobo area.

developments. Such example of vertical closed condominiums is shown in **Figure 33** Wigatar Estate in Sabo Quarters area in Ibadan.

4.5.2 Summary for the types and defensive physical characteristics of GCs

Classifying GCs through a variety of enclosure in Ibadan nine deferent types of gated GCs were identified this include type A (Ornamental gating), type B (Walled subdivisions, type C (Faux-gated entries), type D (Barricaded streets), type E (partially gated roads), type F (Full gated roads), type G (Restricted entry, bounded

Figure 28.
Akala gate, Akobo Ojurin Ibadan.

Figure 29.
View of Oyo-state, housing corporation; Basorun estate zone 4, GATE 1.

Figure 30.
Ogo-Oluwa gate, Agbo area Ibadan.

area), type H (Restricted entry, guarded area), type I (Condominiums). To summarise the physical characteristics of the GCs, it appears OBS has the strongest of these characteristics with an average index of 3.58 as shown in **Table 15** while Agodi GRA appears weakest with an index of 2.63.

4.5.3 Relationship between perception of safety and physical characteristics of the study areas

To discuss the relationship between the Perception of safety and the physical characteristics of the study areas **Table 16** offers a simple way to go about this.

Figure 31.
View of Oyo-state, housing corporation; Basorun estate zone 4, GATE 2.

Figure 32.
Gate and booms across a road, marking the entrance to an exclusive neighbourhood in Ibadan.

Table 16 combines, the perception of the safety of each of the study areas, and the physical characteristics of the neighbourhoods, namely territoriality, surveillance and milieu. The section discussed the relationship between the perception of safety and the physical characteristics of the neighbourhoods.

Therefore, the result shows that ALGRA seems to have the highest perception of safety categories of 'high' (52.9%). This shows that there appears to be some element of correlation between the perception of safety categories of the neighbourhood and physical characteristics. The result shows that OBS has the highest percentage in the high perception categories (48.0%).

OBS has the highest index in territoriality at 3.50 and also has the highest percentage of the 'high' perception categories (48.0%); this suggests that a good

Figure 33.
A typical vertical closed condominiums; Wigatar Estate in Sabo Quarters.

Factors	Gated Communities				
	Old Bodija Scheme	Agodi GRA	New Bodija Scheme	Kolapo Ishola Scheme	Alalubosa GRA
Territoriality	3.68	3.04	2.28	3.40	4.08
Surveillance	3.45	2.85	3.00	3.2	3.15
Milieu	3.60	2.00	3.30	2.60	1.70
Average	3.58	2.63	2.86	3.07	2.98

Table 15.
Summary of indices for physical characteristics of GCs.

showing on territoriality may translate to a better perception of safety in a neighbourhood. This seems to be further confirmed by the fact that NBS has the lowest territorial index at 1.91 the high percentage of the medium perception categories (53.2%). OBS & NBS seems to have the highest index on surveillance with 3.54 and 3.74; this coupled with the fact that OBS has a high percentage at the 'very high' perception categories of 52.9% seems to confirm the contribution of surveillance to the perception of safety. However, this is not to suggest that surveillance alone leads to high perception as demonstrated by ALGRA which has the lowest index on surveillance 3.17 and at the same time having 4.49 the highest percentage of the 'high' perception categories (59.2%).

In the case of milieu NBS seems to have the highest index (3.10) and that it has the highest percentage at 'medium' perception categories (55.3%) but not the highest perception of safety index. It, therefore, suggests that though milieu might contribute to the perception of safety, this may not be enough to solely determine the perception of safety. In summary, OBS appears to have the highest aggregate

Indicator	Neighbourhood					Average
	Old Bodija Scheme	Agodi GRA	New Bodija Scheme	Kolapo Ishola Scheme	Alalubosa GRA	
Territoriality	3.50	1.97	1.91	3.29	3.40	2.81
Surveillance	3.54	3.31	3.74	2.80	3.17	3.31
Milieu	3.00	2.30	3.10	2.40	1.20	2.4
Image	3.90	3.80	3.00	4.30	4.80	3.96
Average (DPCI)	3.49	2.85	2.94	3.20	3.14	3.12
Low	1.4	6.6	6.5	4.0	4.0	4.5
Medium	34.3	55.3	53.2	40.0	32.0	42.96
High	52.9	35.5	35.5	54.0	59.2	47.42
Very high	11.4	2.6	4.8	2.0	4.8	5.12

Table 16.
Relationship between perception of safety and defensible physical characteristics (DPCI).

defensible physical characteristics index (3.49) further reinforced by Perception of safety index (4.08) with the high percentage of 52.9% of 'high perception categories. This suggests that this neighbourhood enjoys a sort of primacy when all these factors are considered. On the other hand, ALGRA seems to have 3.14 aggregates index of defensible physical characteristics but a relatively highest perception of safety index of 4.49 which was also further reinforced by the highest percentage at 'high' perception categories (59.2%). This suggests that the relatively weak defensible physical characteristics of the neighbourhood may not be enough to weaken the residents' perception of safety. This may mean that there are other factors which include: The physical condition /environmental design, social capital and experiences of safety indices which are at play in determining the strength of elements of physical defensible characteristics.

Table 17 clearly shows the summary of the perception of safety (PSI) and defensible physical characteristics (DPCI) in the selected GCs in Ibadan and it can

Indicator	GCs				
	Old Bodija Scheme	Agodi GRA	New Bodija Scheme	Kolapo Ishola Scheme	Alalubosa GRA
Territoriality	3.50	1.97	1.91	3.29	3.40
Surveillance	3.54	3.31	3.74	2.80	3.17
Milieu	3.00	2.30	3.10	2.40	1.20
Image	3.90	3.80	3.00	4.30	4.80
Perception of safety	4.08	3.96	3.80	4.38	4.49
physical/design condition	4.10	3.94	3.53	4.43	4.40
Social capital	3.75	3.02	3.73	2.92	2.95
Experiences of Safety	3.92	3.95	3.82	4.10	4.22
Average	3.72	3.28	3.33	3.58	3.58

Table 17.
Perception of safety PSI and defensible physical characteristics (DPCI).

be deduced from the **Table 17** that surprisingly OBS has the highest safety profile although the area has a lower perception of safety, image, physical and environmental design condition indices than KIS and ALGRA that have higher indices in these factors. From the study, it could be observed that OBS is having better social capital, territoriality, surveillance, milieu indices than these two GCs. The study has also shown that for safety to be achieved in a residential area all these factors must be followed.

5. Conclusions

Based on the physical observation carried out in the study areas, gating are widely employed in various types of housing development. After physical observation of GCs characteristics of various neighbourhood and communities in the urban area of Ibadan, this study identified a working typology of GCs in Ibadan. This was taken as a starting point for this study. This typology comprises nine main types of GC in Ibadan. These include: Type A (Ornamental gating), Type B (Walled subdivisions) Type C (Faux-gated entries) Type D (Barricaded streets) Type E (Partially gated roads) Type F (Full gated roads) Type G (Restricted entry, bounded area, Type H (Restricted entry, guarded area) and type I (Condominium). Eight (8) key elements of physical and none physical characteristics which are territoriality, surveillance, milieu and image, perception of safety, physical/design condition, social capital, experiences of safety were used as parameters in describing the characteristics of the GCs in the study areas.

The result of analysis of defensible physical characteristics of the study areas concerning safety revealed that the elements when examined at the neighbourhood level shows that OBS have good defensible physical characteristics (territoriality, surveillance, milieu image), it also has a high level of perception of safety, physical/design condition, social capital and experiences of safety at the entire neighbourhood level. The overall picture of the study areas is strengthened by the positive value of commercial activities in and around the neighbourhood which tend to aid surveillance. Measured by the physically fortified character, the number of neighbourhoods that falls into the category of GC is huge within the study areas. The respondents generally hold a positive attitude to gating. Physical gating is frequently taken as an inherent element for residential development, and naturally a must-have item within the study areas. No resident hide his/her appreciation of living in a gated or protected residence. Whilst gating is common in the residential culture of Ibadan cities, gating bears unique characteristics in response to its urban context, which is well known for a high density and high-rise.

Author details

Olusola Oladapo Makinde
Department of Architecture, Ladoke Akintola University of Technology, Ogbomoso, Nigeria

*Address all correspondence to: makindeolusola2012@yahoo.com

References

[1] Aulia, D.N and Suryani, L. (2020). Gated community typology based on growth and development in Medan City, Indonesia; IOP Conf. Ser.: Earth Environ. Sci. 452, 012154

[2] Makinde, O. O. (2014). Neighbourhood safety in selected gated communities in Ibadan: Unpublished MPhil Thesis. Oyo-State, Nigeria: Department of Architecture, Obafemi Awolowo University, Osun State.

[3] Bekleyen, A., and Yilmaz-Ay, I. (2016). Are gated communities indispensable for residents? Urbani Izziv, 27. doi:10.5379/urbani-izziven-2016-27-01-005

[4] El-Ekhteyar, E., and Furlan, R. (2016). Sense of community in gated communities in Doha: The case of Al-ein compound in ein khaled neighbourhood. American Journal of Sociological Research, 6(5), 126–134. doi:10.5923/j.sociology.20160605.02, 2016.

[5] Makinde, O.O., (2020). The correlates of residents' perception of safety in gated communities in Nigeria: Social Sciences & Humanities Open 2 (2020) 100018

[6] Elhadary, Y., and Ali, S. (2017). A new trend in urban housing: Gated communities in Khartoum, Sudan. American Journal of Sociological Research, 7(1), 45–55. doi:10.5923/j.sociology.20170701.07, 2017.

[7] Landman, K. (2000). An overview of enclosed Neighbourhoods in South Africa; *Pretoria CSIR*

[8] Makinde, O.O. (2019). Neighbourhood quality in gated communities in Ibadan, Oyo State, Nigeria: Unpublished PhD. Ile Ife: (Arch) Thesis at Obafemi Awolowo University.

[9] Ajibola, M. O., Oloke, O. C., Ogungbemi, A. O. (2010). Impacts of Gated Communities on Residential Property Values: A Comparison of Onietesi Estate and Its Neighbourhoods in Ikeja, Lagos State, Nigeria; *Journal of Sustainable Development Vol. 4, No. 2; April 2011*

[10] Fabiyi O. (2004). Gated Neighbourhoods and Privatisation of urban security in Ibadan metropolis; Occasional Publication No 16 IFRA. Ibadan

[11] Low, S. (2003). Behind the gates: life, security and the pursuit of happiness in fortress America (First Edition). New York: Routledge.

[12] Blakely E. J, Snyder, M. G. (1998). Fortress America: Gated communities in the United States; *Brookings Institution Press.*

[13] Wang, J Lau, S.Y. (2013). Hierarchical Production of Privacy: Gating in Compact Living in Hong Kong; Current Urban Studies. Vol. 1, No. 2, 11–18. *Scientific Research;* (http://www.scirp.org/journal/cus)

[14] Landman, K. (2003). A National Survey of Gated Communities in South Africa; CSIR Building and Construction Technology (Boutek); Programme for Sustainable Human Settlements

[15] García de Alba, E. R. M. (2016). Middle-class gatedness: A practice-based analysis of middle-class gated communities in Mexico. Unpublished thesis submitted in partial fulfilment of the requirements for the degree of doctor of philosophy. Faculty of Social Sciences, Department of Urban Studies and Planning, the University of Sheffield

[16] Grant, J. and Mittelsteadt, L. (2004). Types of Gated Communities:

Journal of Environment and Planning B: Planning and Design, *Vol. 31, pp. 913–930.*

[17] Kohn, M. (2004). Brave New Neighbourhoods: The Privatization of Public Space. *New York: Routledge.*

[18] Muiga, J.G. and Rukwaro, R.W. (2016). The satisfaction of residents with gated community lifestyle: The case of Nairobi County; Kenya. BEST: International Journal of Humanities, Arts, Medicine and Sciences, 4 (12), 85–104. ISSN (P): 2348–0521, ISSN (E): 2454–4728.

[19] Musyoka, M.C., Gakuu, C., and Kyalo, D.N. (2017). Influence of the technological environment on the performance of gated community housing projects in Nairobi County, Kenya; European Scientific Journal, 13 (11), 1857–7881 (Print) e - ISSN 1857–7431.

[20] Gulumser, A.A. Levent, B.T. (2007) Through the Sky: Vertical Istanbul Gated Communities.Location; www.bidoun.com/bdn/ ... /istanbuls-gated-communities.

[21] Grant J. (2003). Planning Responses to Gated Communities in Canada: Paper Presented at Gated Communities: Building Social Division or Safer Communities? Glasgow,

[22] Górczyńska, M. (2012). The specificity of gated neighbourhoods in the Bielany district (Warsaw); Journal of Urban Research: URL: http://articulo.revues.org/2022

[23] Brabec, T. Sýkora, L. (2009). Gated Communities in Prague, in Smigiel C (ed.) Gated and Guarded Housing in Eastern Europe. Forum ifl, Heft 11, Leipzig, Leibniz-Institut für Länderkunde: 83–89

[24] Hegedűs, G. (2009). A Review of Gated Communities in Some Hungarian Cities: Geographica Pannonica 13(3): 85–96.

[25] Glasze, G. (2001). Geschlossene Wohnkomplexe (gated communities): «Enklaven des Wohlbefindens» in der wirtschaftsliberalen Stadt, in Roggenthin R (ed.), Stadt – der Lebensraum der Zukunft? Gegenwärtige raumbezogene Prozesse in Verdichtungsräumen der Erde, Mainz, Mainzer Kontaktstudium Geographie 7: 39–55.

[26] Tobiasz-Lis, P. (2011). Osiedla grodzone w Łodzi. Przyczyny i konsekwencje zjawiska, Space - society - economy, Współczesne przemiany środowiska mieszkaniowego – wybrane problemy 10: 99–114

[27] Gąsior-Niemiec A. Glasze G, Lippok, D, Pütz, R, (2007). Grodzenie miasta: casus Warszawy. Studia Regionalne Lokalne 4(30): 5–30.

[28] Chabowski, R. (2007). Kwestia klasyfikacji i nazewnictwa osiedli zamkniętych w Warszawie, in Jałowiecki B, Łukowski W (ed.) Gettoizacja polskiej przestrzeni miejskiej. Warszawa, Wydawnictwo SWPS Academica, Wydawnictwo Naukowe Scholar: 121–136.

Section 4

Economy and Sustainable Housing

Chapter 13

UK House Prices – Connectedness or Ripple Effect?

Taufiq Choudhry, Syed S. Hassan and Sarosh Shabi

Abstract

The performance of the housing market is currently considered a measure of economic activity. This research explores the *connectedness* vs. *the ripple effect* hypothesis in the current house pricing literature. Using linear causality and nonlinear causality tests we show significant bidirectional dependence between the London house prices and other UK regions' house prices except for Northern Ireland and Wales in contrast to the existing literature where more evidence of ripple effect is reported. Furthermore, linear and non-linear forecasting tests back these results. This result has important implications for policymakers and investors.

Keywords: Connectedness, House Prices, Nonlinearity, Ripple Effect

1. Introduction

The housing market is closely associated with consumer spending, implying that an increase in house prices boosts homeowners' confidence. Similarly, a decline in house prices raises concerns for the homeowners due to the risk of losing the value of their property resulting in a reduction in spending and holding off personal investments. Thus, house prices have become an indicator of the economic performance of a country [1, 2]. Gallin [3] and Costello et al. [4] further show the importance of the role of housing in the economy and the effects of the underlying economic factors on house prices.

Transmission of regional housing prices within one single country has been researched widely over the years [5–11]. Regarding the UK, most of the literature focuses on the ripple effect – i.e. house prices initially originate from London and the South East and are then transmitted to the rest of the country [5, 9].[1] This implies that the housing market in London is the main transmitter of shocks, but developments in other regions have no impact on London. Geographical proximity to London appears to be a decisive factor in relation to the ripple effect [5]. Holly et al. [12] and Cook and Watson [9] report that it takes more time for a shock in the housing market of London to propagate another UK region when this region is relatively distant from London. Further according to Holly et al. [12] any other UK region may have an impact on London prices; however, this impact is relatively short-term.

[1] Antonakakis et al. [5] and Cook and Watson [9] provide an excellent survey of the papers in literature that investigate the ripple effect in the UK housing market. Cook and Watson [9] also provide a good discussion of the different empirical methods applied in these papers.

IntechOpen

Connectedness is defined as the inter-linkage or dependence between two or more-time series [13]. The key differences between connectedness and ripple effect relate to implications in terms of Granger causality. Ripple effect only shows unidirectional shock transmission whereas Connecteness implies bidirectional Granger causality among the underlying variables. Zhu et al. [14] show that rising connectedness may be due to the information spillover considering investment aspects of the housing market which may come from geographically adjacent or economically linked regions. Our study contributes to the literature by empirically investigating the *connectedness vs the ripple effect* between the house prices in London and 13 other regions of the UK.

This chapter studies the price transmission mechanism driving the UK regional house prices using linear causality and the nonlinear Granger causality model proposed by Hiemstra and Jones [15], and the impulse response. Application of the non-linear causality test and impulse response on the UK housing market makes this study unique in the literature. Linear and non-linear forecasting tests are further conducted as a robustness check. This chapter, thus investigates the *connectedness vs the ripple effect* between the house prices in London and other regions of the UK. The key focus is to show that the changes in house prices in the UK are transmitted in a bidirectional manner between London and most of the country. According to Antonakakis et al. [5] this may have implications for investors seeking efficient diversification of investment across mortgage backed assets across various regions in the country. Further, identification of regional disparities can help policymakers and investors to achieve more balanced growth across the regions under study. This chapter is motivated by Antonakakis et al. [5] who claim that recent empirical evidence is rather inconclusive about the actual manifestation of the ripple effect and further by Cook and Watson [9] who advocate further research in this field.

Our results show bidirectional dependence between the London house prices and other regions' prices except for Northern Ireland and Wales. Thus, we provide evidence of connectedness among the house prices in London and other regions of the UK. This result is confirmed by linear causality, the nonlinear causality and impulse response tests. Further empirical examination applying linear and non-linear forecasting tests back the linear and non-linear causality results.

The format of the chapter is the following. Section 2 describes the data and the empirical methodology employed. Section 3 discusses the results, and Section 4 presents the conclusion and implications.

2. Data description and methodology

2.1 Data

This study is based on UK regional housing quarterly prices' data ranging from Q4–1973 to Q2–2018 obtained from the Nationwide website[2] for 13 regions; these are London, East Anglia, East Midlands, North, North West, Northern Ireland, Outer Metropolitan, Outer South East, Scotland, South East, South West, Wales, West Midlands, and Yorkshire and Humberside. The different regions of the UK exhibit certain and unique characteristics. Northern Ireland is unique in the sense it has achieved an increased level of independence from the UK government since the late 1990s when it comes to social security provisions and the taxation of its housing market [16]. Further there is a strong link between the housing markets of the

[2] https://www.nationwide.co.uk/about/house-price-index/download-data

Figure 1.
Regional areas of the UK.

Northern Ireland and the Republic of Ireland [17]. Similarly Wales also achieved increased level of independence from the UK government after the 1997 devolution. East Midlands is a region with a very strong manufacturing in the country and is considered important when it comes to the production sector of the economy [5]. In contrast the West Midlands has a relatively poor economic conditions [18]. According to ONS [19], the South West is one of the UK regions with the highest rates of employment and economic activity. Given the uniqueness, strength and location relative to London of these other regions, it is possible for house prices in these regions to impact London house prices. **Figure 1** shows the approximate location of the 13 regions relative to London. **Figure 2** presents all the prices normalised to one at the start, including the UK. The close movements of all the indices are clearly visible. The rising prices across all regions during the mid-1980s and falling prices during the financial crisis of the late 2000s are clearly prominent.[3]

This figure shows the regional house prices of the UK normalised to one at the start of the sample period; i.e. 1973Q3 – 2018Q2.

2.2 Bivariate and multivariate linear causality

In order to examine the linear relationship between various UK regional house prices with the London house prices, we consider the widely accepted vector autoregression (VAR) specification and the corresponding Granger causality test [20]. This approach enables us to assess whether there is a causal relationship between the variables in terms of time precedence and in which direction the causality flows. This will help us to test whether there is a ripple effect or connectedness between the prices UK regional house prices. The specification of the applied bivariate VAR model can be expressed as follows:

[3] ADF test [20] and KPSS test [21] show that the changes in the house prices (first difference series) are stationary. Results for these tests are available from the authors on request.

$$x_t = \varphi_1 + \sum_{i=1}^{n} \alpha_i x_{t-i} + \sum_{i=1}^{n} \beta_i y_{t-i} + \varepsilon_{1t} \tag{1}$$

$$y_t = \varphi_2 + \sum_{i=1}^{n} \gamma_i x_{t-i} + \sum_{i=1}^{n} \delta_i y_{t-i} + \varepsilon_{2t} \tag{2}$$

where, in our case, x_t represents house prices in London in first differences, y_t is the log-difference of the respective UK regional house prices. φ_1 and φ_2 are the constants, whereas α_i, β_i, γ_i and δ_i, i = 1,...,n, are the parameters for linear relationships between the underlying variables. Ripple effect hypothesis can be tested if only London house prices affect other regions, but not vice versa. On the other hand, connectedness can be shown if bidirectional causality exists between London house prices and other regions. In the next sections, we present the nonlinear approach adopted in our study and describe the relevant tests employed.

2.3 Bivariate nonlinear causality

Arrival of new information and dynamics of economic fluctuations cause changes in the security prices. Campbell et al. [21] describe these processes as nonlinear. Furthermore, many other researchers have highlighted the existence of nonlinear features in macroeconomic variables and models [22–27]. Hiemstra and Jones [15] reported nonlinear causality in financial variables using a correlation integral based approach. Subsequent research papers have provided more evidence on nonlinear modelling of various financial variables [28–34]. Market frictions such transaction cots and information asymmetries could be associated with the nonlinear dynamics and can cause non-convergence towards the long-term equilibrium. Anderson [35] reports that the transaction costs in the asset pricing literature could be one of the factor for disequilibrium error. He further demonstrates that nonlinear models which consider the transaction costs often outperform the parametric models. Some of the studies have identified heterogenous investors' beliefs as one of the sources for nonlinearities in macro-financial time series [36]. This heterogeneity exisit mainly

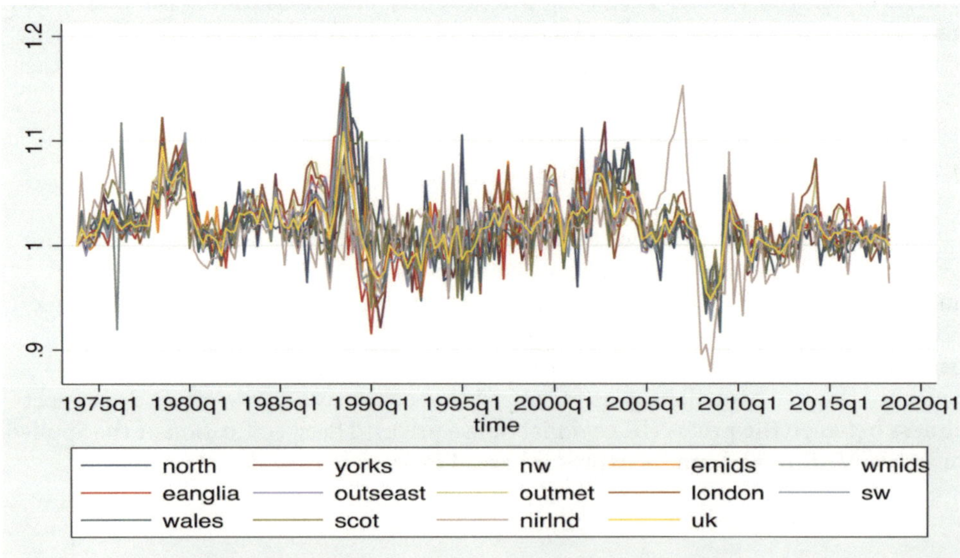

Figure 2.
Regional house prices.

due to differences in investor horizonds, risk profiles [37] and herding behaviour [38]. Due to the above, we study the Granger causality in using nonlinear framework.

Correlation integral based nonlinear Granger causality was introduced by Baek and Brock [39] and was further developed by Hiemstra and Jones [15]. This research studies nonlinear causality between the UK regional house prices, using the Hiemstra and Jones [15] test statistic.

Consider two stationary time series $\{X_t\}$ and $\{Y_t\}$, for t = 1,2,… … An m-length lead vector of X_t is denoted by X_t^m wheras X_{t-Lx}^{Lx} and Y_{t-Ly}^{Ly} are lag vectors of X_t and Y_t as shown below:

$$\begin{aligned} X_t^m &\equiv (X_t, X_{t+1}, \ldots X_{t+m-1}),\ \ m = 1,2\ldots, t = 1,2,\ \ldots, \\ X_{t-Lx}^{Lx} &\equiv (X_{t-Lx}, X_{t-Lx+1},\ \ldots, X_{t-1}), \\ &\quad Lx = 1,2,\ \ldots, t = Lx+1, Lx+2,\ \ldots \\ Y_{t-Ly}^{Ly} &\equiv (Y_{t-Ly}, Y_{t-Ly+1},\ \ldots, Y_{t-1}), \\ &\quad\quad Ly = 1,2,\ \ldots, t = Ly+1, Ly+2,\ \ldots \end{aligned} \tag{3}$$

Using the Hiemstra and Jones [15] framework, Y does not strictly Granger cause X if:

$$\begin{aligned} &Pr\left(\|X_t^m - X_s^m\| < e \,|\, \|X_{t-Lx}^{Lx} - X_{s-Lx}^{Lx}\| < e, \left\|Y_{t-Ly}^{Ly} - Y_{s-Ly}^{Ly}\right\| < e\right) \\ &= Pr(\|X_t^m - X_s^m\| < e \,|\, \|X_{t-Lx}^{Lx} - X_{s-Lx}^{Lx}\| < e) \end{aligned} \tag{4}$$

Porbablity and maximum norm in Eq. (4) are denoted by Pr(.) and ||•||, respectively. The conditional probability that the deviation between two arbitrary lead vectors of $\{X_t\}$ of m-length is less than e, while deviation between the corresponding lag vectors of X_{t-Lx}^{Lx} and Y_{t-Ly}^{Ly} is also less then e, is shown on the left hand side of the Eq. (4). The right hand side represent the conditional probability that two arbitrary m-length lead vectors of $\{X_t\}$ are with a distance of e of each other, assuming that the corresponding lag vectors i.e. X_{t-Lx}^{Lx} and X_{s-Lx}^{Lx} are also within a distance e of each other. For all regions, X_t represents the changes in the London housing prices and Y_t represents the changes in the housing prices in other regions. Therefore, if Eq. (4) is true, this implies that the changes in the London housing prices do not affect the respective changes in regional housing prices. Nonlinear causality test proposed by Hiemstra and Jones [15] is based on the conditional probabilities using corresponding ratios of joint probabilities:

$$\frac{C1(m+Lx, Ly, e)}{C2(Lx, Ly, e)} = \frac{C3(m+Lx, e)}{C4(Lx, e)} \tag{5}$$

where joint probabilities are denoted as *C1, C2, C3 and C4.*[4] Assuming $\{X_t\}$ *and* $\{Y_t\}$ are strictly stationary and weakly dependent, if $\{Y_t\}$ does not strictly Granger cause $\{X_t\}$ then,

$$\sqrt{n}\left(\frac{C1(m+Lx, Ly, e, n)}{C2(Lx, Ly, e, n)} - \frac{C3(m+Lx, e, n)}{C4(Lx, e, n)}\right) \to N(0, \sigma^2(m, Lx, Ly, e)) \tag{6}$$

[4] See Hiemstra and Jones [15] for further details on correlation integrals and joint probabilities.

Details on the definition and the estimator of the variance $\sigma^2(m, Lx, Ly, e)$ are provided in an appendix of Hiemstra and Jones [15].

3. Results

Table 1 shows the results for linear Granger causality based on Eqs. (1) and (2). The results show that London predominantly affects the regional house prices except for Northern Ireland, Outer Metropolitan and Outer East, and Wales. Similarly, regional house prices affect London house prices in seven out of 13 regions with some of these showing a feedback effect from or connectedness to the changes in the London house prices. No evidence of price feedback is found in any direction for Northern Ireland and Wales. The Northern Ireland and Wales results may be due to the increased independence of these regions from the UK government and the far distance location from London. Scotland prices are affected by London but not vice versa. Surprisingly Outer Metropolitan and Outer East affect the London prices but not vice versa. These results confirming connectedness between the house prices may be due to geographically adjacent or economically linked regions.

Table 2 shows results for the nonlinear Granger causality. This test is applied to the standardised residuals obtained from the VAR models after filtering any linear dependence among the underlying variables. The null hypothesis of no nonlinear Granger causality has been rejected in most of the cases except for Northern Ireland and Wales. This shows significant evidence of nonlinear interdependence among the housing prices of London and other regions in the UK. We report bidirectional dependence between London and the other regions except for Northern Ireland and Wales. These results evince the nonlinear feedback effect or connectedness. No evidence of any causality in any direction is found between London and Wales/ Northern Ireland.

Regions	London ➔ Region	Region ➔ London
East Anglia	16.85**	30.12***
East Midlands	27.69***	0.91
North	35.27***	4.49
North West	28.53***	4.52
Northern Ireland	11.30	9.87
Outer Metropolitan	12.33	94.37***
Outer East	5.24	49.02***
Scotland	35.77***	11.41
South East	20.68***	56.11***
South West	24.33***	43.19***
Wales	9.68	5.72
West Midlands	50.03***	24.85***
Yorkshire and Humberside	14.65*	33.97***

*Notes: **Table** 1 shows linear Granger causality results based on Eqs. (1) and (2). ***, ** and * imply significant causality at the 1%, 5% and 10% levels, respectively.*

Table 1.
Linear causality results.

Regions	London ➔ Region	Region ➔ London
East Anglia	12.651***	13.598***
East Midlands	11.193***	12.679***
North	13.697***	14.805***
North West	13.492***	13.734***
Northern Ireland	0.20	0.926
Outer Metropolitan	13.04***	12.817***
Outer East	11.151***	13.617***
Scotland	12.306***	12.575***
South East	14.329***	14.621***
South West	12.397***	13.339***
Wales	0.895	0.157
West Midlands	12.749***	13.915***
Yorkshire and Humberside	13.231***	13.422***

*Notes: **Table** 1 shows test-statistic proposed by Hiemstra and Jones [27] using Eq. (2). ***, ** and * imply significant causality at the 1%, 5% and 10% levels, respectively.*

Table 2.
Nonlinear causality results.

Further evidence is presented by means of the impulse response function. **Figure 3** shows the impulse response function to one-standard-deviation innovations to the housing prices originating in London and other regions, respectively. These graphs can be interpreted into two categories – i.e. i) house prices in other regions responding to the shocks to London house prices and ii) London house prices responding to the shocks occurring in other regions in the UK. Firstly, a one per cent shock to the London house prices shows an immediate impact on house prices in most of the regions within a range from 1.5–3% – e.g., East Anglia, East Midlands, West Midlands, Outer Metropolitan, Outer Southeast, South West and Yorkshire. Geographically speaking, with the exception of Yorkshire, these regions are close to London. In other regions, although the shocks are statistically significant, they are smaller in magnitude. Interestingly, in the second category, innovations that originate in regions like East Anglia, Outer Metropolitan, Outer Southeast, South West, West Midlands and Yorkshire affect the London house prices with shocks in the range of 1% to 2.5%. This shows that London remains the central focus in the overall UK housing market and any shocks occurring here transmit to most of the regions. However, local shocks in other regions also show a spillover effect on London house prices. Antonakakis et al. [5] also report that East Anglia, Outer South East and South West are the major transmitters of regional shocks.

The evidence of connectedness presented here implies that although London is important from the housing market perspective, other regions also transmit the shocks back to the London market. This may be due to the information spillover (investor expectations) between different regions [14] although this research does not explicitly test the information hypothesis. By taking into consideration the impact of the bidirectional spillover effect of price, appropriate regulations and policies for the UK housing sector should be formulated. The results further imply the importance of house prices in other regions when investing in houses in London, and vice versa.

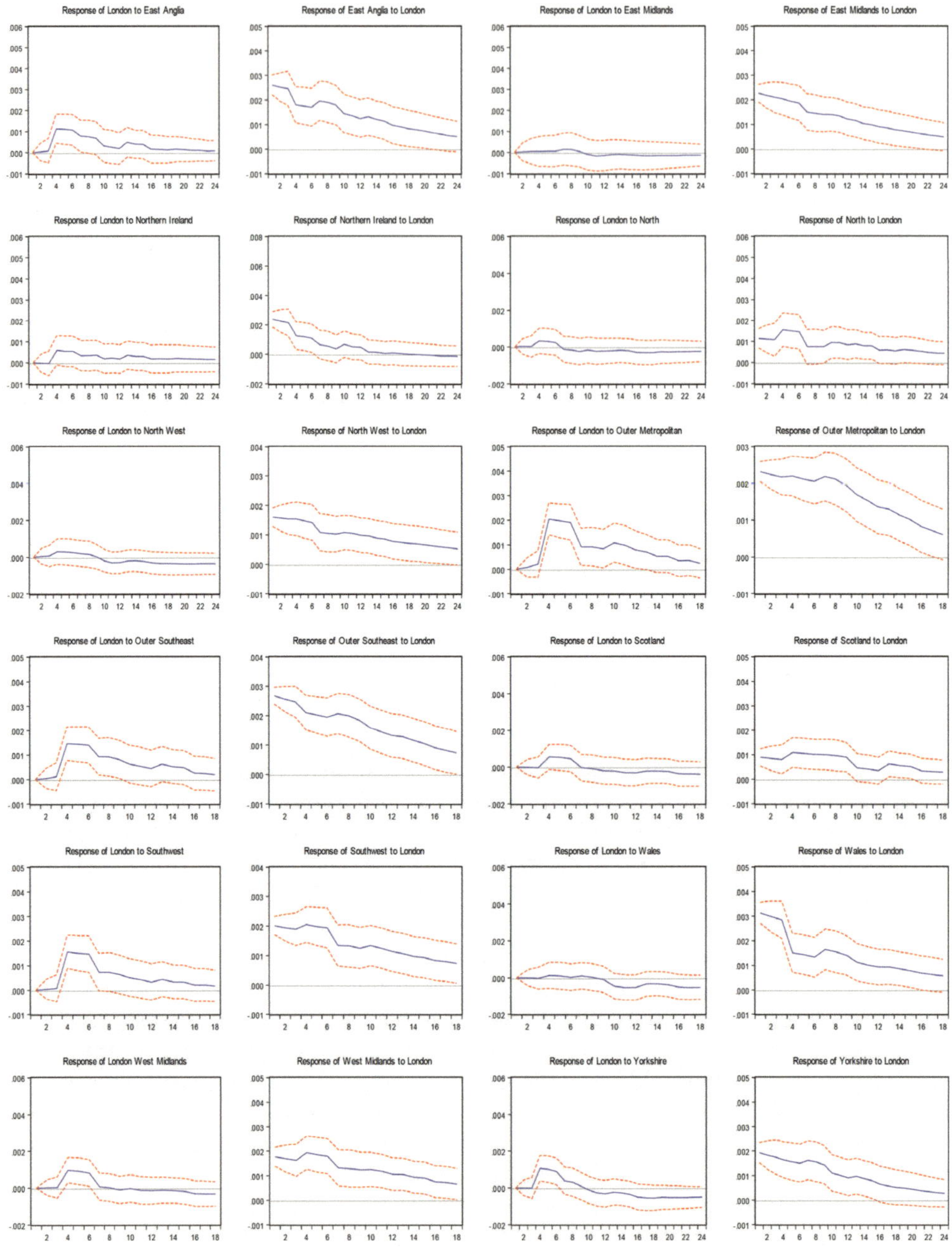

Figure 3.
Impulse response functions (response to Cholesky one S.D. innovations).

4. Robustness checks and further empirical evidence

4.1 Linear and nonlinear forecasting regressions

This section provides additional empirical evidence and explores the nature of the relationship between London and other UK regional house prices. Therefore, it complements the results of Granger causality and serves as a useful robustness check. To this end, we initially focus on the following forecasting regression:

$$y_{t+h} = \alpha + \beta x_t + \sum\nolimits_{i=0}^{\gamma} \gamma_i y_{t-i} + \epsilon_{t+h}, \tag{7}$$

where y_{t+h} refers to the changes in the London house prices, $y_{t+h} = \frac{400}{h+1} ln \left(\frac{y_{t+h}}{Y_t}\right)$, with forecast horizon, $h > 0$ and x represents the changes in the regional house prices. The null hypothesis of $\beta = 0$ is tested here to observe the predictability of changes in the London house prices using the other regional house prices. The corresponding results for h = 1 are presented in **Table 3**.

We report that the other regional house prices are a significant short-term predictor of the changes in the London house prices in most of the cases, with the exception of Northern Ireland, Scotland and Wales. These forecasting results reaffirm and strengthen the evidence against the 'ripple effect' hypothesis in the literature.

We further extend the forecasting model to show more evidence of nolinear relationship between the regional house prices. For this purpose, we use smooth-transition threshold (STR) models [40–44]. Simple threshold model can trigger an abrupt change in the parameter values, however, STR models are capable to allow smooth transition between different regime states. Following Smooth Transition Threshold model is used:

$$y_{t+h} = \alpha + \beta x_t + \sum_{i=0}^{p} \gamma_i y_{t-i} + \left(\varphi_0 + \varphi_1 x_t + \sum_{i=0}^{p} \theta_i y_{t-i}\right) F\left(y_{t-d}\right) + \varepsilon_{t+h}, \tag{8}$$

$F\left(y_{t-d}\right)$ is the transition function and y_{t-d} is the transition variable, whereas remaining variables are as defined in Eq. (7). Based on the existing literature, we first consider the logistic form of transition function (LSTR) as shown in Eq. (9) [40, 42–44]:

Regions	**London ➔ Region**		**Region ➔ London**	
		Adj. R^2		**Adj. R^2**
East Anglia	13.75**	0.48	17.32***	0.45
East Midlands	19.40***	0.51	10.62*	0.43
North	13.35**	0.33	18.37***	0.46
North West	12.91**	0.58	10.69*	0.44
Northern Ireland	13.04**	0.36	5.22	0.45
Outer Metropolitan	4.41	0.68	41.91***	0.52
Outer East	8.79	0.63	24.39***	0.47
Scotland	11.19*	0.31	9.93	0.44
South East	13.79**	0.51	19.63***	0.43
South West	14.41**	0.58	22.91***	0.47
Wales	9.52	0.37	7.30	0.43
West Midlands	23.66***	0.41	13.67**	0.44
Yorkshire and Humberside	12.32*	0.46	17.62***	0.46

*Notes: This table presents the results from the linear forecasting regressions described in Section 4.1 (Eq. (7)). ***, ** and * imply significant causality at the 1%, 5% and 10% levels, respectively.*

Table 3.
Linear forecasting results.

$$F\left(y_{t-d}\right) = \left(1 + \exp\left(-\lambda\left(y_{t-d} - c\right)\right)\right)^{-1}, \lambda > 0, \tag{9}$$

Where λ, d and c are the smoothing, delay and transition parameters, respectively. This function is monotonically increasing in y_{t-d}. Note that when $\lambda \rightarrow +\infty$, $F\left(y_{t-d}\right)$ becomes a Heaviside function: $F\left(y_{t-d}\right) = 0$ when $y_{t-d} \leq c$ and $F\left(y_{t-d}\right) = 1$ when $y_{t-d} > c$.

Monotonic transition may not always be successful in empirical applications. Therefore, we consider exponential transition function (ESTR) [42–44]:

$$F\left(y_{t-d}\right) = 1 - \exp\left(-\lambda\left(y_{t-d} - c\right)^2\right), \lambda > 0. \tag{10}$$

Here, the transition function is symmetric around c. This model emplies that expansion and contraction have similar dynamics while the these vary for the middle ground [44]. STR module can have some issues involving the smoothing parameter λ, therefore, we follow the literature and using variation of the transition varible λ is scaled in both of the models [44].

In this case, the transition function is symmetric around c. The ESTR model implies that contraction and expansion have similar dynamic structures while the dynamics of the middle ground differ [43, 44]. Hence, we have the obtain the following versions of transition functions, respectively:

$$F\left(y_{t-d}\right) = \left(1 + \exp\left(-\lambda\left(y_{t-d} - c\right)/\sigma\left(y_{t-d}\right)\right)\right)^{-1}, \lambda > 0, \tag{11}$$

$$F\left(y_{t-d}\right) = 1 - \exp\left(-\lambda\left(y_{t-d} - c\right)^2/\sigma^2\left(y_{t-d}\right)\right), \lambda > 0. \tag{12}$$

Table 4 presents the results of the LSTR and the ESTR models testing the changes in London house prices as a predictor for changes in the regional house prices. In the LSTR model results, the estimated transition parameter c, which

Country	α	ß	φ_0	φ_1	λ	c	Adj. R^2
Panel – I: Exponential Smooth Transition Threshold Model (ESTR)							
East Anglia	32.09***	11.53***	45.11***	−9.21***	0.071**	−9.23***	0.485
East Midlands	21.79**	3.17**	55.17***	23.98***	0.97	21.13***	0.467
North	35.16	31.19	−16.4	37.61	0.0016	−21.16	0.513
North West	13.31***	19.69***	−11.49***	18.83***	9.71***	−19.71***	0.482
Northern Ireland	4.63	3.51*	5.09	2.45	0.089	−9.28***	0.435
Outer Metropolitan	51.21**	8.6**	57.34**	8.26*	0.046**	−49.28	0.534
Outer East	0.04	0.99	11.45	0.05	0.006	44.63	0.486
Scotland	11.93	5.09	−3.75	11.73	4.70	−6.74	0.45
South East	40.84	−55.72***	40.72***	55.54***	0.10***	−60.74**	0.447
South West	7.31	5.30**	−25.71**	−5.41**	0.035**	13.06**	0.491
Wales	22.55	16.37**	−22.22	15.14**	0.144**	−20.34	0.458
West Midlands	5.67	4.21***	6.52	3.77**	0.068**	−7.30**	0.469
Yorkshire and Humberside	11.31	9.12***	5.17	4.11**	0.533*	−14.30**	0.478
Panel – II: Logistic Smooth Transition Threshold Model (LSTR)							
East Anglia	−5.17**	6.34***	12.03*	2.24**	0.091**	2.38**	0.485
East Midlands	11.27*	8.17***	−25.35***	−16.52	0.012***	23.52***	0.476
North	4.71	3.43	23.71	28.01	0.568	−3.775	0.495

Country	α	ß	φ_0	φ_1	λ	c	Adj. R^2
North West	2.19***	0.89**	−11.5***	−6.07***	0.03***	3.73***	0.489
Northern Ireland	0.91	0.13	−2.19	−2.67	0.049	15.47***	0.462
Outer Metropolitan	4.15**	8.95**	0.37	6.93*	0.048**	1.94**	0.61
Outer East	5.19	3.82	−2.79	5.14	0.013	−1.009	0.548
Scotland	−11.29	21.09	14.81	35.08	0.063	19.05**	0.499
South East	6.84	8.9**	−6.02	−5.14**	0.064**	1.66***	0.473
South West	−18.64***	7.14***	16.97*	−4.27**	0.015***	−42.33***	0.482
Wales	0.98	0.11	−1.61	−6.43***	0.003*	−1.61**	0.49
West Midlands	4.1**	7.88***	−1.22	0.31***	0.013***	39.89**	0.48
Yorkshire and Humberside	3.5	12.14*	2.18	0.77**	0.002***	11.17***	0.479

Table 4.
Nonlinear forecasting results (regional house prices ➔ London house prices).

Country	α	ß	φ_0	φ_1	λ	c	Adj. R^2
Panel – I: Exponential Smooth Transition Threshold Model (ESTR)							
East Anglia	−4.75**	34.06*	9.30***	−5.6*	27.17*	3.19***	0.2559
East Midlands	−6.59***	−31.49**	11.58***	41.47***	21.57**	−3.60**	0.397
North	9.99***	4.51*	9.79***	−63.9***	−66.74	9.6***	0.2449
North West	−6.09***	−86.09***	10.5***	10.68***	11.65	−10.87	0.411
Northern Ireland	−0.064***	18.88	67.44*	11.42***	−23.39	56.83	0.397
Outer Metropolitan	0.0152***	35.54***	−0.061***	−0.056***	85.12	−49.28	0.264
Outer East	−0.071***	10.42	15.07	12.83***	−14.65	24.66	0.408
Scotland	−6.29***	−36.01	84.69**	42.41	78.12	22.31	0.4005
South East	−8.13***	−4.68	9.72***	21.13	15.47	−5.68***	0.4107
South West	−6.94***	−5.26	12.21***	11.66	22.46	−3.49	39.70
Wales	−0.66	−0.079	45.91	17.92	30.18	−8.48**	39.87
West Midlands	10.40***	22.31	−11.27***	−45.63	22.33	−8.76***	40.67
Yorkshire and Humberside	−7.05***	13.41	73.05*	−34.14	19.42	−0.037**	40.35
Panel – II: Logistic Smooth Transition Threshold Model (LSTR)							
East Anglia	−0.0025	0.17	0.0493***	−0.28	27.17*	0.031***	0.4028
East Midlands	−0.0063	−0.087	0.054***	0.173	17.86*	0.0299***	0.3985
North	−0.0057	−0.0864	0.055***	0.066	18.30*	0.0301***	0.401
North West	−0.0064	−0.257*	0.0522***	0.557*	15.70*	0.027**	0.4123
Northern Ireland	−0.0044	0.0705	0.0513***	−0.1169	23.905*	0.0303***	0.3999
Outer Metropolitan	−0.0056	0.0508	0.056***	−0.263	20.128**	0.0306***	0.40
Outer East	−0.0034	0.295	0.056***	−0.67	23.52**	0.032***	0.408
Scotland	−0.0056	−0.132	0.054***	0.168	17.466*	0.029***	0.402
South East	−0.0035	−0.0117	0.072***	−0.61	27.13*	0.059***	0.33

Country	α	ß	φ_0	φ_1	λ	c	Adj. R^2
South West	−0.0061	−0.0395	0.057^{***}	−0.047	18.62^{**}	0.031^{***}	0.398
Wales	−0.0058	−0.071	0.0544^{***}	0.0824	18.79^{*}	0.0299^{***}	0.3987
West Midlands	−0.0081	−0.22	0.05^{***}	0.22	15.34^{*}	0.0284^{***}	0.408
Yorkshire and Humberside	−0.006	−0.021	0.05	−0.20	17.05^{*}	0.0315^{***}	0.405

Table 5.
Nonlinear forecasting results (London house prices ➔ regional house prices).

marks the half-way point between the two regimes, is significantly different from zero in most of the cases. Moreover, we observe that most of the estimated betas are positive and significant (at 1% and 5% levels, depending on the case), suggesting that higher regional house prices boost London house prices in the following quarter. Further, the estimates of φ_1 in the upper regime significance are found in nine out of 13 regions, revealing the importance of regional house prices as an explanatory variable for changes in the London house prices. Insignificant results are found for Northern Ireland, Scotland and Wales. **Table 5** shows the results based on the LSTR and ESTR models confirming as expected that changes in the London house prices are a significant predictor of house price changes in other regions in the UK.

Results for the estimated ESTR models are very similar to the LSTR results. This reaffirms the significance of the regional house prices as a short-term predictor of future changes in the London house prices in a nonlinear context and complements the previously reported results under the linear and nonlinear frameworks. Thus, ESTR and LSTR results reinforce the idea that the regional house prices have a feedback effect or connectedness to the London house prices. This shows evidence against the ripple effect where a unidirectional impact of changes in the London house prices on other regions is reported.

5. Conclusion and implications

This chapter investigates the transmission mechanism driving the UK regional house prices using the linear causality model, the nonlinear Granger causality model, and the impulse response process. We employ quarterly housing prices data ranging from Q4–1973 to Q2–2018 from 13 regions from the UK. Results show bidirectional dependence between the London prices and other regions' prices except for Northern Ireland and Wales. This result is confirmed by the linear causality, the nonlinear causality and the impulse response tests. Further empirical examination applying linear and non-linear forecasting tests support the linear and non-linear causality results. Thus, we provide that London is not always important for the other UK regions over time, as well as that London itself may also receive shocks from other regions. Impulse response shows that London remains the central focus in the overall UK housing market and any shocks occurring here transmit to most of the regions. However, local shocks in other regions also show a spill over effect on London house prices. Identification of regional disparities can help policymakers to achieve a more balanced growth across the country. These results underline the importance of establishing appropriate regulations and stabilisation policies in the housing sector of the economy. Further, the interdependence between regional housing prices might provide significant insight regarding efficient diversification of investments across mortgage-backed securities.

JEL Classification: R2, R21, R31

Author details

Taufiq Choudhry[1]*, Syed S. Hassan[2] and Sarosh Shabi[2]

1 School of Business, University of Southampton, UK

2 School of Management, University of Swansea, UK

*Address all correspondence to: t.choudhry@soton.ac.uk

References

[1] Attanasio, O., A. Leicester, and M. Wakefield, *Do house prices drive consumption growth? The coincident cycles of house prices and consumption in the UK.* Journal of the European Economic Association, 2011. **9**(3): p. 399-435.

[2] Mishkin, F.S., *Financial markets, institutions, and money*. 2007: HarperCollins Publishers.

[3] Gallin, J., *The long-run relationship between house prices and income: evidence from local housing markets.* Real Estate Economics, 2006. **34**(3): p. 417-438.

[4] Costello, G., P. Fraser, and N. Groenewold, *House prices, non-fundamental components and interstate spillovers: The Australian experience.* Journal of Banking & Finance, 2011. **35** (3): p. 653-669.

[5] Antonakakis, N., et al., *The dynamic connectedness of UK regional property returns.* Urban Studies, 2018. **55**(14): p. 3110-3134.

[6] Churchill, S.A., J. Inekwe, and K. Ivanovski, *House price convergence: Evidence from Australian cities.* Economics Letters, 2018. **170**: p. 88-90.

[7] Cook, S., *The convergence of regional house prices in the UK.* Urban studies, 2003. **40**(11): p. 2285-2294.

[8] Cook, S., *Detecting long-run relationships in regional house prices in the UK.* International Review of Applied Economics, 2005. **19**(1): p. 107-118.

[9] Cook, S. and D. Watson, *A new perspective on the ripple effect in the UK housing market: Comovement, cyclical subsamples and alternative indices.* Urban Studies, 2016. **53**(14): p. 3048-3062.

[10] Gupta, R., C. Andre, and L. Gil-Alana, *Comovement in Euro area housing prices: A fractional cointegration approach.* Urban Studies, 2015. **52**(16): p. 3123-3143.

[11] Gupta, R. and S.M. Miller, *The time-series properties of house prices: A case study of the Southern California market.* The Journal of Real Estate Finance and Economics, 2012. **44**(3): p. 339-361.

[12] Holly, S., M.H. Pesaran, and T. Yamagata, *The spatial and temporal diffusion of house prices in the UK.* Journal of urban economics, 2011. **69**(1): p. 2-23.

[13] Billio, M., et al., *Econometric measures of connectedness and systemic risk in the finance and insurance sectors.* Journal of financial economics, 2012. **104**(3): p. 535-559.

[14] Zhu, B., R. Füss, and N.B. Rottke, *Spatial linkages in returns and volatilities among US regional housing markets.* Real Estate Economics, 2013. **41**(1): p. 29-64.

[15] Hiemstra, C. and J.D. Jones, *Testing for linear and nonlinear Granger causality in the stock price-volume relation.* The Journal of Finance, 1994. **49**(5): p. 1639-1664.

[16] McKee, K., J. Muir, and T. Moore, *Housing policy in the UK: The importance of spatial nuance.* Housing Studies, 2017. **32**(1): p. 60-72.

[17] Stevenson, S., *House price diffusion and inter-regional and cross-border house price dynamics.* Journal of Property Research, 2004. **21**(4): p. 301-320.

[18] Bailey, D. and N. Berkeley, *Regional responses to recession: The role of the West Midlands Regional Taskforce.* Regional Studies, 2014. **48**(11): p. 1797-1812.

[19] ONS, *Regional Labour Market Statistics in the UK.* 2017, Office for National Statistics (ONS).

[20] Granger, C.W., *Investigating causal relations by econometric models and cross-spectral methods.* Econometrica: journal of the Econometric Society, 1969: p. 424-438.

[21] Campbell, J.Y., A.W. Lo, and A.C. MacKinlay, *The econometrics of financial markets*. 2012: princeton University press.

[22] Barnett, W.A., et al., *A single-blind controlled competition among tests for nonlinearity and chaos.* Journal of econometrics, 1997. **82**(1): p. 157-192.

[23] Hsieh, D.A., *Chaos and nonlinear dynamics: application to financial markets.* The journal of finance, 1991. **46** (5): p. 1839-1877.

[24] Kahneman, D. and A. Tversky, *Prospect theory: An analysis of decision under risk.* Econometrica, 1979. **47**(2): p. 363-391.

[25] Keynes, J.M., *The general theory of employment, interest, and money*. 1936: Springer.

[26] Shiller, R.J., *Macro markets: creating institutions for managing society's largest economic risks*. 1994: OUP Oxford.

[27] Shiller, R.J., *Irrational Exuberance*. 2005: Princeton University Press.

[28] Bekiros, S.D., *Exchange rates and fundamentals: Co-movement, long-run relationships and short-run dynamics.* Journal of Banking & Finance, 2014. **39**: p. 117-134.

[29] Bekiros, S.D. and C.G. Diks, *The nonlinear dynamic relationship of exchange rates: Parametric and nonparametric causality testing.* Journal of macroeconomics, 2008. **30**(4): p. 1641-1650.

[30] Bekiros, S.D. and C.G. Diks, *The relationship between crude oil spot and futures prices: Cointegration, linear and nonlinear causality.* Energy Economics, 2008. **30**(5): p. 2673-2685.

[31] Chen, A.-S. and J. Wuh Lin, *Cointegration and detectable linear and nonlinear causality: analysis using the London Metal Exchange lead contract.* Applied Economics, 2004. **36**(11): p. 1157-1167.

[32] Diks, C. and V. Panchenko, *A new statistic and practical guidelines for nonparametric Granger causality testing.* Journal of Economic Dynamics and Control, 2006. **30**(9-10): p. 1647-1669.

[33] Shin, Y., B. Yu, and M. Greenwood-Nimmo, *Modelling Asymmetric Cointegration and Dynamic Multipliers in a Nonlinear ARDL Framework*, in *Festschrift in Honor of Peter Schmidt: Econometric Methods and Applications*, R. C. Sickles and W.C. Horrace, Editors. 2014, Springer New York: New York, NY. p. 281-314.

[34] Silvapulle, P. and J.-S. Choi, *Testing for linear and nonlinear Granger causality in the stock price-volume relation: Korean evidence.* The Quarterly Review of Economics and Finance, 1999. **39**(1): p. 59-76.

[35] Anderson, H.M., *Transaction costs and nonlinear adjustment towards equilibrium in the US Treasury Bill market.* Oxford Bulletin of Economics and Statistics, 1997. **59**(4): p. 465-484.

[36] Brock, W. and B.D. LeBaron, *A dynamic structural model for stock return volatility and trading volume*. 1995, National Bureau of Economic Research Cambridge, Mass., USA.

[37] Peter, E.E., *Fractal market analysis: applying chaos theory to investment and economics*. Vol. 24. 1994: John Wiley & Sons.

[38] Lux, T., *Herd behaviour, bubbles and crashes.* The economic journal, 1995. **105** (431): p. 881-896.

[39] Baek, E. and W. Brock, *A general test for nonlinear Granger causality: Bivariate model.* Iowa State University and University of Wisconsin at Madison Working Paper, 1992.

[40] Chan, K.S. and H. Tong, *On estimating thresholds in autoregressive models.* Journal of time series analysis, 1986. 7(3): p. 179-190.

[41] Granger, C.W. and T. Terasvirta, *Modelling non-linear economic relationships.* OUP Catalogue, 1993.

[42] McMillan, D.G., *Non-linear predictability of UK stock market returns.* Oxford Bulletin of Economics and Statistics, 2003. **65**(5): p. 557-573.

[43] Teräsvirta, T., *Specification, estimation, and evaluation of smooth transition autoregressive models.* Journal of the american Statistical association, 1994. **89**(425): p. 208-218.

[44] Terasvirta, T. and H.M. Anderson, *Characterizing nonlinearities in business cycles using smooth transition autoregressive models.* Journal of applied econometrics, 1992. 7(S1): p. S119-S136.

Chapter 14

Housing Cost Dependence on Transport Accessibility Territory

Sharov Maksim Igorevich

Abstract

The study the housing cost dependence on the transport accessibility of the territory of the city will improve the efficiency of the route network and will lead to a reduction in overall transportation costs. The research algorithm consisted of the following stages: data on transport accessibility were systematized; calculation of time costs assessed transport accessibility by city zones; the dependence of the cost per square meter on the time cost of movement is obtained. When assessing the impact of transport accessibility on the housing cost it is important to pay attention to the remoteness from the central part of the city, the proximity of highways, the system of access roads; the proximity of public transport stops with a large number of routes connecting different zones.

Keywords: urban planning, transport accessibility, Cost of Housing

1. Introduction

The development of urban areas requires integrated and sustainable territorial development focused on ensuring a safe and favorable environment for human life, Limiting the negative impact of various activities on the environment and ensuring the protection and rational use of natural resources in order to safeguard the interests of present and future generations. The development of urban passenger transport and green mobility is therefore an important issue.

Priority development of urban passenger transport is ensured both by urban planning solutions and by technical solutions. In modern urban development theory and practice, the most effective solutions for the development of urban passenger transport systems include:

- urban planning for urban passenger transport, including the creation of an attractive and convenient environment for public passenger transport infrastructure;

- development of areas adjacent to high-capacity passenger transport corridors;

- multifunctional use of land to reduce the need to travel long distances.

As the accessibility of urban passenger transport increases, so does the attractiveness of the city as a whole, resulting in a greater concentration of service enterprises and other places of work.

Urban population, thousand people	Travel times, min
2000	45
1000	40
500	37
250	35
100	30

Table 1.
Standard travel times to work places in cities.

In the general practice of urban planning and transport, accessibility is determined by various factors, including:

- accessibility of transport;
- availability of public services;
- creating the necessary number of public spaces.

Transport accessibility refers to the normative measure of travel time between different points within cities and agglomerations. In the design of the road network, one of the important parameters determining the required quality of the urban road network is the time between any two points in the city [1].

Transport accessibility of urban and regional areas is one of the most important economic characteristics.

In this connection, it is necessary to point out the need to develop up-to-date criteria and standards for assessing accessibility. Since Soviet times, the Russian Federation has had building regulations and norms (SNIP 2.07.01–89, SP 42.13330.2016) [2], which establish time-consuming requirements for travel to work. Even in cities with more than one million inhabitants, 90 per cent of working people are not required to spend more than 45 minutes traveling from their place of residence to their place of work (one way) (**Table 1**). The established travel time standards need to be calculated to obtain intermediate values based on the number of urban residents. In addition, the standard states that daily commuting from other locations to a downtown place of work is permitted, but not more than two times.

The improvement of transport infrastructure is an essential tool in the implementation of regional policies for sustainable development, it also makes it possible to reduce disparities in the spatial development of different urban centers and agglomerations and to increase the attractiveness of remote and sparsely populated areas [3, 4].

2. Methodology

However, improvements in transport accessibility influences the cost of housing in the area. For example, the price of an apartment in the city of Moscow near metro stations may differ by an order of magnitude from apartments located at a sufficient distance, with a analogous type of residential development. The recent opening of the Central Moscow Diameter has already led to an average increase in housing prices of 13 per cent. In this case, it is important to have tools for calculating and forecasting the cost of housing on the basis of the accessibility of the territory. This chapter presents a study of the impact of transport accessibility on the cost of a square meter of housing on two different types of cities in the Irkutsk region of the

Russian Federation. Also, two different methods were used to estimate transport accessibility of the territory, in one case - data of the questionnaire population, in the second - the method of transport modeling using the software of PTV "VISUM".

The city of Angarsk plays an important role in the system of interregional transport corridors linking the European part of Russia and Western Siberia with the regions of the Far East. Angarsk is an industrial city. The territory of Angarsk is heterogeneous, with different socio-economic, infrastructural and environmental conditions determining the basic directions and functional priorities of the territory's development. In general, Angarsk plays a large role in the economy of the Irkutsk region and contributes greatly to the development of the region's competitive advantages. Underestimation of the problem of determining transport demand and the relevance of transport infrastructure to the social and economic needs of city residents is one of the causes of economic difficulties and negative processes, such as out-migration due to poor quality of life. The development of transport accessibility in Angarsk is an important condition for improving the quality of life of the inhabitants.

The spatial planning of the city makes it possible to form a promising residential development in the south-western part of the city of Angarsk, as well as new quarters of medium, low-rise and individual residential development in the southern and western parts of the city. The construction involves, first and foremost, the development of land through the construction of multi-floor housing units in the western part of the city in two neighborhoods and in the eastern part of the city in three quarters. Therefore, ensuring the required level of transport accessibility for the developing area of the city is a very urgent task.

The main residential development of the city of Angarsk is concentrated in its center, while the industrial zones are located along its south-eastern, southern, south-western and western periphery. At the same time, new green areas of great breadth are maintained and formed between them and the residential territory, providing ecological and esthetic comfort of the residential territory. The road network of the city of Angarsk is based on the need to provide stable, as safe and as short as possible communication points, places of work and rest.

In assessing transport accessibility, the most important task is the transport zoning of the territory. To achieve this, the following data are needed:

- population in the zone;
- maps of the terrain;
- number of places of residence and places of work
- availability and location of basic cultural and amenities facilities
- list of streets and characteristics of the carriageway
- traffic patterns on the road network;
- transport and housing costs.

Transport zoning has two main characteristics: the number of zones into which the territory is divided and the size of those zones. The number and size of zones are determined on the basis of the objectives of the study and the required accuracy of the results. Due to the fact that the assessment of accessibility in the city of Angarsk was carried out using computer simulations, the division into transport zones was more detailed than in the city of Irkutsk (**Figure 1**).

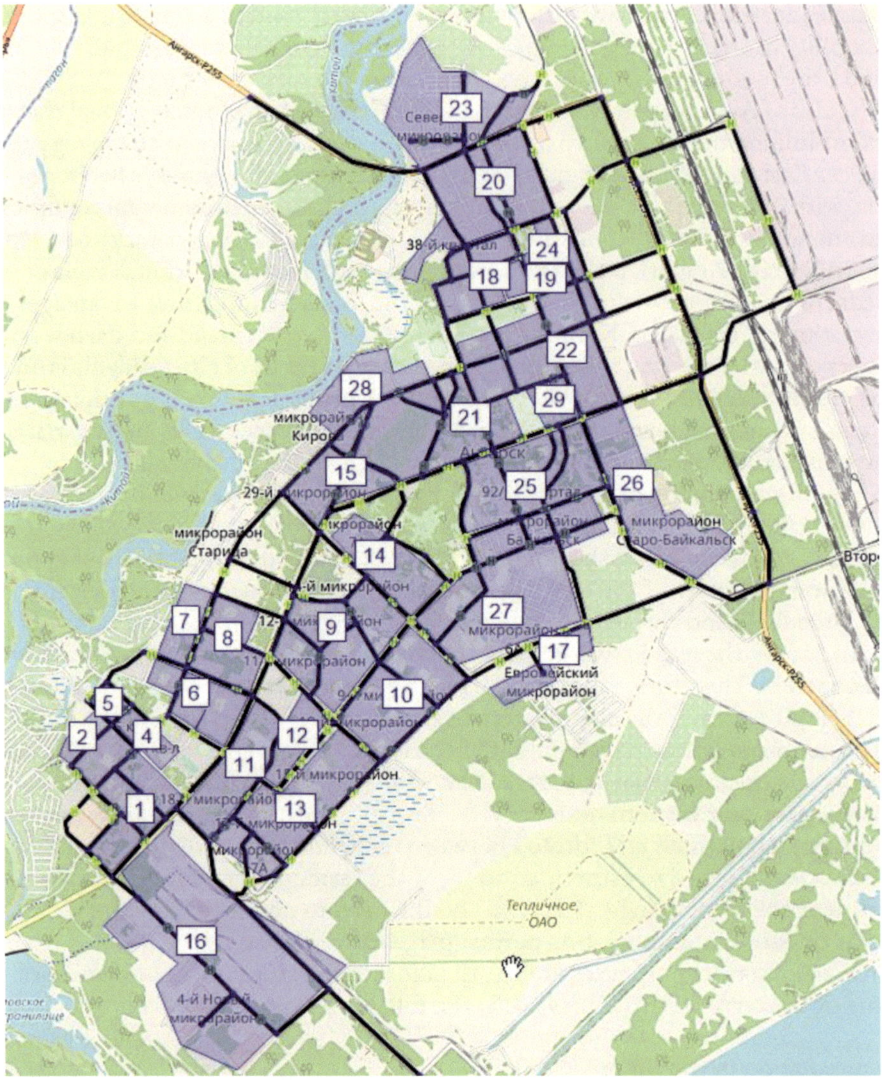

Figure 1.
Transport zones of the city of Angarsk.

When carrying out transport zoning, the following rules should be observed:

1. Traffic and pedestrian accessibility inside the zone should be taken into account.

2. In defining the boundaries of the zone, it is necessary to combine the choice with administrative boundaries (if any). Available statistical information can then be used.

3. The selection of zones should take into account territories of the same purpose: (residential, industrial, recreational).

In this study, in order to achieve the objectives of assessing transport accessibility and its impact on housing costs, we select areas in such a way that the development covers a sufficiently large area, for example, a few neighborhoods. When dividing into zones we try to take into account the type of construction (lots with a large housing stock, lots with two- or three- floor houses with wooden floors, single- floor houses).

One possible method of processing the data obtained in the study is the calculation of a time-use matrix, which made it possible to determine the accessibility of urban areas. The creation of the transport model and the assessment of transport accessibility based on it were carried out with the help of the software product "VISUM" [5].

Then we analyze each area separately and get the average time spent moving around the area. The following groups were identified: high availability to 13 minutes; average availability from 13 to 17 minutes; low availability of more than 17 mines (**Figure 2**).

The data obtained make it possible to estimate the impact of transport accessibility on the cost of 1 sq. m. dwellings. There is virtually no alternative definition of the link between transport affordability and the cost of sq. m. as this is one of the main parameters of housing value formation. Analyzed the real estate market of the city of Angarsk, examining the value of sq. M / p on sites avito.ru and domclick.ru, which are currently one of the most popular electronic sites for housing sales, average cost of 1 sq. m for 2020. The dependence of price policy on zone division is illustrated in **Figure 3**.

In assessing the impact of transport affordability on housing costs, it is important to pay attention to a number of aspects [6–11]:

- Distance from the city center;
- Proximity to motorways;
- An advanced system of access roads;
- Proximity of public transport stops with a large number of routes connecting different zones;
- Availability of parking spaces, parking areas or garage cooperatives;

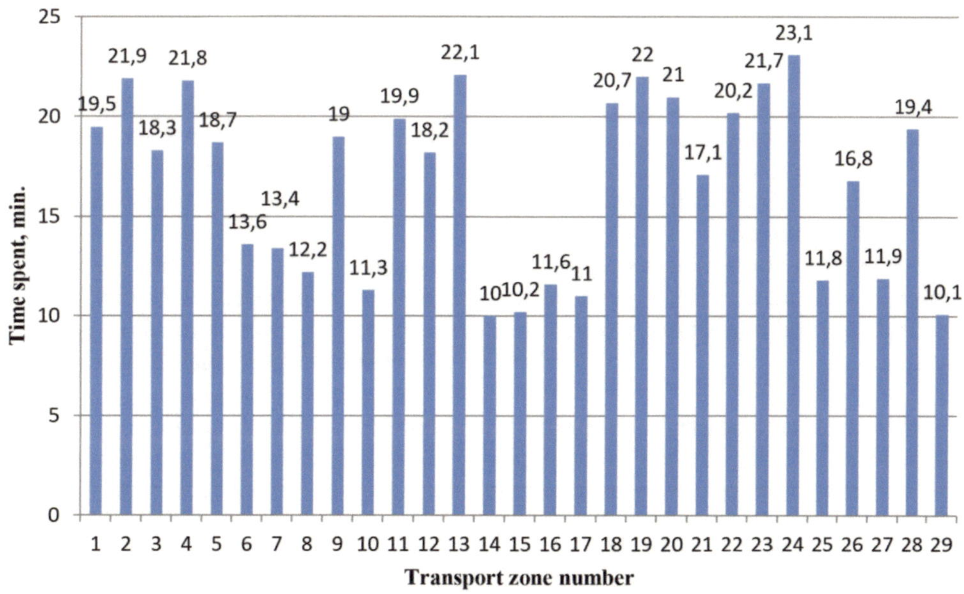

Figure 2.
Distribution of travel time by transport zones in Angarsk.

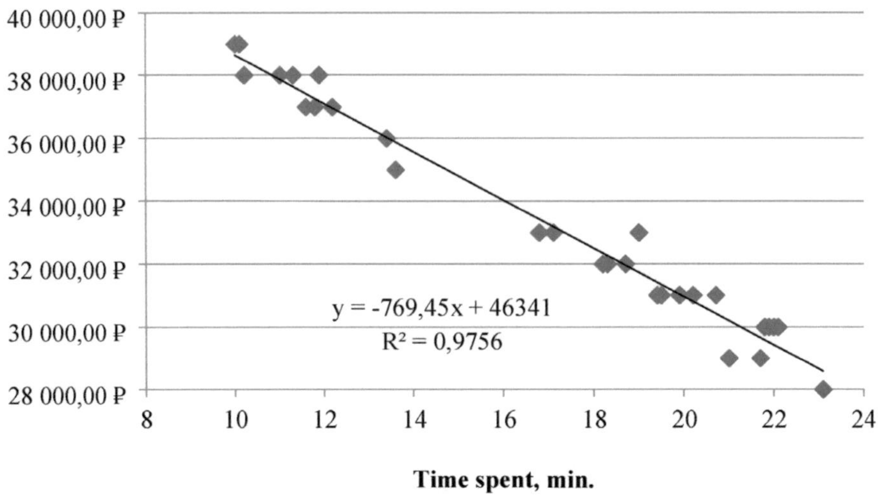

Figure 3.
Dependence of the cost per square meter on the time cost of movement.

- Proximity to cultural and social places (kindergartens, schools, health centers, supermarkets);
- Exposure to noise, vibrations and exhaust.

Most often in studies on this topic, authors focus on estimating the distance from the city center, as it is there that all the main objects of gravity are concentrated. Accessibility is measured by the time needed to travel from a given point in a city to the external borders of the central region. The accessibility rating makes it possible to create a reasonable zoning of the city.

As the level of accessibility decreases, the cost of travel to the central part of the city increases and the cost of a square meter decreases gradually. The analysis showed that dwellings located in zones 8, 10, 14, 15, 16, 17, 25, 27, 29 will have the highest level of comfort in terms of transport accessibility, the least comfortable dwellings in areas 1–5, 9, 11–13, 18–24, 28 zones (**Figure 4**).

However, it is worth noting that although transport zones 6, 7, 26 have average transport accessibility, the cost per square meter is over 33,000 rubles, this is due to the fact that the buildings are in good condition and people living in these areas, use private transport. On the basis of an analysis of the average market value of a square meter, it can be concluded that dwellings located in the central part of the city are the most attractive.

On the basis of the analysis, a correlation has been established between the time costs of travel and the cost of a square metre of residential property in the city of Angarsk. A trend line was drawn and an equation of this line was obtained (**Figure 3**) to explain the functional relationship between the cost of a square meter and the time spent on labor moves.

With the increasing remoteness of the residential building from the geographical center of the city, the level of transport accessibility of the territory decreases, and as a result, the cost of one square meter decreases. So, in particular, the difference in the housing costs in the area with the best transport accessibility is higher by 28% than in the remote territory.

In the city of Irkutsk, a different method of assessing transport accessibility was applied, based on a questionnaire survey of the population.

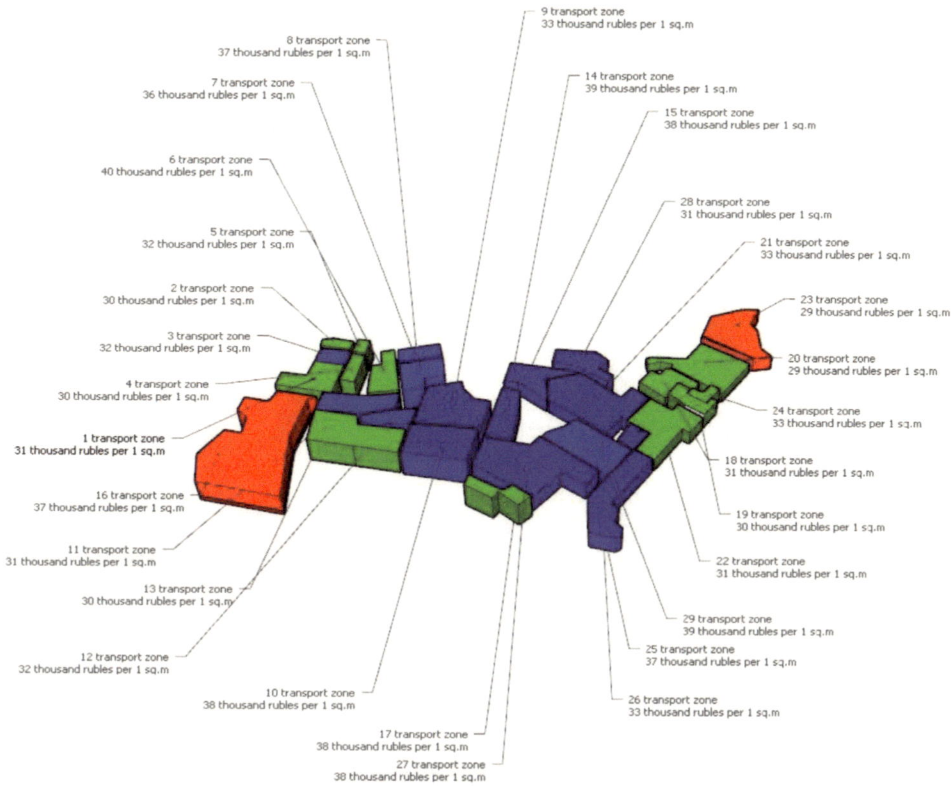

Figure 4.
Cost per square meter for transport zones in Angarsk.

Irkutsk is a city located in Eastern Siberia in Russia, the administrative center of Irkutsk Oblast, one of the largest cities in Siberia. Founded in 1661, Irkutsk is included in the list of historical settlements of Russia. The population of Irkutsk is about 623,000 (2019), the area - 277 sq. km.

Today's Irkutsk originates from a fortified settlement laid by the Russian explorer Yakov Pokhabov in the summer of 1661. The territory on the bank of the Angara River at its confluence with the Irkut River (hence the name of the settlement) was favorable for agriculture and cattle breeding. The waterway provided communication with the Yenisei River and Lake Baikal.

Unlike modern Angarsk, Irkutsk has a historically developed road network, and it is also divided by the Angara River. The Angara River divides Irkutsk into the right-bank and left-bank parts. The length of the river within the city is 29 km, the width under the Old Angara Bridge in the city center is about 300 meters.

In such conditions, changes in the parameters of transport accessibility can lead to rather complex and expensive solutions.

The results of the questionnaire survey carried out by the IRNITU Transport Laboratory in Irkutsk showed [5, 12–14] that even the existing town planning standard of 38 minutes per one travel is realized only by 50%. In general, the results of the estimation of the enlarged zones of the city of Irkutsk are presented in **Figure 5**.

Having analyzed the housing market of the city of Irkutsk, the average per one square meter price in newly built houses and at the pre-owned market is defined (**Figure 6**).

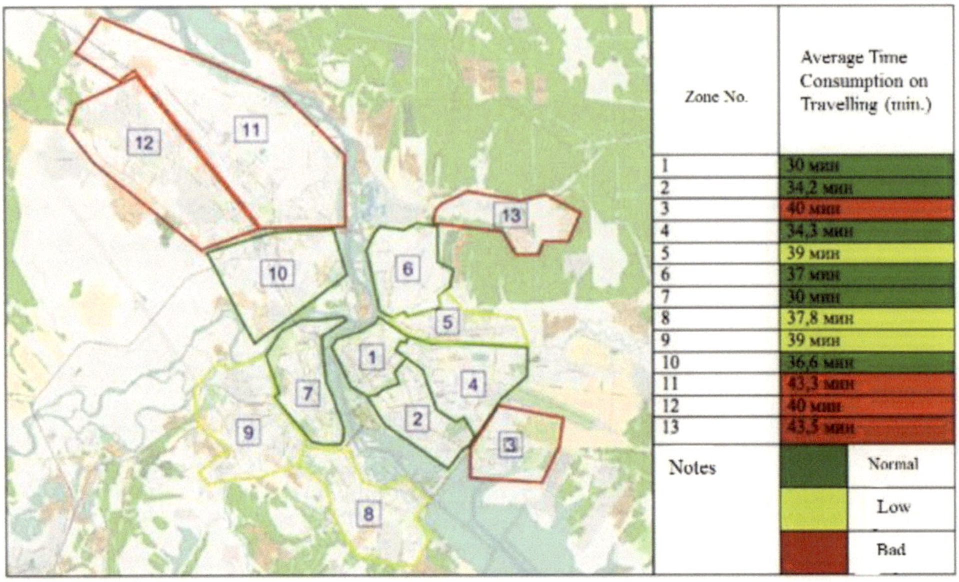

Figure 5.
Average time spent on a one-way work trip in Irkutsk.

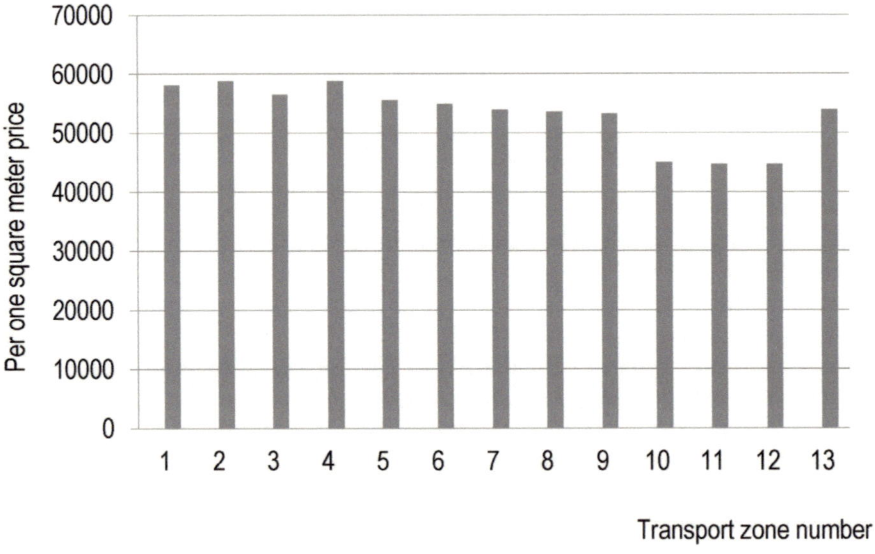

Figure 6.
Average price per one square meter according to transport zones in Irkutsk.

The analysis showed that residential buildings located in 1, 2, 4, 6, 7, 8 and 10 zones will have the highest level of comfort in terms of transport accessibility, residential buildings located in 3, 5, 9, 11,12 and13 zones (**Figure** 7) – the least one. Accordingly, the location of the object will determine their cost.

However, it should be noted that, despite the fact that the transport accessibility is low in the third transport zone, per one square meter price amounts to RUB 56 533.82; it is explained by the fact that the building is mostly elite, and people living in the area use individual means of transportation.

Based on the analysis of the average market per one square meter price in a newly-built and pre-owned residential fund, it can be concluded that residential buildings located in the central part of the city are the most attractive.

Based on the analysis, the time consumption on traveling and the per square meter price of residential real estate in the city of Irkutsk was determined. For a functional explanation of the dependence of the per square meter price on the time consumption on traveling for labor purposes, a trend line was drawn up and the equation of this line was defined (**Figure 8**).

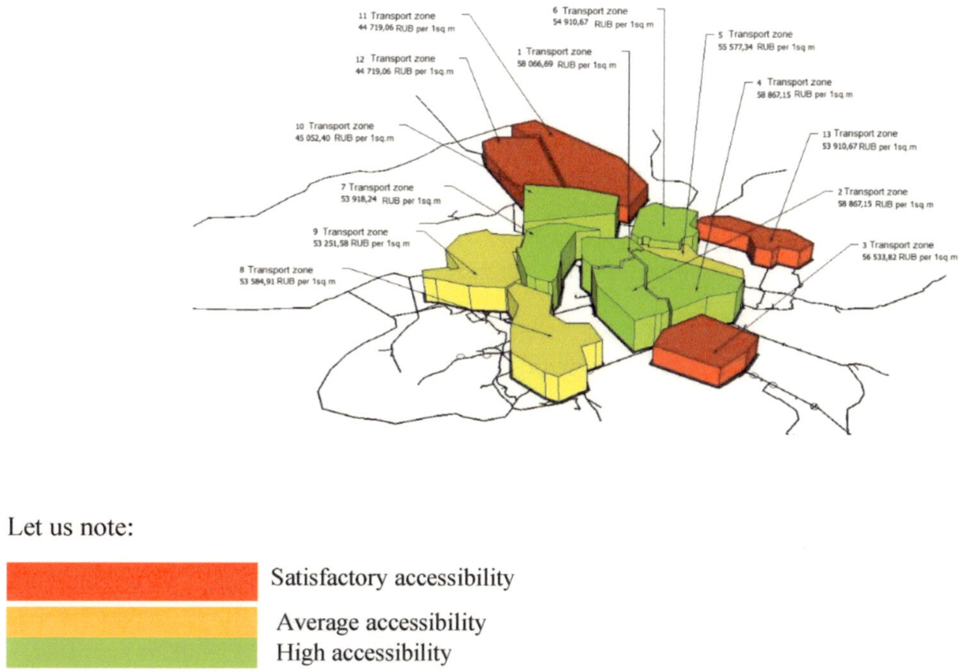

Figure 7.
Per square meter price according to transport zones in Irkutsk.

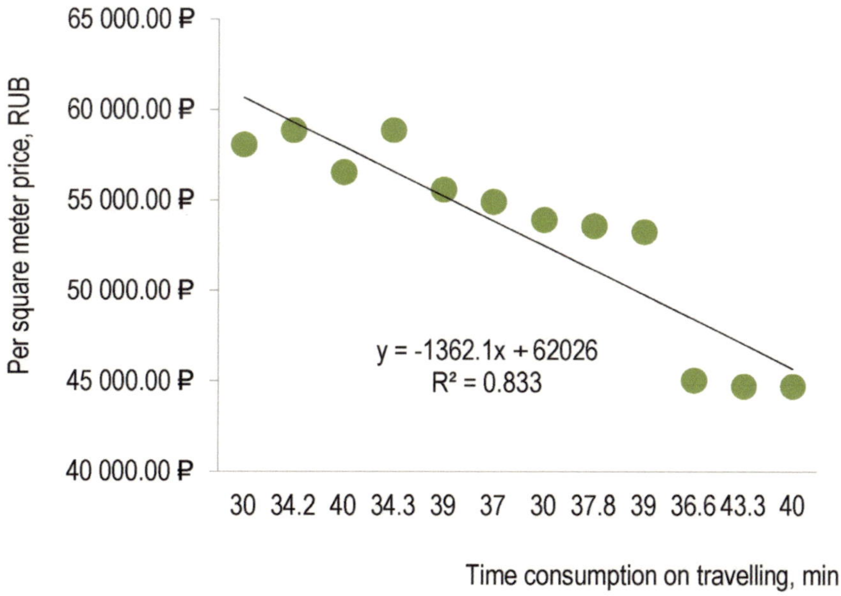

Figure 8.
Dependence of the per square meter price on the time consumption on traveling.

Going beyond the boundaries of the geometric center decreases the value of the property. However, the value of the object increases due to the benefits of location (transportation hubs, objects of social attraction) and causes an increase in its value.

The validity and reliability of the results can be confirmed by representative sample sizes, verification of experimental results by generally accepted statistical criteria.

3. Conclusions

In conclusion, it can be said that studies on the assessment of urban transport accessibility, which improve the quality of service of the transport network and make use of these data for operational management and transport planning, are vital.

The study identified the rules to be followed in transport zoning, indicators to be taken into account when assessing transport accessibility. The more data available, the higher the quality of research. The level of detail is determined on its own; Most often, the zone includes several neighborhoods of the same type and series of developments.

The need to ensure stable functional connections of the central residential territory with peripheral zones (centripetal connections), as well as the need to organize peripheral connections, create a spatial basis for the development and transformation of the existing transport and architectural-planning structure of cities.

Depending on the availability of data on income and expenditure of the population, the relevance of information on the organization of the road network increases, and the possibility of obtaining the most realistic assessment of transport accessibility with the definition of areas, requiring the development of a route network to be covered by urban passenger transport.

As a result, it can be concluded that increasing the accessibility of the territory is a key factor in economic growth. The availability of tools for detailed and accurate accessibility assessment is currently one of the most important research studies in transport planning.

Author details

Sharov Maksim Igorevich
Irkutsk National Research Technical University, Irkutsk, Russia

*Address all correspondence to: sharov.maksim@gmail.com

References

[1] Guidelines for the design of city streets and roads Under the editorship of Yu S Lanzberg Yu and A Stavnichego (Moscow: Stroyizdat) p 324

[2] SP 42.13330 «SNIP 2.07.01-89 * URBAN PLANNING. PLANNING AND BUILDING OF URBAN AND RURAL SETTLEMENTS»

[3] Sharov, M. I., & Lebedeva, O. A. (2020). Housing cost dependence on transport accessibility territory of industrial city. Paper presented at the IOP Conference Series: Materials Science and Engineering, 880(1) doi:10.1088/1757-899X/880/1/012073

[4] Sharov M I, Lebedeva O A 2019 Modern technologies. System analysis. Influence of transport zoning on the operation of the city route network No 2 (62) (Irkutsk) pp 196-202

[5] PTV PARTNER, http://ptv-vision.ru/

[6] Koryagin M 2018 Urban planning: A game theory application for the travel demand management Periodica Polytechnica Transportation Engineering 46(4) (Budapesti Muszaki es Gazdasagtudomanyi Egyetem/ Budapest University of Technology and Economics) pp 171-178

[7] Gil J 2016 Urban modality: Modelling and evaluating the sustainable mobility of urban areas in the city–region A+BE Architecture and the Built Environment 1 (Portugal) pp 1-436

[8] Lai L.W.C and Davies S.N.G 2011 Government transport land – use planning and development by implicit contract for franchised buses and ferries in Hong Kong, 1933-1972 Planning Practice and Research 26(4) (United Kingdom) pp 435-466

[9] Rasay, Hasan; Golmohammadi, Amir Mohammad 2020 Modeling and analyzing incremental quantity discounts in transportation costs for a joint economic lot sizing problem Iranian journal of management studies (Iran) vol. 13 Issue 1 pp. 23-49

[10] Arbib Claudio and Pinar Mustafa C 2020 Competitive location and pricing on a line with metric transportation costs European journal of operational research vol. 282 Issue 1 (Netherlands) pp 188-200

[11] Laurent Achille-B and Vallerand Steve 2020 International journal of sustainable transportation vol. 14 Issue 3 (United States) pp 205-214

[12] Blumenberg, E. 2003. Transportation Costs and Economic Opportunity Among the Poor. Access, 23:40-41.

[13] Carruthers, R., Dick, M. & Saurkar, A. 2005. Affordability of Public Transport in Developing Countries. Transport Papers. TP-3. World Bank: Washington DC

[14] PROGRAM of integrated development of transport infrastructure of the Angarsk urban district for 2017-2036 p58

Chapter 15

Low-Cost Single-Family House through The Use of Precast Reinforced Concrete Elements

Guillermo Yorel Noriega Aquise

Abstract

A technical design is developed to attend and assist populations in need of single-family housing and for populations in post-emergency situations. It exposes a production process of precast reinforced concrete elements, to be produced in a small production plant or at the site, with a minimum of equipment and tools. It is intended to establish a low-cost single-family house construction system with prefabricated reinforced concrete elements, which will become a technological alternative to traditional confined masonry construction. It presents a production line of six types of houses. For the comparison, a methodological process is followed, the comparison is made between the process of building houses with precast concrete elements and with the building process by confined masonry, the traditional process most used in Peru. The dominant principle of comparison is the equality of the useful surface of the rooms. The outstanding and visual difference is in the thickness of the walls, in the prefabricated house it is 0.10 meters and in the houses with confined masonry it is 0.15 m. The costs, production times and assembly, of the building with prefabricated elements, is low, compared to the building process by confined masonry. An in-line production process is established, of prefabricated elements with minimal equipment. The basic criterion is to manufacture that does not exceed the capacity of the size of the manufacturing, transport and assembly equipment. A simple process for the assembly was examined, a minimum period of construction of a prefabricated house of 2 hours was determined at any time of the year. The lowest cost, the direct cost has been achieved in VUF 04 at $ 264.50 US dollars per square meter and in total costs of $ 374.54 US dollars per square meter.

Keywords: manufactured houses, costs, prefabricated elements, construction system

1. Introduction

A technological alternative is sought that responds to the need for the construction of a low-cost single-family basic house and is implemented agilely in the short term.

When analyzing the housing situation in Peru, common problems are found, as in all Latin American countries. There is a very large housing deficit both because of the famine and because of the physical deficiencies of the existing houses. According to different estimates, this deficit is equivalent to just over half of all existing houses. The severity of the housing problem varies greatly within the region and within each country and even within the same city [1].

Nine out of 10 houses in Latin America and the Caribbean are of low quality. "*Currently, more than 75% of the inhabitants of Latin America and the Caribbean reside in urban areas. The challenge is how to eradicate the poverty cords and the deterioration that this generates and that does not allow good quality of housing in our hemisphere*," Luis Alberto Moreno, president of the BID, explained to EL PAÍS. In the last 20 years Brazil, Colombia, Peru, Chile, Paraguay and Argentina, delivered more than six million housing units to low-income populations. This policy has not prevented the emergence of low-quality housing clusters on the outskirts of cities, far from work centers [2].

Peru is the third country in Latin America with the highest housing deficit, according to the report of the Ministry of Housing, Construction and Sanitation, of the Peruvian government. They state that there is a deficit of 1800,000 houses, among families that do not have a house or live in a precarious house. Peru ranks third in Latin America as the country with the highest housing deficit. After Nicaragua and Bolivia [2].

The housing deficit in the face of a low supply leads to an increase in the price of houses. According to the Peruvian Chamber of Construction (CAPECO), in recent years there has been an increase in the average price of departments in Lima and Callao. The price per square meter (m^2) during 2015 was 4,623 soles and increased by 5.5% for 2016; by zone, the largest hike was presented in Lima Modern that went from 4,794 soles per m^2 in 2015 to 5,187 soles per m^2 in 2016. The high prices make it unattainable to obtain a house for everyone. Additionally, the high informality in construction in low-income districts increases the housing deficit [3].

The need for housing is critical in certain geographic spaces where telluric and catastrophic events occurred. Thus, the mayor of the Provincial Municipality of Ica-Peru, in August 2017 pronounced on the insufficient reconstruction of Pisco and states that after 10 years "*little or nothing has been done*" to rebuild the city that was later devastated of the catastrophe. "*There are families that still live on mats and plastics*." The earthquake of 7.9 degrees, left 595 dead, about 2,291 injured, 76 thousand houses destroyed and 431 thousand people affected [4].

Also, after two years of the earthquake that occurred in Colca (Caylloma – Arequipa – Peru) families from Ichupampa continue to live in plastic modules, the reconstruction did not reach 15% of the houses affected by the earthquake of August 14, 2016. Of the 410 houses, only six were left intact, 234 were found collapsed and the rest suffered several cracks and fissures. The only church in the town was destroyed like the initial school. Two years have passed since the tragedy and the streets express that the earthquake had occurred last week. According to the Commerce of August 15, 2018, only 132 were qualified to be rebuilt and each house will be served with the S/43,348 bond and to date, only one house has been rebuilt [5].

2. Methodology

Science shows that the practice of comparison has been and continues to be an essential resource for responding to problems of natural and social knowledge. But we must not forget the important differences that exist between comparison as a way of thinking and as a scientific procedure. The first compares simple operations; the second compares complex operations, although the difference does not lie in the complexity of the logical structure of the comparisons, does not present significant contrasts in science and in everyday life, but rather in the selection and definition of the objects and properties that are compared, as well as in the care and systematicity of the production procedures and data analysis from which the comparisons are made [6].

The comparative method consists of empirical generalization and hypothesis verification. The advantages offered by the comparative method include understanding unknown things from known ones, the possibility of explaining and interpreting them, profiling new knowledge, highlighting the peculiarity of known phenomena, systematizing information, distinguishing differences with similar phenomena or cases [7].

The comparative method is inherent in any scientific procedure, it is expected that whenever it is compared following scientific procedures, it will be possible to compare; in aspects that are comparable and follow the analysis strategy to reach conclusions. It is not understood any type of unconscious comparison, that is not premeditated, rather this comparison is based on established objectives [6].

> *For the comparison process, on the one hand, there are single-family houses designed and built with pre-manufactured elements of reinforced concrete and on the other hand, there is a confined masonry design and construction.*

The methodological process to follow is defined in three phases:

2.1 Phase 1

Architectural design of basic single-family houses: It comprises the architectural design of twelve types of single-family houses with one story, which in turn will be the basis for building a second story in the future. The design corresponds to six houses with prefabricated elements and six houses with confined masonry, in both cases the useful surface is the same.

2.2 Phase 2

Structural Design of basic single-family houses and building process: Includes the structural design of basic single-family houses, using the calculation process. For both manufactured houses and confined masonry designed houses. Building processes are also defined.

2.3 Phase 3

Analysis of costs, budgets and times for the construction of basic single-family houses: The designs and processes are analyzed according to the specified and delimited approaches. In order to define the costs, the budget and the times that the building demands. The analysis of results must achieve:

- Design construction processes for the building.
- Establish the shortest construction time for the building.
- Establish the lowest cost of manufacture, construction and building.

3. Results

3.1 Design and development of basic housing

The design was carried out according to the requirements of the comparative method, in order to observe the data of the proposed variables and indicators. Architectural and structural designs and construction processes are analyzed.

In order to define the costs, the budget and the times required for the construction of the houses. The houses designed is the result of an exploration of the housing need of the City of Arequipa and nearby cities affected by telluric processes.

The dominant design principle is the useful surface of the environments of the designs are equal, that is, the useful areas of a bedroom, is the same in both types of design (precast and confined masonry), as well as in all the components of the living place. The outstanding and visual difference is in the thickness of the walls, in the prefabricated house it is 0.10 meters and in the houses with confined masonry it is 0.15 m. The total area that a manufactured house occupies is less than a house with confined masonry.

3.1.1 Prefabricated basic single-family housing (VUF)

Six types of basic single-family housing of various lengths and widths were designed, whose representative product on the surface are: 25.83 m^2, 33.39 m^2, 39.06 m^2, 42.21 m^2, 51.03 m^2 and 59.85 m^2 (**Tables 1–3**; **Figures 1–6**).

3.1.2 Basic single-family house-confined masonry (VAC)

The six types of basic single-family housing built by confined masonry of various dimensions of length and width whose representative product on the constructed area are: 27.09 m^2, 35.15 m^2, 41.28 m^2, 44.51 m^2, 53.54 m^2, 63.86 m^2 (**Tables 4–6**).

Items	Basic Housing	Long m	Wide m	T. Area m^2	Code
1	Prefabricated Single Family Housing 01	6.30	4.10	25.83 m^2	VUF 01
2	Prefabricated Single Family Housing 02	6.30	5.30	33.39 m^2	VUF 02
3	Prefabricated Single Family Housing 03	6.30	6.20	39.06 m^2	VUF 03
4	Prefabricated Single Family Housing 04	6.70	6.30	42.21 m^2	VUF 04
5	Prefabricated Single Family Housing 05	8.10	6.30	51.03 m^2	VUF 05
6	Prefabricated Single Family Housing 06	9.50	6.30	59.85 m^2	VUF 06

Table 1.
Areas and codes of prefabricated single-family housing.

Environments in square meters (m^2)	VUF 01	VUF 02	VUF 03	VUF 04	VUF 05	VUF 06
Kitchen, dining room and living room.	13.00	12.36	14.2	18.9	23.7	21.4
Bedroom: a 2/seater bed	7.50	7.5	7.5	7.5	8.76	8.76
Bedroom: two 1/square beds		7.5	7.5	7.5	7.5	7.5
Bedroom: two 1/square beds						7.5
Bathroom, toilet, sink and shower.	2.60	2.6	2.6	2.6	2.6	2.6
Bathroom, toilet, sink and shower.					2.6	2.6
Passage			3.06	1.2	1.26	1.26
Walls	2.73	3.43	4.20	4.51	4.61	6.97
Total Area m^2	25.83	33.39	39.06	42.21	51.03	58.59

Table 2.
Detail of environments by surface in square meters – VUF.

Environments in percentage (%)	VUF 01	VUF 02	VUF 03	VUF 04	VUF 05	VUF 06
Kitchen, dining room and living room.	50.33	37.02	36.35	44.78	46.44	36.53
Bedroom: a 2/seater bed	29.04	22.46	19.20	17.77	17.17	14.95
Bedroom: two 1/square beds		22.46%	19.20	17.77	14.70	12.80
Bedroom: two 1/square beds						12.80
Bathroom, toilet, sink and shower.	10.07	7.79	6.66	6.16	5.10	4.44
Bathroom, toilet, sink and shower.					5.10	4.44
Passage			7.83	2.84	2.47	2.15
Walls	10.57	10.27	10.75	10.68	9.03	11.90
Total Area %	100.%	100.%	100.%	100.%	100.%	100.%

Table 3.
Detail of the environments in percentage of occupancy – VUF.

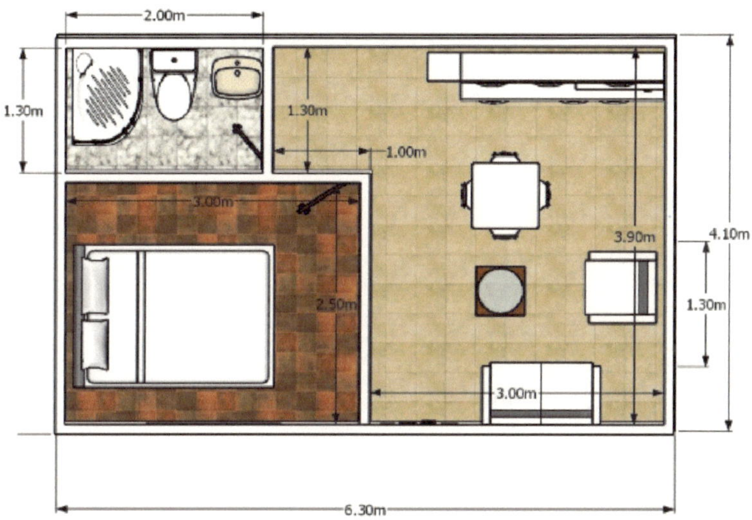

Figure 1.
Plant view VUF 01.

3.2 Isometry of basic single-family houses

Isometric projection allows us to show basic housing in three-axis dimensions (height, width and depth) allows us to understand the desire of the work to achieve. It is practically a cube of variable dimensions, where the basic characteristics of a house are shown.

3.2.1 Isometry in prefabricated houses with elements of reinforced concrete

It has been projected in this way in order to reduce costs to a minimum. Common measures are used. The front view, is 6.30 m long, the free height is 2.40 m, the doors, windows and some rooms have the same dimensions that provide a minimum surface of habitability. This reduces production costs and increases its efficiency on the production line (**Figure 7**).

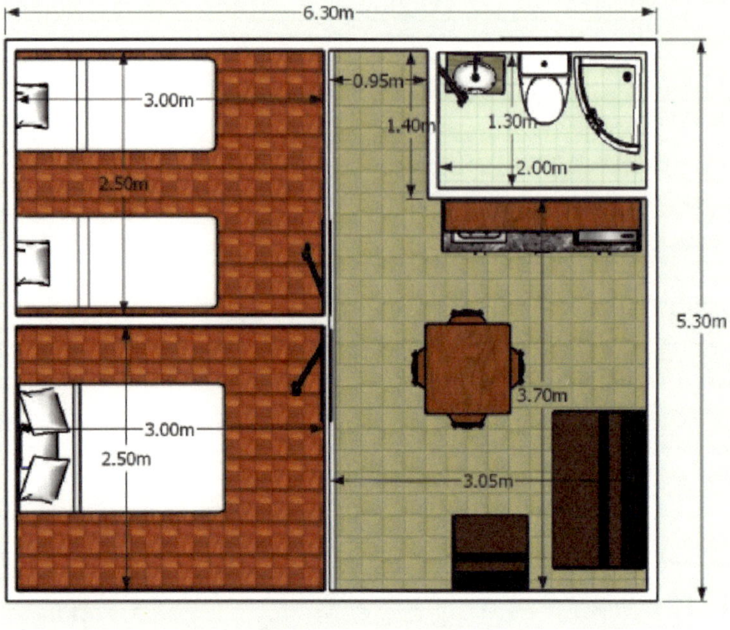

Figure 2.
Plant view VUF 02.

Figure 3.
Plant view VUF 03.

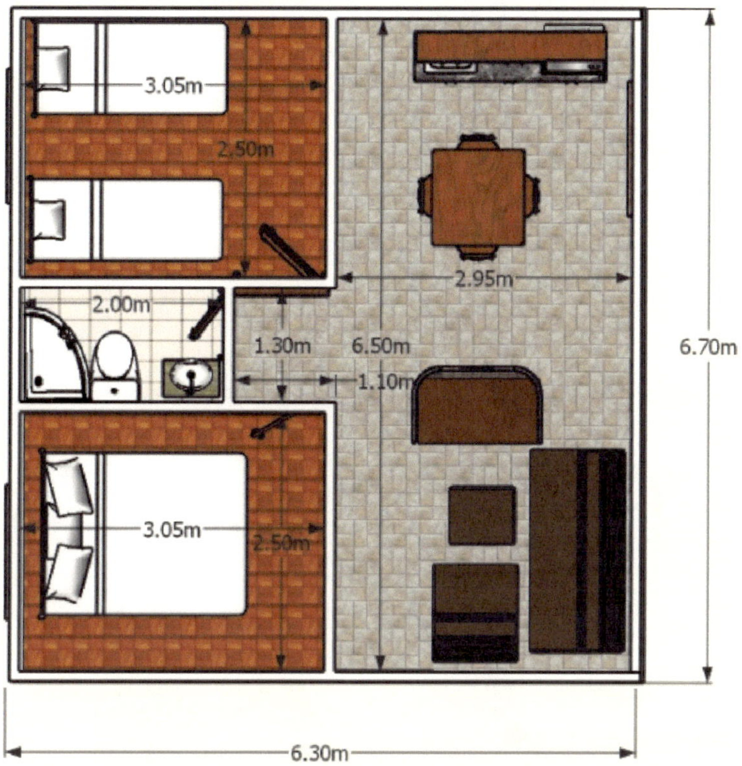

Figure 4.
Plant view VUF 04.

The view is common and its unique perspective, which differs externally between each other are longitudinal dimensions and the number of windows, the details are specified in the floor views.

3.2.2 Isometry in prefabricated houses with confined masonry

The isometry of houses with confined masonry is as similar as those built with prefabricated elements with reinforced concrete element s. The view is common and the singular perspective, which differs externally between them are the different dimensions of length and width, which implies a larger surface area, expressed in different values.

3.3 Process of elaboration of precast reinforced concrete elements

In the manufacturing process, equipment and machines are used that must have the property of moving them to a place with minimal conditions and can be established in nearby places where prefabricated houses are assembled.

The machines and equipment to be used are:

- Portable laboratory for design and testing.
- Devices for loading, storage and dosing of supplies.
- Self-propelled concrete mixer or purchase ready-mix concrete.

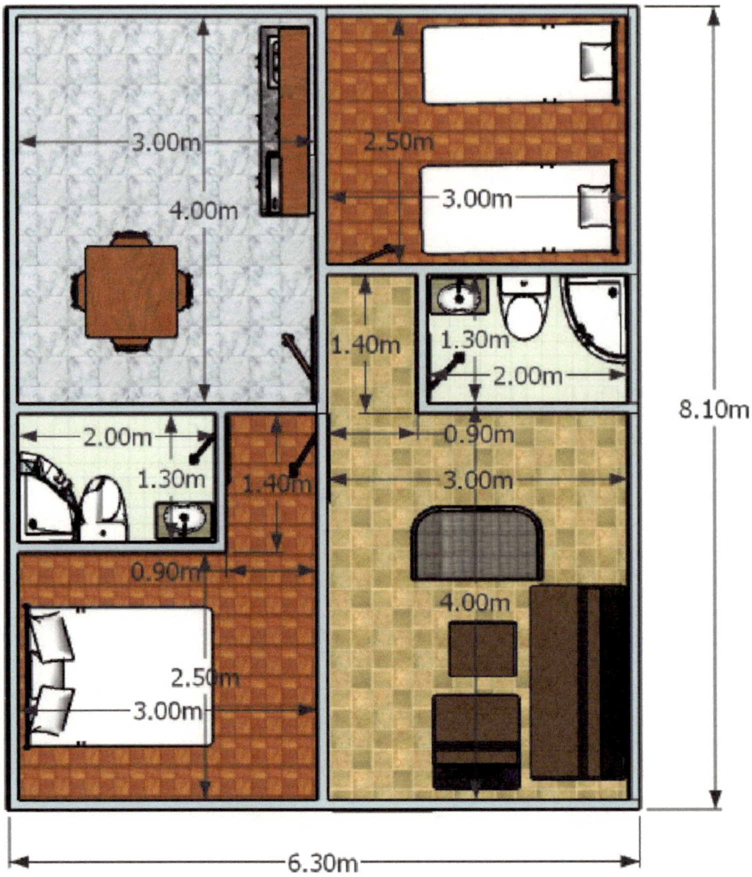

Figure 5.
Plant view VUF 05.

- Independent molds, with horizontal and vertical movement.
- Independent vibration or additive platform.
- 10 TM combustion engine forklifts.
- 10TM truck crane with a 10 m free platform.
- Manual tools.

The manufacturing process requires having molds to make the prefabricated elements. The molds respond directly to the design and must respond to the requirements of the projects. Efficiency is increased when a mold or a small number of molds can be so versatile and can produce a wide variety of precast elements.

A 2.40 m x 9.60 m mold has been designed with a variable mold thickness from 0.10 m to 0.40 m, designed from iron plate, having to have several molds that can give shape to pieces of variable configuration. There must be space for doors and windows, it must allow to place the sanitary and electrical installations. You can also use wooden formwork to expose a glossy finish.

Concrete to achieve the projected strength must be subjected to a vibration process. The designed system consists of placing the molds on rails with a vibration

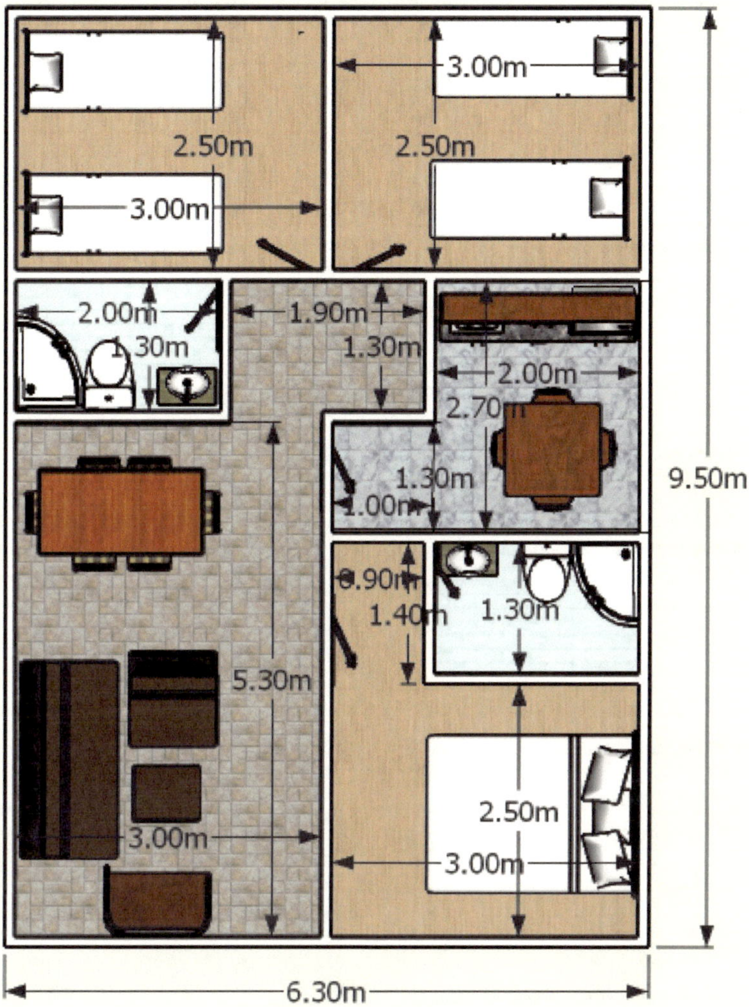

Figure 6.
Plant view VUF 06.

Items	Basic Housing	Long m	Wide m	T. Area m^2	Code
1	Confined Masonry Housing 01	6.45	4.20	27.09 m^2	VAC 01
2	Confined Masonry Housing 02	6.45	5.45	35.15 m^2	VAC 02
3	Confined Masonry Housing 03	6.45	6.40	41.28 m^2	VAC 03
4	Confined Masonry Housing 04	6.90	6.45	44.51 m^2	VAC 04
5	Confined Masonry Housing 05	8.30	6.45	53.54 m^2	VAC 05
6	Confined Masonry Housing 06	9.75	6.55	63.86 m^2	VAC 06

Table 4.
Areas and codes of single-family housing confined masonry VAC.

platform. When the production of precast elements moves to another location the vibration process is replaced with the use of additives, depending on the cost, the use of additives can replace the physical vibration process.

Environments in m^2	VAC 01	VAC 02	VAC 03	VAC 04	VAC 05	VAC 06
Kitchen, dining room and living room.	13.00	12.33	14.35	19.94	23.4	21.72
Bedroom: a 2/square bed	7.50	7.5	7.5	7.63	7.5	7.63
Bedroom: two 1/square beds		7.5	7.5	7.63	7.5	7.63
Bedroom: two 1/square beds						7.63
Bathroom, toilet, sink and shower.	2.60	2.6	2.6	2.6	2.6	2.6
Bathroom, toilet, sink and shower.					2.6	2.6
Passage			2.93	1.37	2.47	5.13
Walls	3.99	5.22	6.40	6.03	7.47	8.92
Total Area in m^2	27.09	35.15	41.28	45.20	53.54	63.86

Table 5.
Detail of the environments by surface in square meters – VAC.

Environments in percentage (%)	VAC 01	VAC 02	VAC 03	VAC 04	VAC 05	VAC 06
Kitchen, dining room and living room.	47.99	35.08	34.76	44.12	43.71	34.01
Bedroom: a 2/square bed	27.69	21.34	18.17	16.88	14.01	11.95
Bedroom: two 1/square beds		21.34	18.17	16.88	14.01	11.95
Bedroom: two 1/square beds						11.95
Bathroom, toilet, sink and shower.	9.60	7.40	6.30	5.75	4.86	4.07
Bathroom, toilet, sink and shower.					4.86	4.07
Passage			7.10	3.03	4.61	8.03
Walls	14.73	14.85	15.50	13.34	13.95	13.97
Total Area	100.%	100.%	100.%	100.%	100.%	100.%

Table 6.
Detail of the environments in occupancy percentage – VAC.

The process for making precast elements is as follows:

- Define the design of the basic single-family house.
- Preparation of elements to be prefabricated.
- Formwork – molding, placement of meshes and pipes.
- Pouring of ready-mixed concrete.
- Physical vibration or with additives.
- Finishes and accumulation of pieces and cataloging.
- Transport and assembly.

3.4 Assembly of a prefabricated house with elements of reinforced concrete

The assembly process, perform the following process (**Figure 8**).

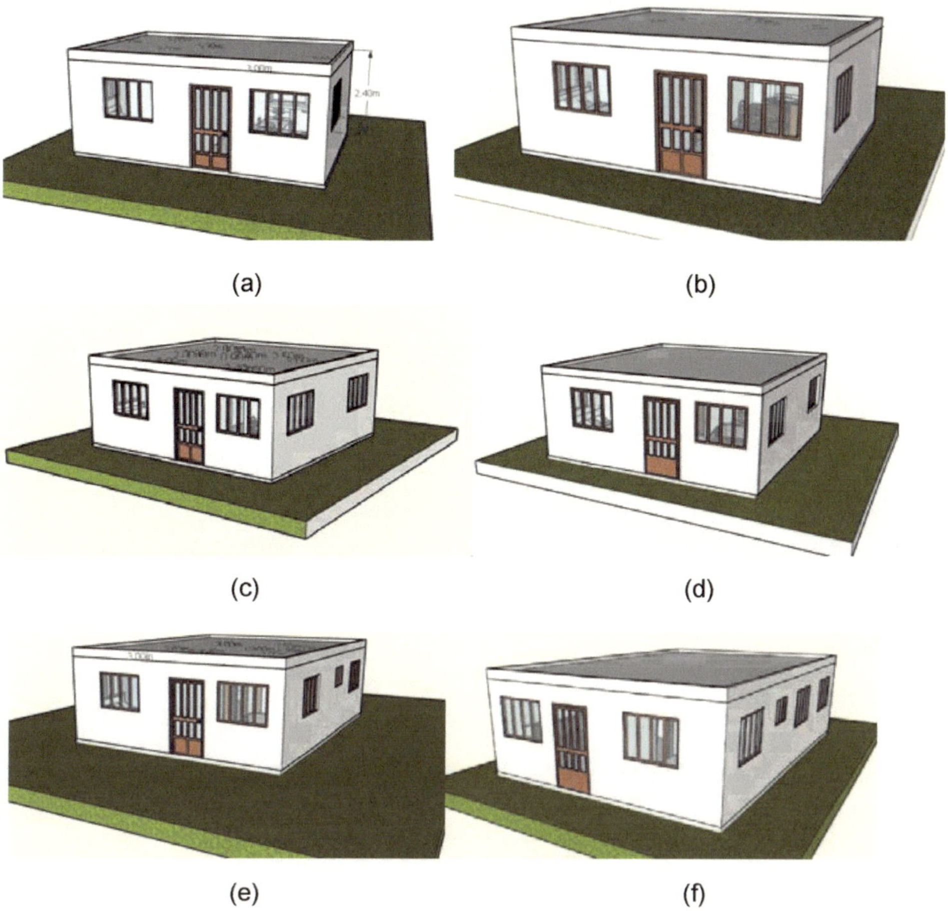

Figure 7.
(a) Isometric view of VUF 01. (b) Isometric view of VUF 02. (c) VUF 03 isometric view. (d) VUF 04 isometric view. (e) VUF 05 isometric view. (f) VUF 06 isometric view.

3.5 Structure of prefabricated elements of reinforced concrete

The structural system of precast concrete elements comprises foundation beams, floor slabs, walls and ceiling, under an approach of articulated structural panels with anchors that support stress requests. The elements must reach a sufficient resistance for their handling and transport to the assembly place. They must withstand the weather and inclement places.

Each element is made of reinforced concrete of 210 Kg/cm^2 with mesh every 0.20 m in both directions of iron with a diameter of 3/8 in diameter and in the perimeter near the edge there is an iron with diameters of 1/2″ in diameter, reinforcements are also applied in corners and critical places. Lifting and transport devices are installed. It is possible to use various anchoring systems, as long as it responds to the stress requests required by the prefabricated parts.

Two pieces are shown with exposed structures, the others show a similar constitution (**Figure 9**).

3.6 Production time and assembly of a prefabricated house

For this analysis, we worked with the value of the unit time, which is the direct relationship between the metering and the unit yield of the item, in relation to the

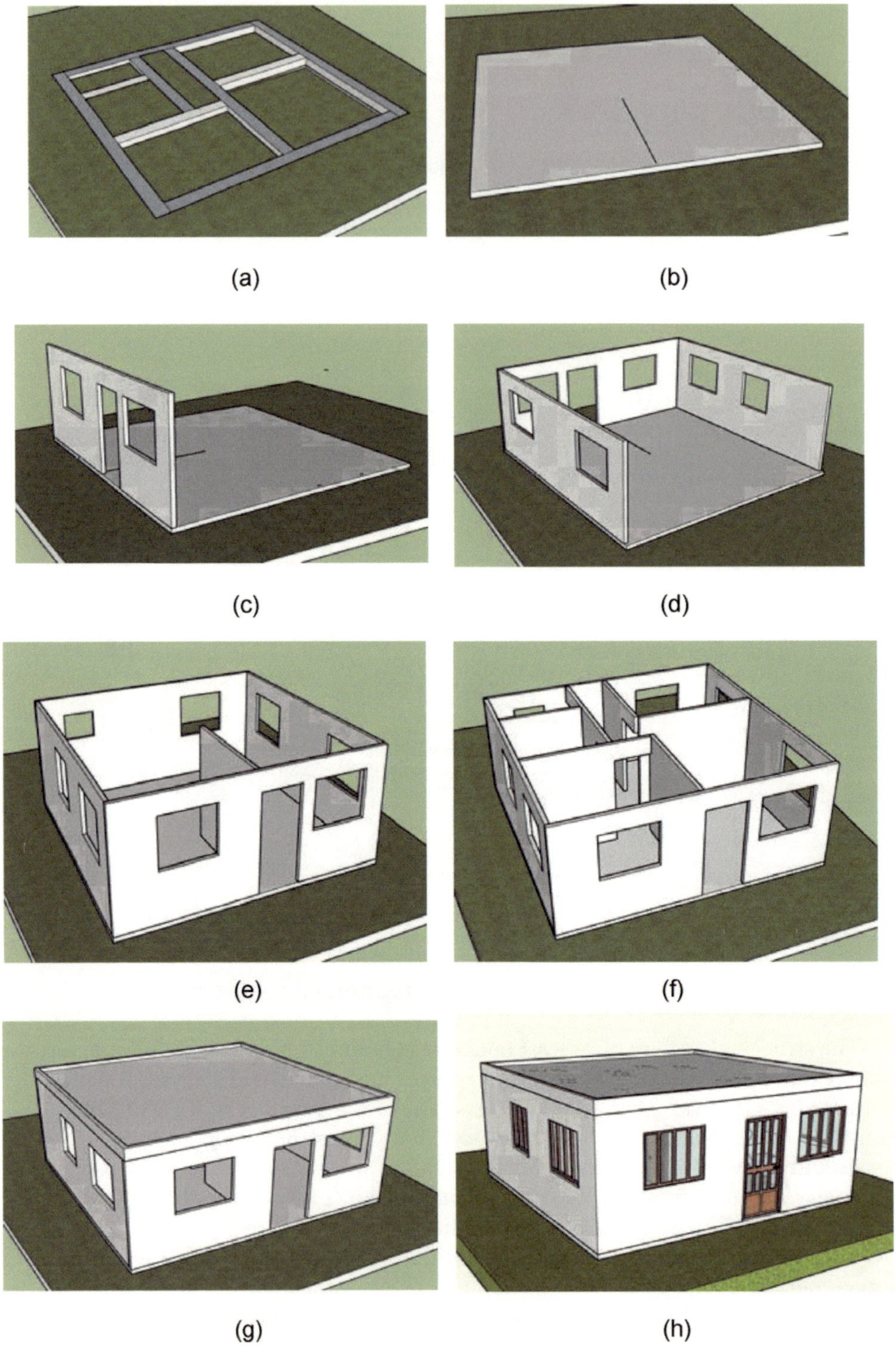

Figure 8.
(a) Foundation installation. (b) Foundation slab. (c) Front wall installation. (d) Installation of sidewalls. (e) Installation perimeter wall. (f) Installation of interior walls. (g) Ceiling installation. (h) Completion of work finishes.

value of the direct cost. The individual values per item and the total per dwelling, were obtained according to the S10 program, which gives us a gross unit time of the need in days and that is required for the execution of the work.

The value obtained is an indicative and orientate parameter that allows us to adjust the necessary times, according to an adjusted time analysis of unit times and the Gantt chart is made. The minimum and effective duration of production and assembly of the prefabricated houses has been determined, which are shown in the following (**Table 7**).

The unit time of VUF Prefabricated Single Family Housing has been determined from 32.07 days in VUF-01 to 65.70 days in VUF 06. Adjusted time with a Gantt diagram of VUF Prefabricated Single Family Housing has been determined from 4.00 days in VUF-01 to 6.00 days in VUF 06 (**Figure 10**).

El Unitary time of VUF houses is adapted to a quadratic polynomial trend of y x $0.005x^2$ + 0.6022x + 12,706 (R2 x 0.9916, 95%). El Adjusted Unit Time in Gantt of

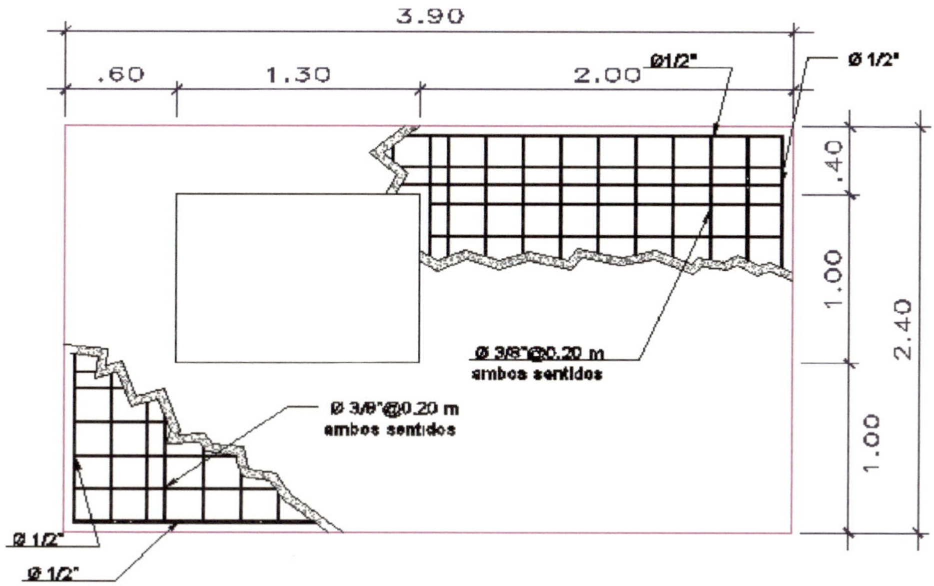

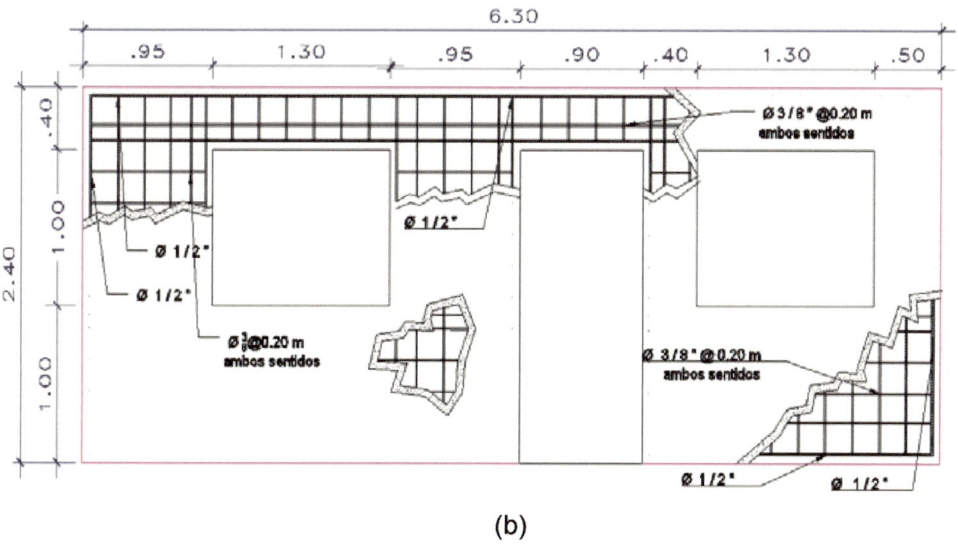

Figure 9.
(a) Structure of a side piece with window of the VUF 01. (b) Structure of a front part of the VUF 01.

VUF houses is adapted to a quadratic polynomial trend of y – $0.0003x^2$ + 0.0876x + 2,004, (R2 x0.9072, 95%).

The unit time on constructed housing area expressed in days/m^2, has been determined, the lowest value in VUF-04 of 1.09 and the highest value found in VUF-01 of 1.24. Performing calculations at set times in the Gannt chart has achieved the lowest value in VUF-06 of 0.10 and the highest value in VUF-01 of 0.15.

These indices are achieved without increasing the efficiency of the machines and equipment, by carrying out a greater analysis with efficiency indices, the production level would increase and achieve cost reduction, which implies producing larger units of prefabricated elements per unit of time. There is a direct relationship between execution times and investment requirements, it is shown in the following (**Table 8**).

It presents a direct relationship between the execution days and the investment requirement, expressing that, the greater the number of execution days, the greater the investment. The largest investment is concentrated in the production of prefabricated parts.

Description	VUF Air m^2	T. U. Metrado/Ru days VUF	T. A. Gantt, VUF days.	T. U. days/ m^2	T. A. Gantt days/m^2
VUF-01:	25.83	32.07	4.00	1.24	0.15
VUF-02:	33.39	37.88	4.88	1.13	0.15
VUF-03:	39.06	43.99	4.66	1.13	0.12
VUF-04:	42.21	45.84	5.00	1.09	0.12
VUF-05:	52.29	59.90	6.00	1.15	0.11
VUF-06:	59.85	65.70	6.00	1.10	0.10

Table 7.
Unit time (TU) and adjusted (TA) in VUF Gantt.

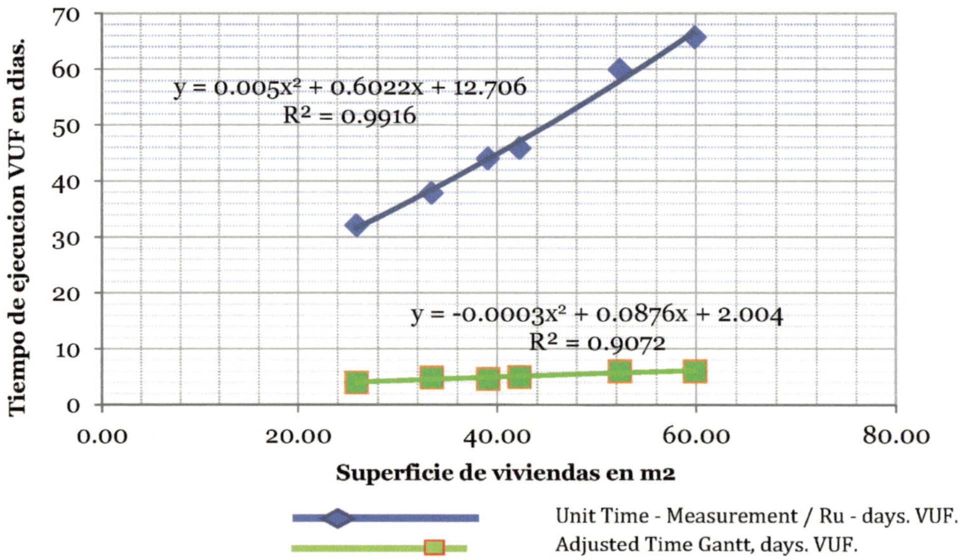

Figure 10.
Unit time and adjusted execution of the VUF.

The investment trend responds to a cubic polynomial function where the cusp is on the second day of execution, there is a decrease on the fourth day. However, it must be understood that the work can be executed within four to six days, therefore, before starting the execution of works all available investment must be available (**Table 9**).

Of the trends of the polynomial functions, there are three subtypes in the relationship of execution time and investment requirement, which is expressed in three values close to 1 of the coefficients of determination. Which tells us that according to the sub types of prefabricated houses, the cost of the work and the investment requirements can be adjusted (**Figure 11**).

3.7 Costs and budgets of VUF prefabricated single-family houses

3.7.1 Direct costs VUF

The direct cost are all those expenses that are directly related to the construction of a work, specifically it will be expressed in the amount of labor, materials and equipment involved in the execution of a work, which will be expressed in national currency and in US dollars at the change of budgeting.

Six types of houses have been designed for a production system of prefabricated elements of single-family houses in order to optimize resources and adapt to a minimum production line at scale, be it on a production line in plant or at the foot of the play.

Architectural and structural plans have been developed for each type of house designed, from which the corresponding measurements have been made. These

Description	Monday	Tuesday	Wednesday	Thursday	Friday	Saturday	TOTAL
	1.dia	2.dias	3.dias	4.dias	5.dias	6.dias	
VUF-01	8,428	13,994	9,558	6,710			38,690
VUF-02	8,929	17,542	11,884	4,083	4,308		46,745
VUF-03	11,605	18,043	6,753	12,889	5,101		54,391
VUF-04	11,079	19,258	7,402	12,063	5,672		55,475
VUF-05	15,217	25,297	8,893	11,942	6,612	4,647	72,608
VUF-06	19,602	25,437	9,392	14,877	5,590	6,346	81,244

Table 8.
VUF daily investment requirement.

Code	Inversion polynomial functions	R^2
VUF 01	$y = 1931.4x^3 - 16589x^2 + 41813x - 18727y$	R^2 = perfecta.
VUF 02	$y = 1858x^3 - 18074x^2 + 49687x - 24514$	$R^2 = 0.9996$
VUF 03	$y = 316.89x^3 - 3639.7x^2 + 10388x + 5489.6$	$R^2 = 0.4056$
VUF 04	$y = 748.51x^3 - 7638.2x^2 + 21273x - 2388.3$	$R^2 = 0.469$
VUF 05	$y = 608.77x^3 - 6676.6x^2 + 18265x + 4691$	$R^2 = 0.6687$
VUF 06	$y = 469.56x^3 - 4901.2x^2 + 11242x - 14014$	$R^2 = 0.7084$

Table 9.
Inversion polynomial functions VUF.

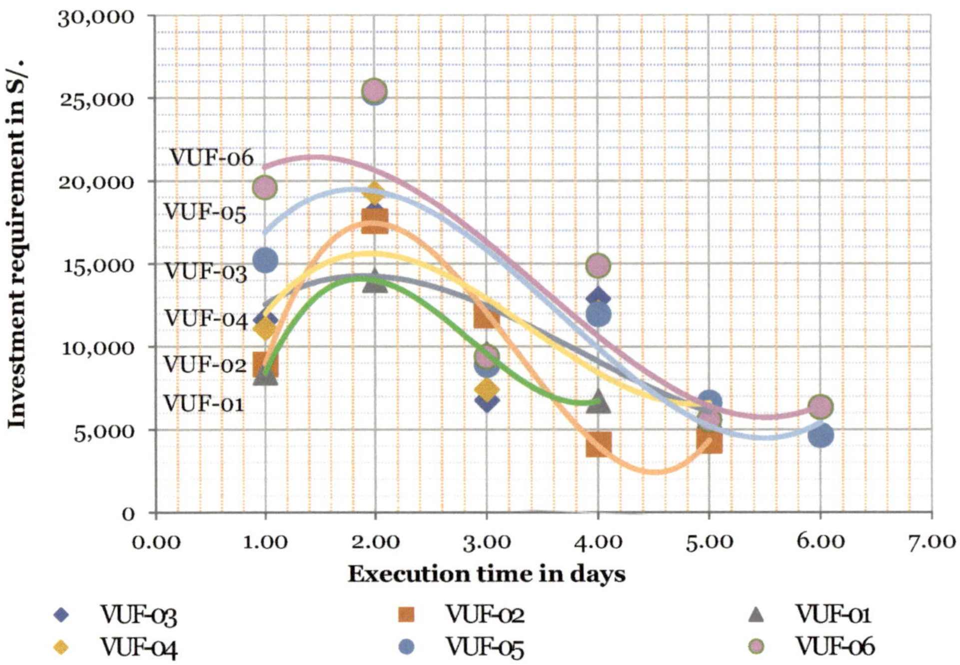

Figure 11.
Polynomial lines trend reversal time in execution VUF.

measurements were processed in the cost and budget program S10. A summary of direct costs is presented in the **Table 10** below.

Description	Cost in Soles	Cost in Dollars	Cost in $/m^2
Direct Cost VUF-01	27,323.20	7,786.61	301.46
Direct Cost VUF-02	33,012.08	9,407.83	281.76
Direct Cost VUF-03	38,411.45	10,946.55	280.25
Direct Cost VUF-04	39,176.99	11,164.72	264.50
Direct Cost VUF-05	51,277.09	14,613.02	279.46
Direct Cost VUF-06	57,375.35	16,350.91	273.20
T/C: 3,509 to 26 June 2020, SUNAT-PERU [8].			

Table 10.
Direct cost of a VUF in US suns and dollars.

The costs of the manufactured houses shown in the table are in Peruvian currency in soles, in US dollars and the cost per square meter, according to the type of VUF house. For each item, a unit cost analysis has been carried out indicating the amount of materials used and considers the equipment and machines necessary for its manufacture and assembly (**Figure 12**).

The direct costs in dollars of basic housing range from US $ 7,786.61 to US $ 16,350.91 and are accessible to the economies of populations in need of housing. These prices are competitive with the costs offered in the real estate industry.

The bond that the Peruvian government assigns to the victims of collapsed or uninhabitable houses due to disasters according to RM No. 012 2018 Housing [9] establishes a housing reconstruction bond for Ichupampa, Lari, Tuti and others in Arequipa, of S/43,497 y its value in US dollars is $ 13,171.

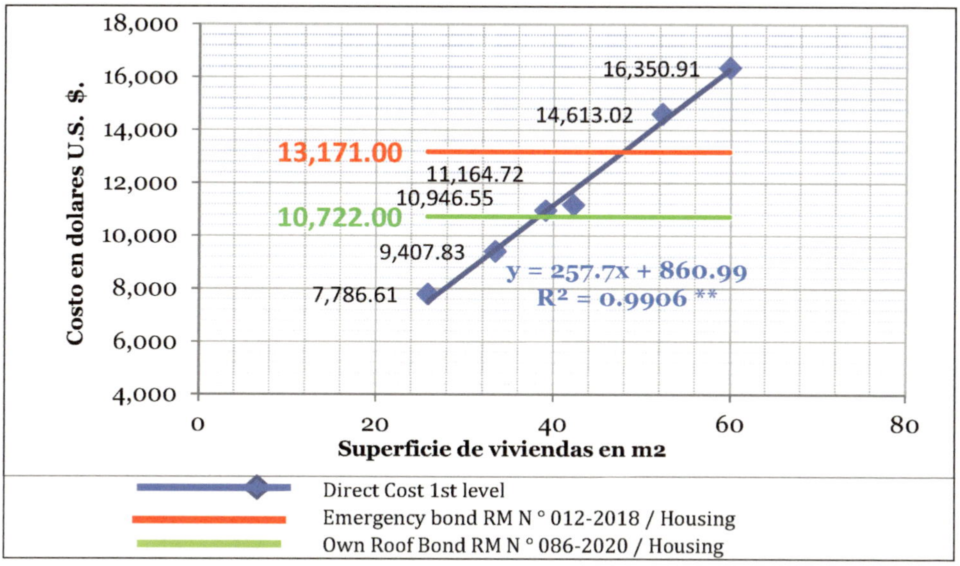

Figure 12.
Direct cost of VUF and its relationship to emergency bonds.

The value of the house VUF 01, VUF 02, VUF 03 and VUF 04 are below the bonus assigned by the Peruvian Government. The proposal to serve in situations of need for housing and post-emergency is framed for direct care, effectively and in a short period of time. The correlation analysis between the surface of the VUF and the cost in dollars of its building has a linear trend and the coefficient of determination is very high.

The direct cost per square meter of manufactured housing, in US dollars, ranges from $ 273.23 to $ 301.46, these amounts are one of the lowest compared to traditional construction costs.

According to Ministerial Resolution No. 086–2020-Housing [10] it provides a family housing bonus for families in need of S/. 37,625 soles, in US dollars it is equivalent to US $ 10,722.43 and VUF 01 and VUF 02 houses are below this value.

3.7.2 Total costs. VUF

The total cost of a work is defined as the sum of direct cost-plus indirect costs. For our case, indirect expenses are covered by overheads of 10% of direct cost plus 10% profit; from which a subtotal is obtained and from this value is added 18% corresponding to the IGV, resulting in the total cost. For a production line it is common for indirect cost to be a fixed and non-alterable proportional value in costing processes.

The value of the total cost has been obtained in soles and then it has been converted to US dollars according to the exchange rate and the cost per square meter is established in US dollars (**Table 11**).

The total costs in dollars of basic housing range from US $ 11,025.83 to US $ 23,152.89 and are accessible values for the economies of populations in need of housing (**Figure 13**).

The bond that the Peruvian government assigns to the victims of collapsed or uninhabitable houses due to emergencies or disasters according to RM No. 012 2018 Housing The housing reconstruction bond of US $ 13,171 would only allow VUF 01 and VUF 02 to be reached.

Description	Cost Soles	Cost Dollars	VUF Area	Cost in $/m^2
C.T. VUF-01	38,689.65	11,025.83	25.83	426.86
C.T. VUF-02	46,745.11	13,321.49	33.39	398.97
C.T. VUF-03	54,390.61	15,500.32	39.06	396.83
C.T. VUF-04	55,474.62	15,809.24	42.21	374.54
C.T. VUF-05	72,608.36	20,692.04	52.29	395.72
C.T. VUF-06	81,243.50	23,152.89	59.85	386.85
T/C 3,509 to 26 June 2020, SUNAT-PERU.				

Table 11.
Total cost of prefabricated single-family housing (VUF).

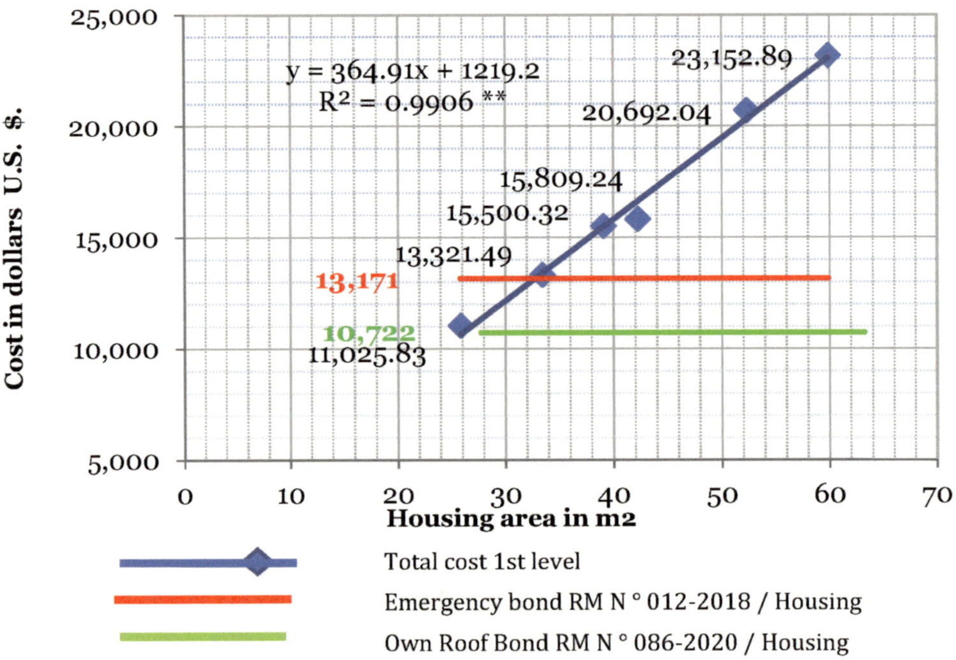

Figure 13.
Total cost of VUF and its relationship to emergency bonds.

The bonus that the Peruvian government assigns for its own roof according to Ministerial Resolution No. 086–2020-Housing of $ 10,722.43 US dollars would only be enough for VUF 01.

It is necessary to highlight that the bonds assigned to the population in need with the amounts assigned are incorporating 100% of the amount in the construction of a house, and under these conditions the recipient population can have a house without any particular investment contribution from the beneficiary. At present, it is usual for the beneficiary who receives these bonds to make a larger contribution to the received bonus, in some cases it can exceed up to five times the amount, and in many cases the desired house is not completed.

The correlation analysis of the total cost and the construction area of the VUF has a parallel, increasing linear trend and the coefficient of determination is very high and is the same value as that obtained in the direct cost.

The trend line of the direct cost, the coefficient of determination, is y = 257.70x + 860.99 R^2 = 0.9906. The trend line of the Total cost and the coefficient of determination is y = 364.91x + 1219.2 R^2 = 0.9906.

It is usual to calculate the cost trend in a production line from the direct cost, and on this amount an additional percentage is assigned to the value of the production line, which does not alter the cost trend. In our case, the value added to the specific direct cost is proportional and unalterable during the production and building processes, which is reflected in the trend lines and in the determination coefficient.

The direct cost per square meter of prefabricated housing, in US dollars ranges from $ 386.85 to $ 426.86, these amounts are low in comparison and are competitive costs with the real estate industry that exists in the city of Arequipa – Peru.

3.8 Cost separation and building times prefabricated VUF single-family housing and houses with VAC confined masonry

A comparison has been made between the design of a building with precast reinforced concrete elements and a traditional construction design that is known as confined masonry, it is a system that is traditionally used in Peru and Latin America.

The confined masonry is defined as that which is entirely bordered by elements of reinforced concrete (except for the foundations that can be made of cyclopean concrete and in other cases it is made of reinforced concrete), emptied after the masonry wall has been built and with a distance between columns that does not exceed more than 2 times the height of the floor. It is important to follow the construction sequence indicated so that the confinements adhere to the masonry and form a whole that acts in an integral way [11].

For this reason, six single-family houses built with confined masonry have been designed that have similar characteristics to those of single-family houses with prefabricated elements. That is to say, the architectural plans are similar with regard to the space of the architectural environments, the difference is found in the width of the walls in confined masonry, the width of the wall is 15 cm and in the precast it is 10 cm.

From which the anti-seismic structural calculations with response of equal magnitude have been modeled. Once this similarity was achieved, the metric analysis was then developed, and it was processed in the S10 cost and budget program, in the Microsoft Project and in the SPSS.

3.8.1 Comparison of building times VUF and VAC

The unit time of VUF Prefabricated Single Family Housing has been determined from 32.07 days in VUF-01 to 65.70 days in VUF 06 and the Time Adjusted with a Gantt chart has been determined from 4.00 days in VUF-01 to 6.00 days in VUF 06.

The unit time of the House built with VAC confined masonry has been determined from 61.11 days in VAC-01 to 136.39 days in VAC-06 and the Adjusted time with a Gantt chart has been determined from 35.38 days in VAC-01 to 79.20 days in VAC 06 (**Table 12**).

The shortest time required for the construction of houses is with prefabricated elements, it is due to the simplicity of the production and assembly process, whereas in confined masonry a dependent, time-consuming and stationary entangled process is required, which leads to lengthening the processes of completion of works.

Description	VAC areas	T. U. (Metrado/Ru) VAC days	T. A. Gantt VAC, days.	T. U. days/m^2	T. A. Gantt days/m^2
VAC-01	27.74	61.11	35.38	2.20	1.28
VAC-02	35.15	76.87	47.26	2.19	1.34
VAC-03	40.96	92.50	56.53	2.26	1.38
VAC-04	44.51	90.88	55.84	2.04	1.25
VAC-05	52.59	117.69	68.32	2.24	1.30
VAC-06	63.21	136.39	79.20	2.16	1.25

Table 12.
Unit time (TU) and adjusted time (TA) in VAC Gantt.

Houses with prefabricated elements require less time for their construction, in unit time values, it takes from 1.10 days/m^2 to 1.24 days/m^2. With time adjusted in Gantt, values lower than 0.10 days/m^2 to 0.15 days/m^2 are achieved.

On the other hand, in a confined masonry building process the time requirements/m^2 are higher, so in unit time 2.04 days/m^2 have been found to 2.24 days/m^2, in adjusted time it has been found from 2.04 days/m^2 to 2.26 days/m^2.

In the relationship between execution time and building surfaces for times adjusted by Gantt, for VUF, there is a polynomial trend of y = −0.0003x2 + 0.0876x + 2.004 and (R^2 = 0.9072) this trend is low in comparison to the VAC trend. For VAC, there is a polynomial trend of y = −0.0075x2 + 1.8927x – 10.777 (R^2 = 0.9879), this trend is very high compared to the VUF trend (**Figure 14**).

The trend of the construction processes of houses with confined masonry requires a longer time than for the processes of building houses with premanufactured elements. The VAC, the greater the construction area, the longer time is required for the building, whereas the VUF the building process in the

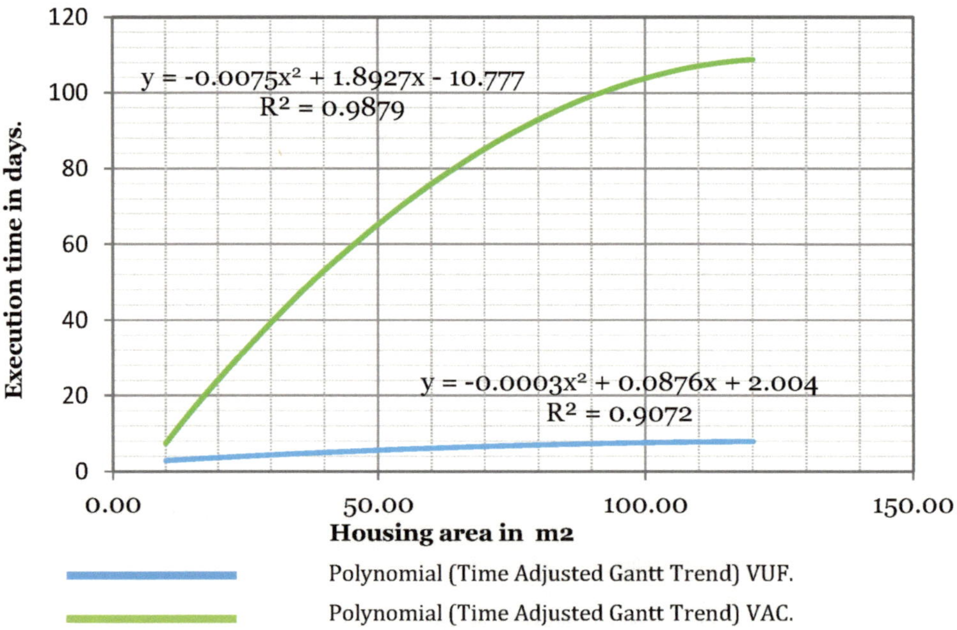

Figure 14.
Projected trend lines time/surface ratio of VUF and VAC.

surface variable there are no significant differences, this item is more dependent on the production capacity of machines and equipment.

3.8.2 Total costs VUF and VAC

From a common architectural design, the structural plans were developed and from which the metrics were measured and from which the housing costs were established, according to each type of housing designed, that is, 12 budgets were developed, with which the comparative analysis is carried out. By the comparative method, the principle of homogeneity and similarity was taken into account in order to make a comparison according to the proposed methodological design.

The design of items and costs of both building processes was developed from the same architectural design, expressed in a production line and trying to avoid falling into a comparative analysis of efficiencies, an aspect that is not considered in the research project. As a result of this process, the costs of the VUF and VAC single-family houses are presented (**Tables 13** and **14**).

Carrying out a conceptual contrast of the building processes by VUF precast elements in comparison with the construction processes by confined VAC masonry, the building with precast concrete elements exceeds in the reduction of time and costs, which has been reconfirmed in the calculations made (**Figure 15**).

According to the graphical representation of the costs of the VUF and the VAC, the cost difference is notable having the lower cost of the houses built with precast

Description	Cost in Soles	Cost in Dollars	VUF areas	Cost per m^2
C.T. VUF-01	38,689.65	11,025.83	25.83	426.86
C. T. VUF-02	46,745.11	13,321.49	33.39	398.97
C. T. VUF-03	54,390.61	15,500.32	39.06	396.83
C. T. VUF-04	55,474.62	15,809.24	42.21	374.54
C. T. VUF-05	72,608.36	20,692.04	52.29	395.72
C. T. VUF-06	81,243.50	23,152.89	59.85	386.85
T/C: 3,509 to 26 June 2020, SUNAT-PERU.				

Table 13.
Costs in suns and dollars of one (VUF).

Description	Cost in Soles	Cost in Dollars	VAC areas	Cost per m^2
C.T. VAC-01	56,959.17	16,232.31	27.74	585.16
C. T. VAC-02	70,700.67	20,148.38	35.15	573.21
C.T VAC-03	84,752.67	24,152.94	40.96	589.67
C.T. VAC-04	82,244.61	23,438.19	44.51	526.58
C.T. VAC-05	108,023.74	30,784.76	52.59	585.37
C.T. VAC-06	126,326.68	36,000.76	63.21	569.54
T/C: 3,509 to 26 June 2020, SUNAT-PERU.				

Table 14.
Costs in suns and dollars of an VAC.

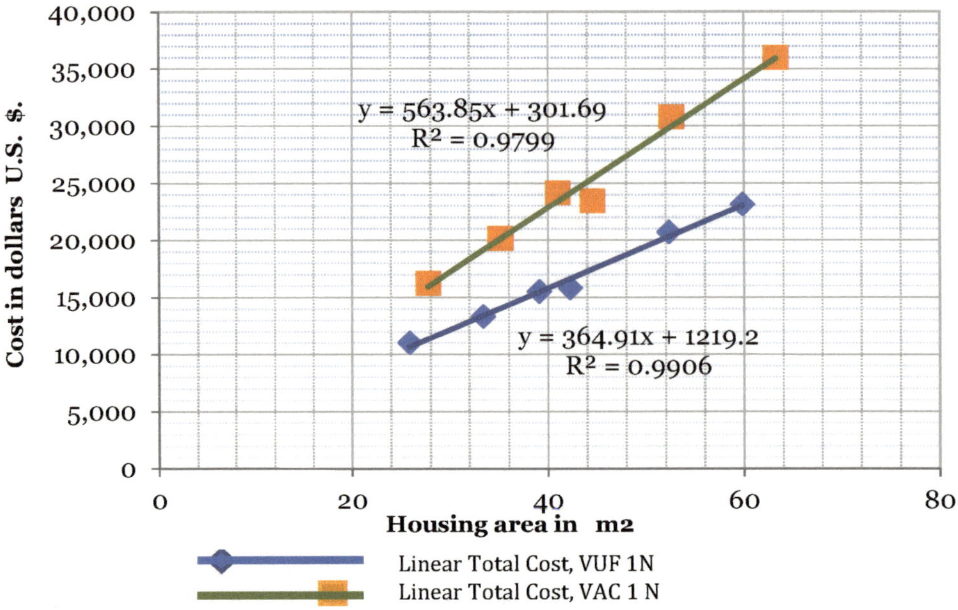

Figure 15.
Linear Total cost of VUF and VAC.

VUF concrete elements. The relation of total costs and surfaces for building processes, presents the following trend.

For VUF, a linear trend of y = 364.91x + 1219.2 (R^2 = 0.9906) is presented. This trend is very low compared to the VAC trend. Which shows that the VUF building costs are below the costs of a confined masonry building process.

For VAC, there is a linear trend of y = 563.85x + 301.69 and (R^2 = 0.9799) this trend is very high compared to the VUF trend. Which implies that VAC costs are high compared to VUF costs (**Figure 16**).

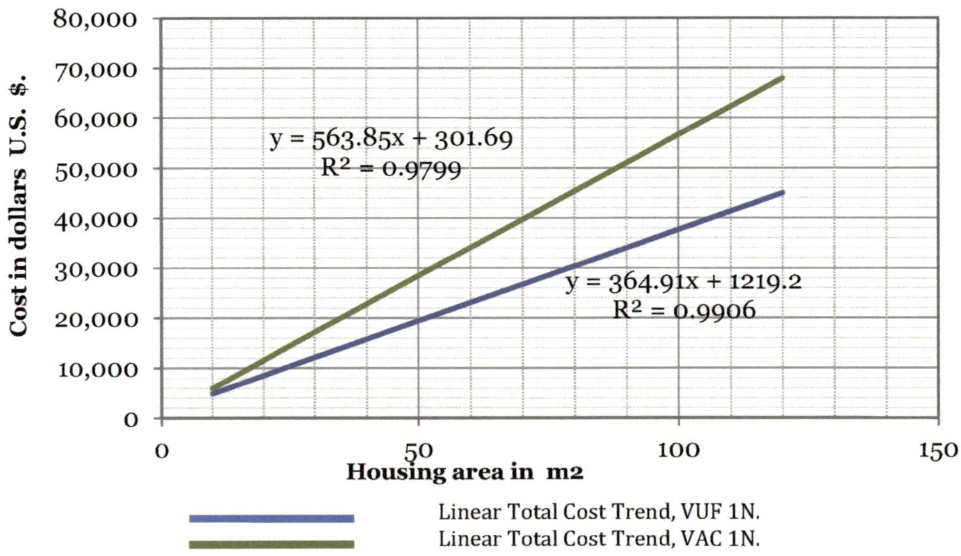

Figure 16.
Projected trend lines of cost/surface of VUF and VAC.

The projected trend of a VAC confined masonry building process exists a direct relationship and is determined by a coefficient of 0.9799 between cost in dollars and housing surface, which implies that the greater the built surface, the greater the investment and the slope of growth are required. is high. The trend of a VAC process significantly exceeds in costs a building process with VUF prefabricated elements, which allows confirming the efficiency of the VUF building process.

4. Conclusions

- Six types of basic single-family houses have been designed and are in accordance with the standards and are accessible to low-income economies. Four types of Housing to be served by the emergency bond granted by the Peruvian government in emergency situations and in need of housing.

- Parts and elements were designed for the assembly of prefabricated houses, seeking efficiency in terms of resistance and investment costs.

- A precast elements production line process was designed with minimal equipment that can be set up anywhere, even in post-emergency situations. The basic and fundamental criterion that the manufactured parts do not exceed the capacity of the size of the manufacturing, transport and assembly equipment.

- A process analysis was examined and a simple process was designed for the assembly of buildings, a minimum period of construction of a prefabricated house of 2 hours was determined at any time of the year.

- Lowest cost of a manufactured house fabrication and assembly system. Direct cost in VUF 04 has been achieved at a cost of US $ 264.50 per square meter and total costs of US $ 374.54 per square meter.

- There is a personal conviction to develop and make available a technology, techniques and processes that can be used by the various entities that assist the population in need of housing. With which families can have access to a basic house or to solve the effects of a post-emergency situation. We hope to be able in the future to disseminate the knowledge achieved, after putting it into practice.

Author details

Guillermo Yorel Noriega Aquise
Catholic University of Santa María, Arequipa, Peru

*Address all correspondence to: gnoriega@ucsm.edu.pe; yornoriegaa@gmail.com; gmonoriegae@gmail.com

References

[1] J. Timaná and M. d. P. Castañeda, Factores Determinantes en la selección de Vivienda Social en el Perú: el Caso de Chincha, Lima Peru; Determining Factors in the Selection of Social Housing in Peru: The Case of Chincha, Lima Peru: ESAN University, November 2019.

[2] M. L. Rossel, "Nueve de cada 10 viviendas de América Latina y el Caribe son de Baja calidad," "nine out of 10 houses in Latin America and the Caribbean are of low quality," El Pais America, October 28, 2018.

[3] J. Zuñiga, "Déficit habitacional en el Perú,", "housing déficit in Perú," Successful Peru News, Lima Peru, June 29, 2019.

[4] Exitosa Noticias, "Alcalde de Ica: Nada se ha hecho por Pisco a 10 años del terremoto," "Mayor of Ica: Nothing has been done for Pisco 10 years after the earthquake," August 15, 2017. [Online]. Available: https://elcomercio.pe/peru/arequipa/dos-anos-sismo-colca-familias-ichupampa-continue-living-modules-news-546874.

[5] Z. Condori, "A dos años del sismo en el Colca familias de Ichupampa siguen viviendo en módulos," "two years after the earthquake in Colca, families from Ichupampa continue to live in modules," Diario el Comercio, Lima Peru, August 15, 2018.

[6] SUNAT TIPO DE CAMBIO, "SUNAT – Tipo de Cambio Oficial," SUNAT TYPE OF EXCHANGE, "SUNAT – Official Exchange Rate 2020. [Online]. Available: https://e-consulta.sunat.gob.pe/cl-at-ittipcam/tcS01Alias. [Accessed 26 junio 2020].

[7] J. J. Fuentes Romero and V. Rodriguez Fernandez, "Una revisión bibliográfica de los estudios comparativos: su evolución y aplicación a la ciencia," "A bibliographic review of comparative studies: its evolution and application to science," Coruña Spain, October 19, 2019.

[8] C. Gomez Diaz de Leon and E. A. De Leon de la Garza, "MÉTODO COMPARATIVO," "COMPARATIVE METHOD," Autonomous University of Nuevo León (UANL) (Monterrey, Mexico), Monterrey Mexico, 2014.

[9] Normas legales El peruano, "Resolucion Ministerial N° 012-2018-VIVIENDA.," Diario Oficial del Gobierno del Peru, Lima Peru, 17 enero 2018.

[10] Normas Legales El peruano, "Resolución Ministerial N° 086-2020-VIVIENDA," "Ministerial Resolution No. 012-2018-HOUSING.," Official Gazette of the Government of Peru, Lima Peru, January 17, 2018.

[11] INSTITUTO CAPECO, Albañilería confinada. Confined Masonry. Maestro builds well, Lima Peru, 15 05 2018.

Chapter 16

Evaluation of Well-Being and Thermal Comfort of the LAD-MA Construction System for Low-Cost Homes

Rodolfo Jiménez Cavieres, Javier Carrasco Eade and Camilo Valdebenito Monsalve

Abstract

This work is part of a research into the state of conservation and behavior of a group of self-built social housing. The construction, which dates from 1990, was carried out with an original low-cost construction system that uses clay and wood bricks called LAD-MA. This was implemented by the NGO Urban Technical Assistance Center "Taller Norte", in the Peñalolén commune, Santiago de Chile, Metropolitan Region. The study focuses on the evaluation of well-being and thermal comfort in these homes, which is determined through environmental monitoring by meteorological stations installed for six months in 4 homes. It is established that the houses do not comply with the parameters set up by the international standards ISO 7730 and ASHRAE 55. For this, constructive solutions are proposed to thermally improve the current houses, and update the LAD-MA construction system to comply with the thermal Insulation standards stipulated for the Sustainable Housing Certification of the Ministry of Housing and Urbanism.

Keywords: social housing, low-cost construction system, environmental comfort assessment

1. Introduction

In the mid-eighties of the twentieth century, in Chile, around 29,373 families were living in 166 registered irregular settlements (camps), located on the outskirts of different cities in the country [1]. The people who lived in these camps, mainly inhabited houses made of light material, popularly known as "mediaguas", with an area of approximately 18 m2 that, due to their size and materiality, presented habitability problems for families, such as overcrowding, thermal and humidity discomfort, lack of lighting and basic hygiene services [2]. This problem led the authorities of the time to implement urbanization programs that consisted of building basic kitchen and bathroom modules called "sanitary booths" that provided the supply of drinking water, sewage service, and electricity. For that, the inhabitants of Camps had to implement other enclosures such as bedrooms and living room, through self-construction. In this context, an NGO called "TALLER NORTE Urban

Technical Assistance Centre", in conjunction with organizations from the Peñalolén commune, develops a program called "Building Together" in which various constructive solutions are implemented, the first experiences being the construction with wood and clay. Towards the end of the '80s, Taller Norte developed a construction system that uses handmade brick from baked clay and wood for the self-construction of houses attached to sanitary huts called "LAD-MA" [3]. Because it is made up of bricks (**Lad**rillo) and wood (**Ma**dera), it allows the implementation of 1-level homes, expandable to 2 levels, depending on the needs and possibilities of each family.

Despite the advancement of housing policies in Chile during the last decades, irregular settlements and their consequent habitability problems are far from disappearing. According to official figures from the cadastre carried out by the Ministry of Housing and Urbanism of Chile [4], there are currently 802 camps nationwide in the country's main cities, with the presence of 47,050 households. Among the reasons that explain the persistence of the camps are the lack of state regulation of the real estate market, the effects of socioeconomic inequality and low salaries, as well as migration [5] from other countries and cities.

The foregoing makes it possible to wonder about the viability of implementing technically assisted self-construction programs, which facilitate efficiently solving the current lack of habitability in the context of irregular settlements. In these programs, there is an opportunity to implement construction systems like LAD-MA, developed 30 years ago, which uses inexpensive materials such as wood and fiscal bricks and is easy to execute.

In this sense, it is worth asking what are the technical gaps in a construction system such as LAD-MA regarding compliance with contemporary standards of habitability, taking into account that in recent years, Chile has presented significant advances in the matter. Specifically, it is interesting to review if the LAD-MA system allows acceptable conditions of thermal comfort according to standards such as ASHRAE 55, as well as, if technically, it is possible to adapt the construction system to achieve compliance with the current Thermal Regulations for homes of Chile [6].

2. Background

2.1 The LAD-MA construction system

The LAD-MA system consists of a set of 4 wall panels (**Figure 1**), which can be arranged in order to solve different architectural solutions. They are made up of a structural framework of upright feet and upper sills of untreated 2x4 "pine wood, braced with 1x4" pine diagonals, which act jointly with a body of prefabricated plates made with 'fiscal brick'.

The brick plates are fixed to the wooden structure and to each other, using cement mortar of 400 [kg/cm3] and with a horizontal reinforcement for the shear stress, with smooth round iron of 6 [mm], fixed at end of the wall panel with a fold and staples, fixed and tensioned at the other end with nuts and washers.

An 8 [mm] slotted steel bar is used to join the wooden structure with the foundation beam, one at each end of the wall panel, with an upper tie-down with clamps and two bars in the central columns as seen in (**Figure 2**).

On-site, the foundations are made directly on excavation and without molds. The foundation beams are precast on-site using molds containing guides to locate the center columns' 8 [mm] *Feith* anchors. The foundation beams are mounted on the foundation dice, perfectly level on their upper face. A wooden floor structure is

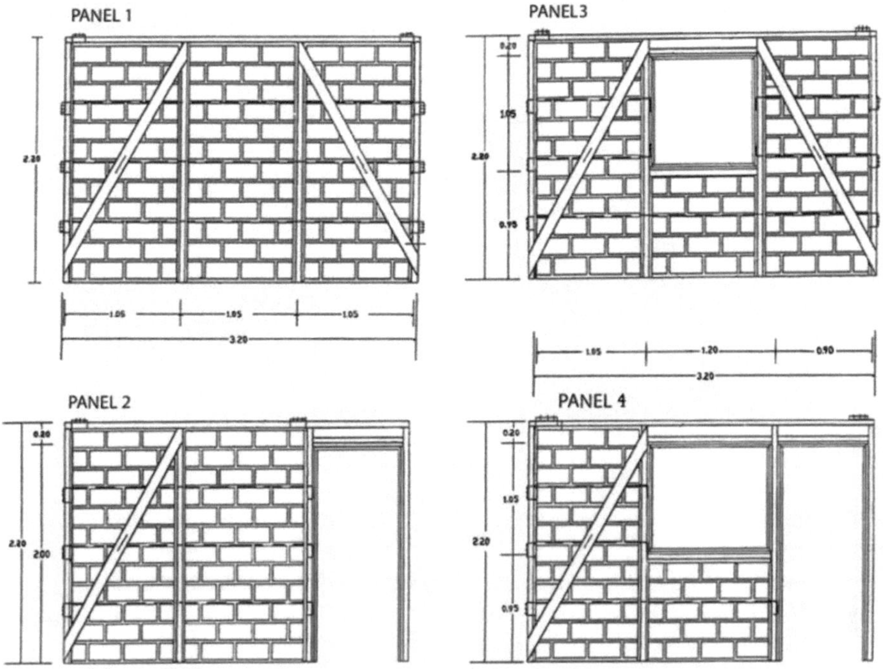

Figure 1.
LAD-MA set of wall panels.

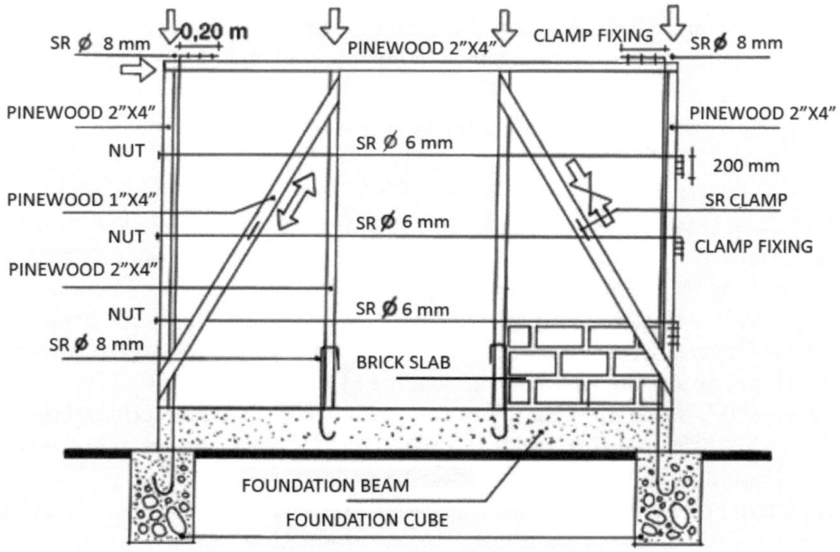

Figure 2.
Schematic detail of the parts and components of the LAD-MA construction system.

arranged on the LAD-MA structure, to enable a second floor, which is entirely resolved with a traditional wooden structure that in many cases comes from the recycling of precarious homes (**Figure 3**).

This construction system was designed with a purely social and non-profit objective, it is low cost, enables and facilitates self-construction work, is based on the prefabrication of components, uses low-cost local materials and recycled

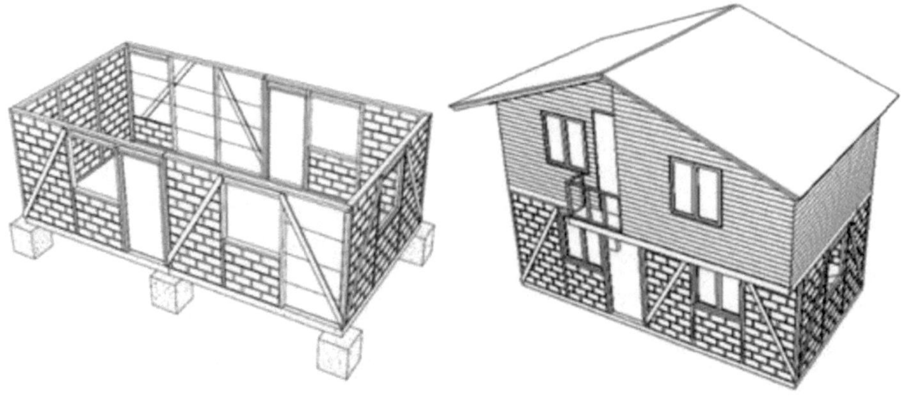

Figure 3.
LAD-MA Architectural Module.

materials, and is intensive in unskilled labor, with technical support that guarantees its proper implementation.

In addition, the LAD-MA construction system was conceived considering that it provides a positive perception in the beneficiaries of a "solid" home, using brick as a material commonly used in permanent homes, giving the feeling of durability, thermal comfort, and structural stability. In contrast to the "mediaguas" units in which the beneficiaries of the construction system previously lived.

It is worth mentioning that for this research, it was analyzed the evolution over time of 4 homes built in the LAD-MA system, which are located in the Peñalolén Commune, in Santiago de Chile. In some of the observed cases, the owners made modifications, making extensions and plastering the interior and exterior of the walls with cement mortar, and painted the outside and the inside.

2.2 The current standards of habitability of housing in Chile

The regulation of building standards in Chile arose in 1992 with Decree No. 47 that created the General Code of Urbanism and Constructions [7]. The first topics that were regularized through this document were the fire resistance standards of buildings and the requirements for sewage and rainwater systems. Then, in the year 2000, the first stage of prescriptive Thermal envelope regulations for housing was added to the General Code, to establish a minimum insulation standard for the roof of the dwelling, according to the climatic zone of the country.

Then, in 2007, the second stage of prescriptive Thermal envelope regulation expanded the requirements to other building components, such as exterior walls, ventilated floors, and windows of the dwellings.

A third stage of the Thermal envelope Regulation is pending, however, it is known that it will consist of an adjustment and reduction of the prescriptive U-Value of some of the building components, for some of the climatic zones of the country. These adjustments have already been applied on a mandatory basis for new homes in cities of Chile that currently have Atmospheric Decontamination Plans (PDA), as well as in a voluntary way, through the Certification System for Sustainable Housing [6].

The following **Table 1** presents a comparison between the current thermal envelope standards of the Chilean Regulation and the thermal envelope standards from the Certification System for Sustainable Housing in Chile:

Table of comparison. Current Chilean Building regulation for thermal insulation standards and Minimum thermal insulation standards complying with the Sustainable Housing Certification				
Roofs				
City of reference	Current Chilean Building regulation for thermal insulation standards		Minimum thermal insulation standards complying with the Sustainable Housing Certification	
	Thermal Zone	U Value (W/m2k)	Thermal Zone	U Value (W/m2k)
Arica	1	0,84	A	0,84
Iquique				
Antofagasta				
Copiapó			B	0,47
La Serena			C	0,47
Valparaíso	2	0,6		
Santiago	3	0,47	D	0,38
Rancagua				
Talca	4	0,38		
Concepción			E	0,33
Temuco	5	0,33	F	0,28
Valdivia			G	0,28
Puerto Montt	6	0,28		
Coyhaique	7	0,25	I	0,25
Punta Arenas				
Walls				
City of reference	Current Chilean Building regulation for thermal insulation standards		Minimum thermal insulation standards complying with the Sustainable Housing Certification	
	Thermal Zone	U Value (W/m2k)	Thermal Zone	U Value (W/m2k)
Arica	1	4	A	2,1
Iquique				
Antofagasta				
Copiapó			B	0,8
La Serena			C	0,8
Valparaiso	2	3		
Santiago	3	1,9	D	0,8
Rancagua				
Talca	4	1,7		
Concepción			E	0,6
Temuco	5	1,6	F	0,45
Valdivia			G	0,4
Puerto Montt	6	1,1		
Coyhaique	7	0,6	I	0,35
Punta Arenas				

Table of comparison. Current Chilean Building regulation for thermal insulation standards and Minimum thermal insulation standards complying with the Sustainable Housing Certification				
Ventilated floors				
City of reference	Current Chilean Building regulation for thermal insulation standards		Minimum thermal insulation standards complying with the Sustainable Housing Certification	
	Thermal Zone	U Value (W/m2k)	Thermal Zone	U Value (W/m2k)
Arica	1	3,6	A	3,6
Iquique				
Antofagasta				
Copiapó			B	0,7
La Serena			C	0,87
Valparaíso	2	0,87		
Santiago	3	0,7	D	0,7
Rancagua				
Talca	4	0,6		
Concepción			E	0,6
Temuco	5	0,5	F	0,5
Valdivia			G	0,39
Puerto Montt	6	0,39		
Coyhaique	7	0,32	I	0,32
Punta Arenas				

Source: Own elaboration based on Art. 4.1.10 of the O.G.U.C and the Sustainable Housing Construction Standards, volume II, Energy [6].

Table 1.
Comparative table of thermal transmittance standards for ventilated roofs, walls, and floors.

2.3 Thermal Comfort

Considering thermal comfort as one of the main variables in building design, standards such as ISO 7730 and ASHRAE 55 have been developed. These Standards are used as a reference to determine the performance of buildings, through measurement tools. In the case of Chile, the ASHRAE standard has been used as a reference for the design of the housing energy-rating tool [8].

2.3.1 Standard UNE-EN ISO 7730: 2006

The purpose of the standard is to predict the thermal sensation and the degree of discomfort within a built environment, by calculating a Predicted Mean Vote (PMV) and an Estimated Percentage of Dissatisfied (PPD), taking into account levels of clothing and metabolic activity of people, as well as wind speed, turbulence percentage, among other parameters.

The PMV is an index that reflects an average of votes cast by a large group of people concerning a 7-level thermal sensation scale, which is expressed in **Table 2**.

The PMV index can be estimated for different combinations of metabolic rate, clothing insulation, air temperature, mean radiant temperature, relative air velocity, and air humidity; for the effect of the purpose of the standard, the following simplified expression is used:

$$PMV = aT + bPv - c$$

Where T is the ambient temperature in [°C] and Pv the pressure of the water vapour in the environment in [kPa].

The constants a, b and c are constants that relate the physical quantities of temperature and pressure to obtain the PMV which is a dimensionless variable and these are obtained from the following table, depending on the time of exposure to the indoor environment and depending on the gender of the subject (**Table 3**).

On the other hand, the PPD index is determined based on the PMV expressed in the following equation:

$$\text{PPD} = 100 - 95\text{e} - \left(0.3353\text{PMV}^4 + 0.2179\text{PMV}^2\right)$$

2.3.2 ASHRAE 55

This standard determines the influence of environmental variables on human comfort. Although these variables are the same used in the ISO 7730, the difference is that ASHRAE 55 seeks to determine the comfort temperature ranges and then determine the PMV-PPD values, thus deriving in two ways of determining the thermal comfort based on De Dear's studies.

The ASHRAE Standard provides two approaches to thermal comfort on buildings. The first approach focuses on buildings with centralized HVAC systems,

+3	**hot**
+2	warm
+1	slightly warm
0	neutral
−1	slightly cool
−2	cold
−3	chill

Table 2.
Scale of 7 levels verbalized according to thermal sensation, PMV (UNE-EN ISO 7730: 2006).

Time	Gender	a	b	c
1 hour	Male	0.220	0.233	5.673
	Female	0.272	0.248	7.245
	Both	0.245	0.248	6.475
2 hour	Male	0.221	0.270	6.024
	Female	0.283	0.210	7.694
	Both	0.252	0.240	6.859
3 hour	Male	0.212	0.293	5.949
	Female	0.275	0.255	8.620
	Both	0.243	0.278	6.802

Table 3.
Values of constants a, b and c to be used in the PMV estimation equation.

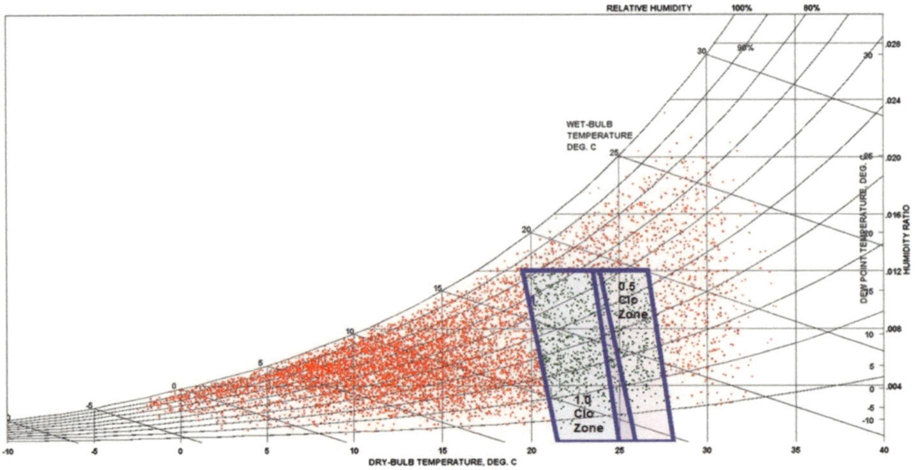

Figure 4.
Thermal Comfort Range for Santiago, Chile, based on ASHRAE 55 Standard. Source: own elaboration using climatic data from Santiago (855740 WMO Station Number) plotted on the Psychrometric Chart provided by the software Climate Consultant.

considering an airspeed of 0.2 [*m/s*], a sedentary metabolic activity between 1 [*met*] and 1.3 [*met*], and the option of insulation of clothing varies between 0.5 [*clo*] and 1.0 [*clo*], similar to that described above, corresponding to summer and winter respectively. Based on the Fanger thermal balance, a hygrothermal comfort range illustrated in **Figure 4** is determined.

The second method focuses on buildings without centralized HVAC systems, and determines a dynamic comfort temperature based on the average ambient temperature outside the buildings, this continues to maintain a requirement that there be a sedentary metabolic activity between 1 [met] and 1.3 [met], and that people can vary the clothing insulation between 0.5 [*clo*] and 1.0 [*clo*]. However this method is valid only if the outside temperature oscillates between 10 [°C] and 33.5 [°C], and that their measurements are greater than 7 days and a maximum of 30 days, thus generating the minimum and maximum comfort temperature equations.

$$Tmín,\ máx = 0.31 \cdot TExt + 17.80 \pm 3.50$$

Regarding humidity, the standard establishes a maximum humidity radius of 0.012, equivalent to the vapor pressure of 1.910 [kPa], and does not determine a minimum. Also, consider an acceptability index of 80%.

3. Methodology

The study set out to determine if the homes built using the LAD-MA system, in the mid-90s, meet the contemporary comfort parameters described in the international standards ISO 7730 and ASHRAE 55. Given the results obtained, possible modifications to the construction system were studied, to optimize its performance and conform to the thermal envelope standards for housing. The above, understanding that the LAD-MA system was conceived as a self-construction method, economically accessible for families in conditions of socioeconomic vulnerability, who require housing solutions that can be extended over time; a problem still in force at the local level.

The study has been composed of two stages. The first stage, of an empirical nature, consisted of measuring environmental parameters in four existing houses built under the LAD-MA construction system and located in the Peñalolén Commune in Santiago. This measurement was carried out by data loggers that assessed internal and external environmental parameters for six months. The external parameters measured were temperature and humidity, while the internal parameters were temperature, humidity, CO2 concentration, and noise. The data obtained through these data loggers allowed us to determine if the homes achieve thermal comfort standards, based on the PMV and PPD indicators, analyzing these according to the parameters established in the international standards ISO 7730 and ASHRAE 55. The measurements were analyzed based on periods of continuous occupation (24hrs), and limited periods of occupation (only the hours of occupation) under both standards. It is worth mentioning that some of the houses studied made use of heating systems during the winter period, so the measurement does not strictly reflect passive environmental conditions of the construction.

In the second stage of this study, possible modifications to the LAD-MA construction system were analyzed, in order to adjust the system to comply with the Chilean Thermal Regulations and thus, theoretically improve its compliance with the PMV and PPD indicators. To carry out this part of the study, possible constructive solutions were proposed to reduce the thermal transmittance of the LAD-MA system, using the static energy simulation tool Therm 7.7. The constructive adjustments were then represented in a dynamic thermal model of the LAD-MA architectural module, using the energy simulation tool DesignBuilder 6.1 and the Energy Plus 8.9 calculation engine, thus determining interior temperatures and the variation of PMV and PPD indicators.

4. Discussion of the results

4.1 LAD-MA housing monitoring

NET ADMO environmental measurement equipment was installed in 4 homes originally built with the LAD-MA system. It is worth mentioning that the 4 LAD-MA homes monitored present differences in their state of conservation and their current architectural configuration due to the different modifications and extensions that users have made since they were delivered. The first and second level extensions that have been adhered to the original LAD-MA modules are wooden structures that have been progressively built by the owners, in general with a low thermal standard. However, they have generated an effect on the environmental conditions of the original module. **Table 4** presents a brief architectural characterization of the 4 monitored dwellings, highlighting in color the location of the original LAD-MA module in each case, the orientation of the dwelling relative north, and the description of the general state of conservation in each a. case.

The results of the environmental measurement of the 4 LAD-MA houses in Peñalolén, showed that they present minimum periods of thermal comfort, which are reduced towards the winter months. The average temperature in the monitored homes during February is 24°C, while the average temperature barely reaches 14°C in July. On the other hand, the indoor relative humidity oscillates within acceptable levels. **Table 5** presents graphs of the average monthly conditions of temperature, and relative humidity, together with some general observations.

Finally, when analyzing the Percentage of People in Thermal Dissatisfaction through the PPD indicator, during the months in which the LAD-MA dwellings were monitored, it is observed that the dwellings present high levels of dissatisfaction. On average, under the parameters of the ISO 7730 standard, the 4 thermally monitored homes have 75% thermal discomfort in the total measurement period

Homes built in the LAD-MA system monitored - Peñalolén, Santiago			
Monitored housing - M1 - Panes Family			
1st level plan	2nd level plan	Exterior view of the house	General description
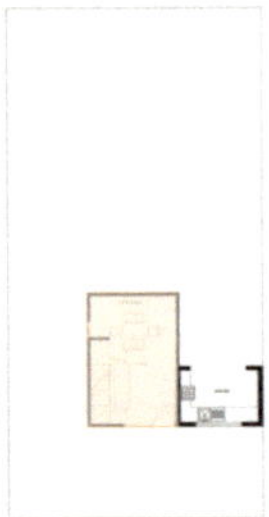	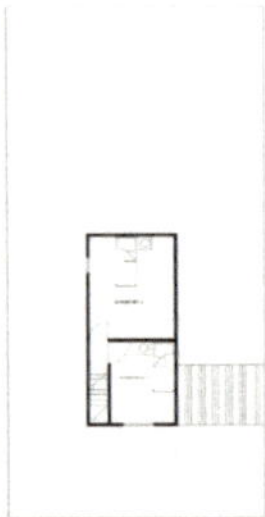		It does not have modifications or maintenance of the LAD-MA panels, even generating deterioration of it. The environment of the module is mainly closed, avoiding good ventilation of the space.
Monitored housing - M2 - Sandoval Family			
1st level plan	2nd level plan	Exterior view of the house	General description
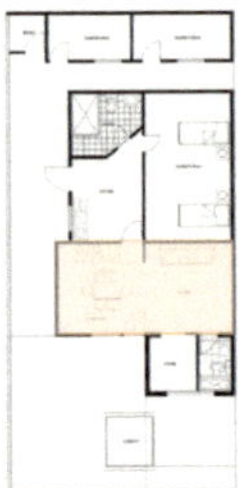	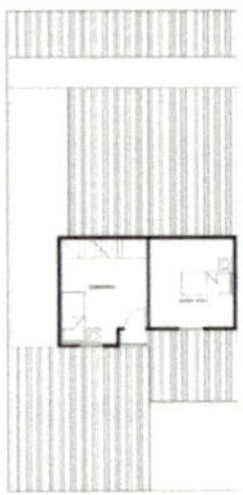		Modifications were made, plastering the inside of the panels with cement, and painting them outside and inside. Maintenance is carried out every 3 to 5 years. Closed environment, avoiding good ventilation Ceramic floor
Monitored housing - M3 – Lopez Family			
1st level plan	2nd level plan	Exterior view of the house	General description
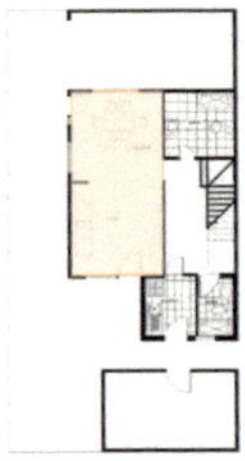			Modifications were made, plastering the inside of the panels with cement, and painting them outside and inside. Maintenance is carried out every 3 to 5 years. The house has a large patio and large windows, they also left space at the entrance, factors that give it very good ventilation. Ceramic floor
Monitored housing - M4 – Vargas Family			
1st level plan	2nd level plan	Exterior view of the house	General description
			Modifications were made, plastering the inside of the panels with cement, and painting them outside and inside. Maintenance is carried out every 3 to 5 years. The house is higher than the others, having a very good inlet for ventilation, however, the outlet to the patio is smaller, reducing the potential for air exchange. Ceramic floor

Table 4.
Characterization of the Homes built in the monitored LAD-MA System, Peñalolén, Santiago. Own elaboration.

Comparison of Average Indoor Temperature and Relative Humidityin Monitored Homes Built with the LAD-MA System		
February		
Temperature	Relative Humidity	Observations
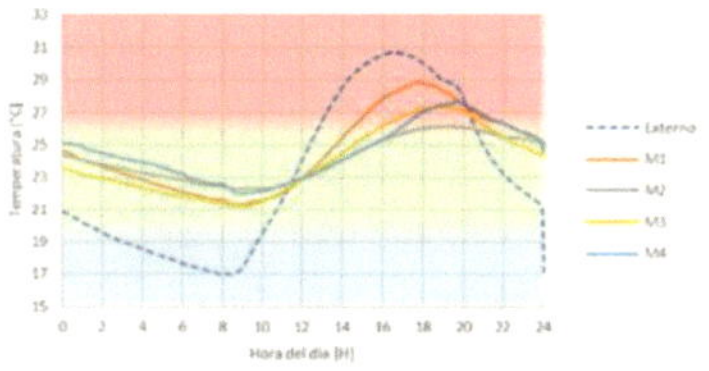	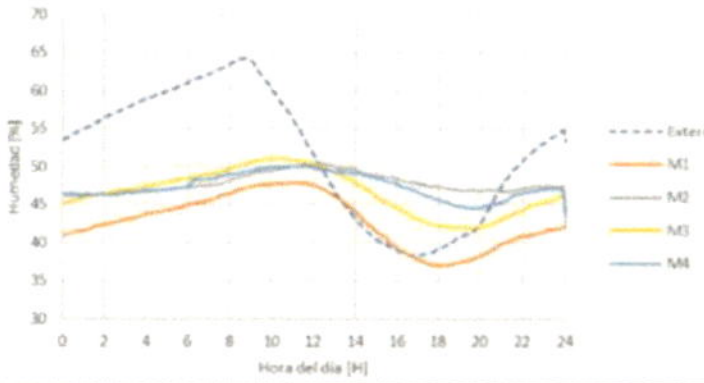	The homes have temperatures slightly below the minimum comfort level, with M1 being the coldest home and the one with the highest indoor temperatures during the day. Indoor humidity in all homes is within acceptable levels.
April		
Temperature	**Relative Humidity**	**Observations**
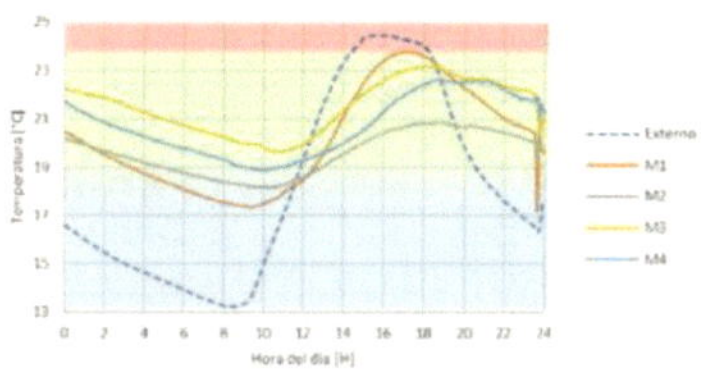	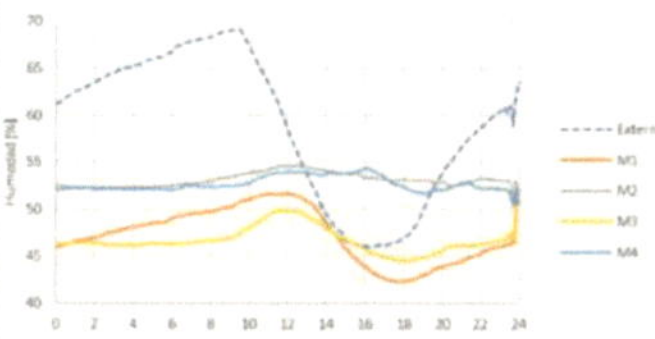	The homes have temperatures slightly below the minimum comfort level, with M1 being the coldest home and the one with the highest indoor temperatures during the day. Indoor humidity in all homes is within acceptable levels.
May		
Temperature	Relative Humidity	Observations
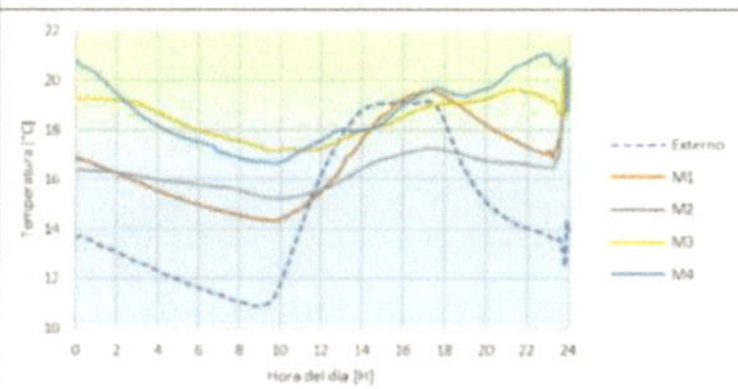	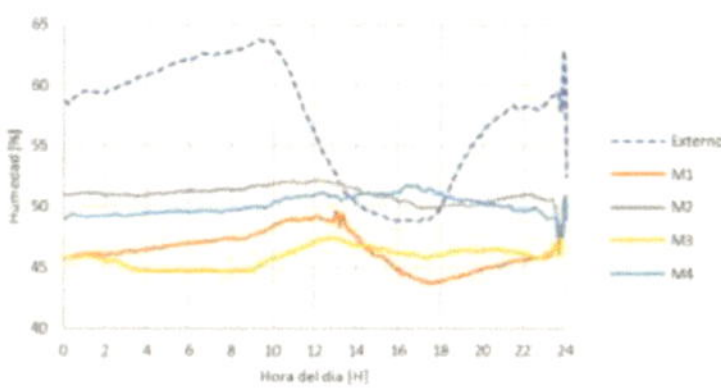	The dwellings have T ° below the minimum comfort level, with M1 and M2 being the coldest dwellings. M1 has higher indoor temperatures during the day. The use of heating is observed in M3 and M4. Indoor humidity is within acceptable levels.
June		
Temperatura	**Humedad relativa**	**Observaciones**
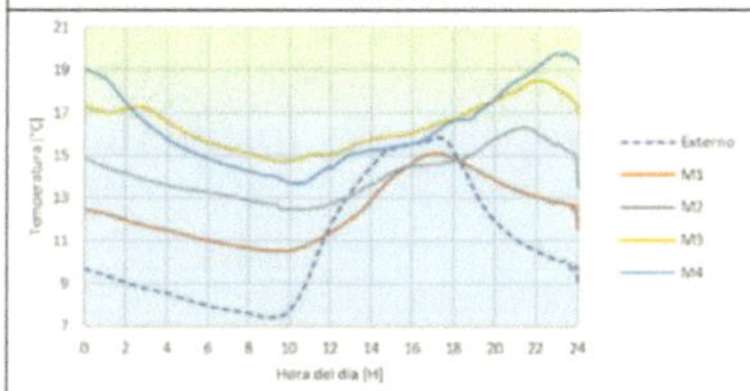	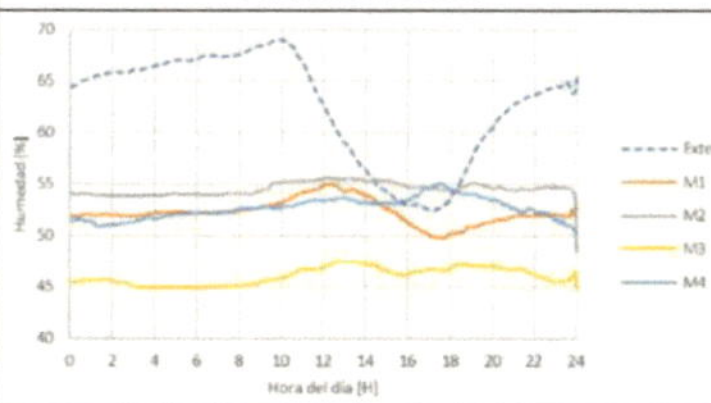	The houses practically do not present T ° within comfort ranges. M1 is the coldest dwelling. The use of heating is observed in M2, M3 and M4. Indoor humidity is within acceptable levels.
July		
Temperatura	**Humedad relativa**	**Observaciones**
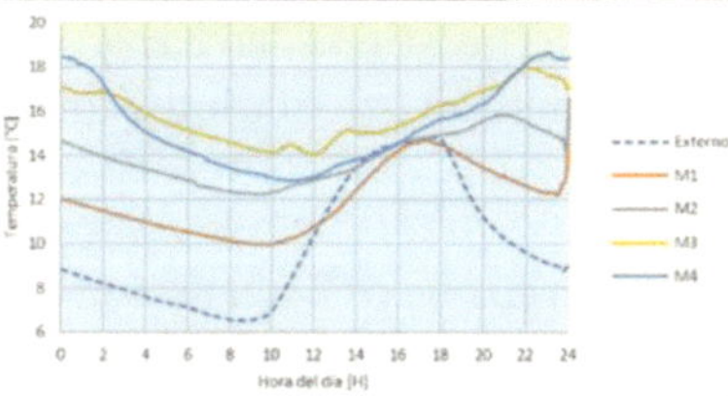	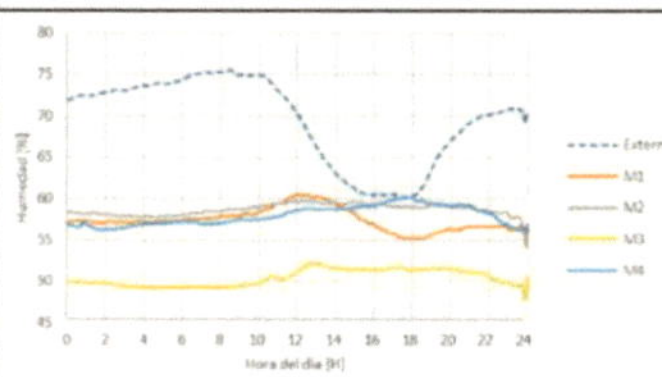	The houses practically do not present T ° within comfort ranges. M1 is the coldest dwelling. The use of heating is observed in M2, M3 and M4. Indoor humidity is within acceptable levels.

Table 5.
Comparison of Average Indoor Temperature and Relative Humidity in monitored LAD-MA homes. Own elaboration.

<table>
<tr><td colspan="2">Analysis of the period of discomfort of the monitored LAD-MA dwellings</td></tr>
<tr><td colspan="2">ISO 7730 Discomfort Model</td></tr>
<tr><td>Total period</td><td>Limited period of occupation</td></tr>
<tr><td>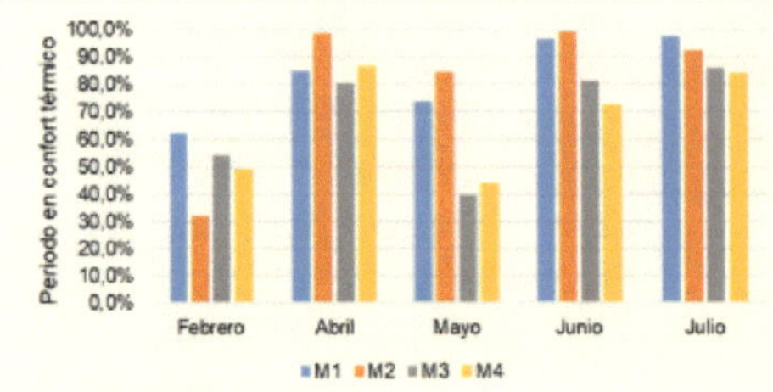
</td><td>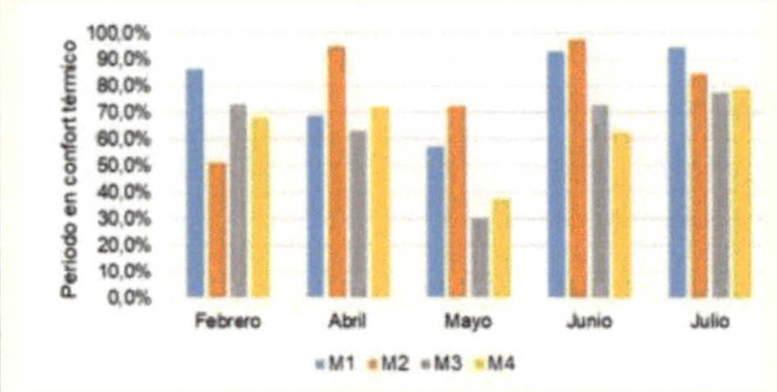
</td></tr>
<tr><td colspan="2">ASHRAE Discomfort Model 55</td></tr>
<tr><td>Total period</td><td>Limited period of occupation</td></tr>
<tr><td>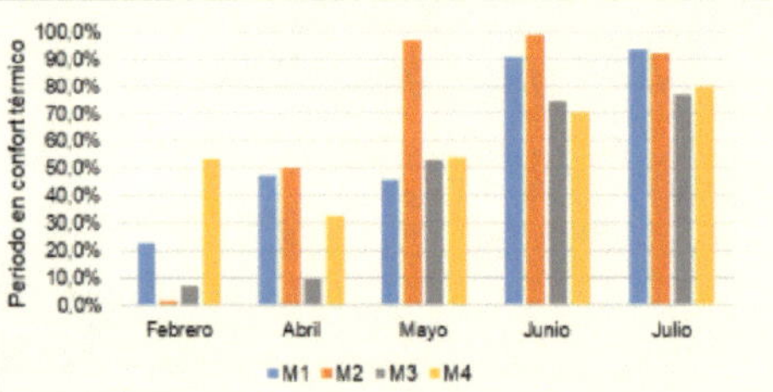
</td><td>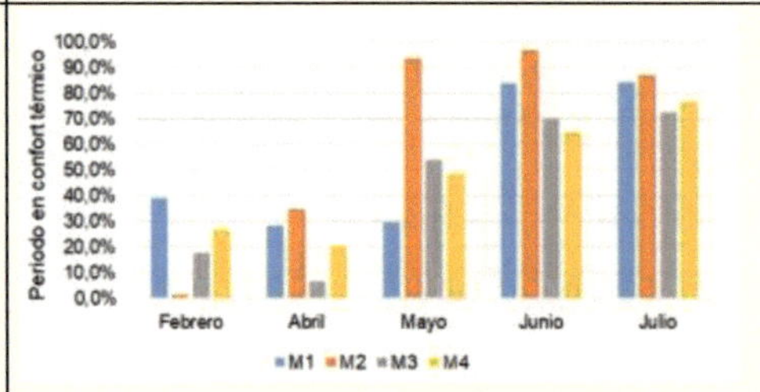
</td></tr>
</table>

Table 6.
Analysis of time of discomfort of the Monitored LAD-MA Homes. Own elaboration.

and 71% during the limited period of occupancy, while under the parameters of the ASHRAE 55 standard, the 4 monitored dwellings have 57.5% of thermal discomfort time in the total measurement period and 54% during the limited period of occupation. Towards the winter months, the period of discomfort increases between 80% and 90% of the occupation time, due to the low temperatures inside the houses. The above is observed in **Table 6**.

4.2 Thermal Transmittance Analysis of the LAD-MA construction system

The next stage of the study analyzes the thermal performance of the wall construction system of the LAD-MA module, together with the adjustments required so that the system can comply with the country's thermal regulations. To carry out this first part of the analysis, the Therm 7.7 thermal transmission calculation simulation tool was used to represent a typical section of the LAD-MA wall, composed of brick plates and wooden studs.

For analysis purposes, some characteristics of materials have been assumed based on the Chilean Standard NCh 853, which presents typical conductivities of materials according to their density. The edge conditions used have been 6.5°C outside and 25 W/m2K of air film coefficient on the outside, and 13°C inside, with 7.69 W/m2K of air film coefficient on the inside.

Thermal analysis of the original construction system was carried out, plus the analysis of 4 variants that represent possible improvements of the thermal envelope. The summary of transmittance and thermal resistance results of the variants studied is presented in **Table 7**.

The analysis of variants allows us to observe the alternatives for improving the thermal quality for the LAD-MA construction system, to the extent that stucco (variant 1 and 2) and thermal insulation could be progressively added to the walls (variant 3 and 4), in the context of a thermal improvement for said dwellings. From the point of view of thermal resistance, it is evident that the incorporation of thermal insulating materials contributes in a better way to the fulfillment of the thermal regulations for homes. The most recommended solution being the installation of thermal insulation on both sides of the wall, thus which the thermal transmittance of the wall would make it possible to comply with the normative requirements in force in 6 of the 7 thermal zones, and with the normative requirements of the third stage of

Constructive analysis of variants LAD - MA system		
LAD-MA Original System		
Transmittance U (W / m2K)	Resistance (m2K / W)	Commentary
3,23	0,31	The original system complies with the Thermal Regulations in force for Thermal Zone 1, in the north of the country. It does not comply with the Third Stage of the Thermal Regulation of Homes, in any of the Thermal Zones of the Country
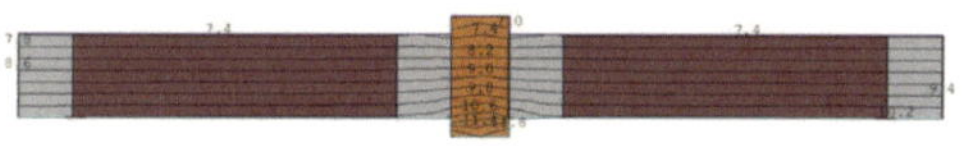		
Variant 1- LAD-MA + plus Internal Stucco		
Transmittance U (W / m2K)	Resistance (m2K / W)	Commentary
2,93	0,34	By adding mortar stucco on the outer face, 15mm thick, the transmittance is reduced by 9%. It complies with the Thermal Regulations in force for Thermal Zones 1 and 2, in the north of the country. It does not comply with the Third Stage of the Thermal
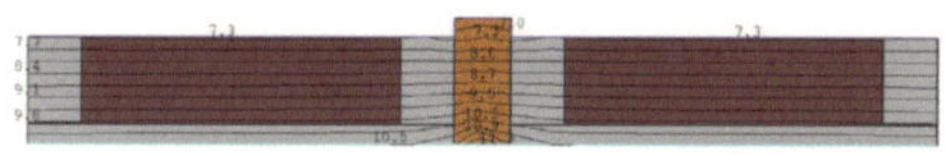		
Variant 2 - LAD-MA + Stucco both sides		
Transmittance U (W / m2K)	Resistance (m2K / W)	Commentary
2,70	0,37	By adding mortar stucco on both sides, 15mm thick, the transmittance is reduced by 16%. It complies with the Thermal Regulations in force for Thermal Zones 1 and 2, in the north of the country. It does not comply with the Third Stage of the Thermal Regulation of Homes, in any of the Thermal Zones of the Country
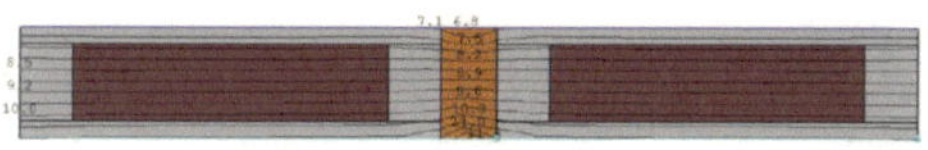		
Variant 3 - LAD-MA External stucco plus interior Mineral Wool + Wood cladding		
Transmittance U (W / m2K)	Resistance (m2K / W)	Commentary
1,38	0,72	By adding a 15mm thick thermal insulator, the transmittance is reduced by 57%. It complies with the current Thermal Regulations for Thermal Zones 1, 2, 3 and 4 of the country. Complies with the Third Stage of the Thermal Regulation of Homes, in Thermal Zone A, in the north of the country
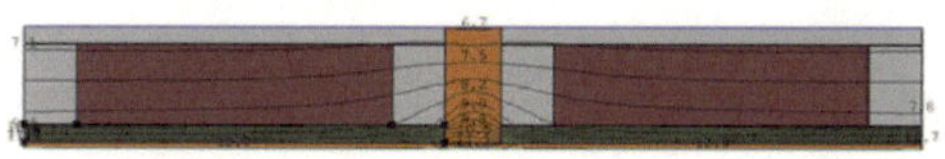		
Variant 4 - LAD-MA Construction System with mineral wool on the outside and inside		
Transmittance U (W / m2K)	Resistance (m2K / W)	Commentary
0,77	1,29	By adding 15mm thick thermal insulation on both sides of the wall, the thermal transmittance is reduced by 76%. It complies with the current Thermal Regulations for Thermal Zones 1, 2, 3, 4, 5 and 6 of the country. Complies with the Third Stage of the Thermal Regulation of Homes, in Thermal Zone A, B, C, and D of the Country
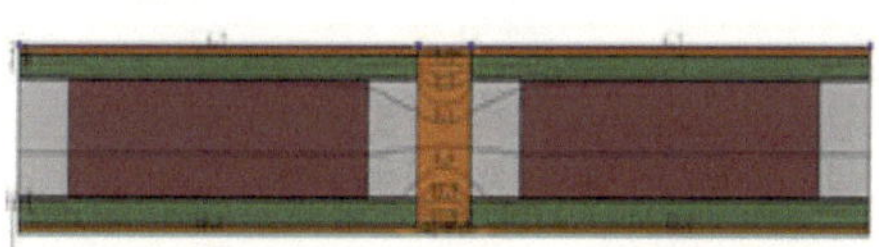		

Table 7.
Constructive analysis of variants System LAD - MA. Own elaboration.

the thermal regulation of houses, in 4 of the 8 thermal zones. It is worth mentioning that this thermal solution can be obtained in several ways, among others by using a solution of the STATE type (Exterior Thermal Insulation System). Better known in Chile by its acronym in English EIFS (Exterior insulation finishing system) that incorporate a plastered plaster on the outside, or failing that, by installing a secondary

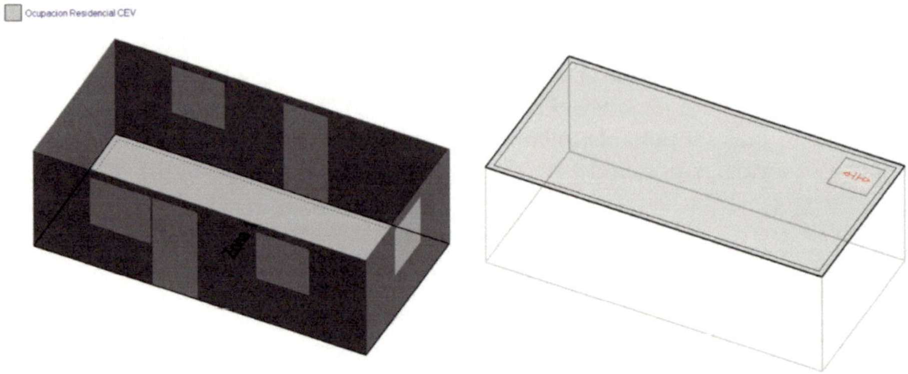

Figure 5.
Computational thermal model in Design Builder program. Own elaboration.

PMV and PPD indicators from the LAD-MA module, under the studied variants						
Variantes	PMV			PPD (%)		
	Winter	Summer	Annual	Winter	Summer	Annual
LAD-MA Original System	-2,4	-1,4	-1,9	78,2	61,1	69,7
Variant 1- LAD-MA + Stucco Internal	-2,4	-1,4	-1,9	78,8	59,9	69,4
Variant 2 - LAD-MA + Stucco both sides	-2,6	-1,7	-2,1	82,6	60	71,3
Variant 3 - LAD-MA External stucco, inner Mineral Wool + Timber finishing	-2,3	-1,6	-2	79,6	57	68,3
Variant 4 - LAD-MA System with mineral wool both sides	-2	-1,4	-1,7	73,3	53,1	63,3

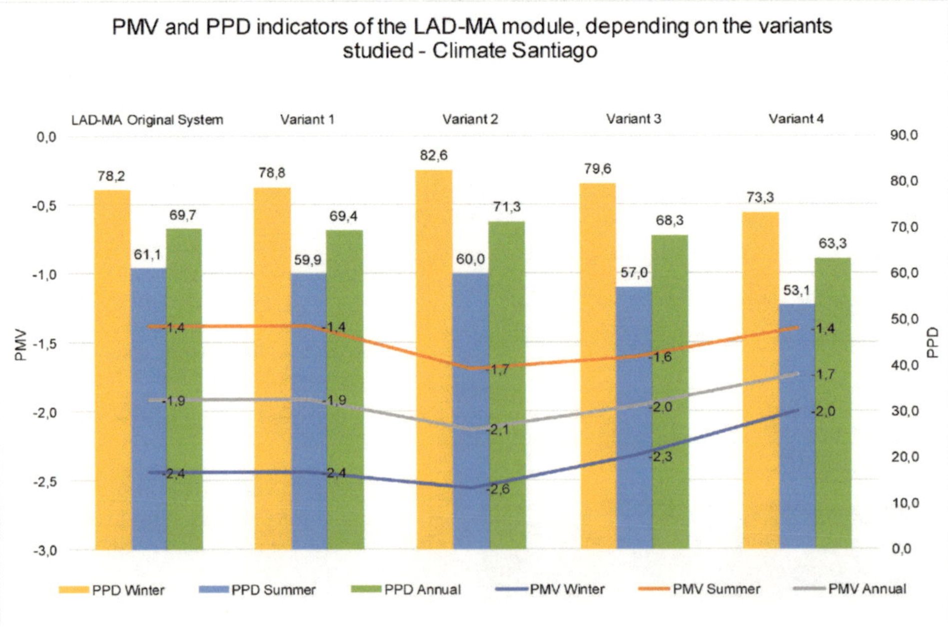

Table 8.
PMV and PPD indicators of the LAD-MA module, depending on the variants studied. Own elaboration.

structure that serves to support some thermal insulator in plate or roll format, on which a wooden or similar coating is installed.

4.3 Analysis of the potential for improvement in the PMV and PPD indicators according to the wall variants LAD-MA

The construction variants are represented below, through a dynamic thermal calculation model developed in the DesignBuilder 6.2 program, using the Energy Plus calculation engine. This representation aims to approach the thermal results obtained empirically in the LAD-MA homes built through a representative digital model of the LAD-MA architectural module. For the analysis, an adiabatic condition was considered for the roof of the module. **Figure 5** shows the model represented.

The analysis is carried out based on the climate of Santiago and focuses on the thermal comfort results of the module and the PPD and PMV indicators for each variant studied. **Table 8** presents the results obtained from these indicators, considering the winter and summer periods and the annual total.

It is observed that the representative thermal model of the original LAD-MA system presents 69.7% of the time in thermal discomfort in the annual period. It should be noted that the discomfort conditions increase when applying stucco on both sides (variant 2). The solution with the greatest impact on reducing annual thermal discomfort is clearly variant 4, consisting of the incorporation of thermal insulation on both sides of the wall, with which the time in thermal discomfort is reduced from 69.7% to 63.3%. It is worth highlighting the positive effect of the thermal insulation solution during the summer period, which reduces the transfer of heat into the home, reducing, in turn, the time in thermal discomfort of 61.1% in the home with the LAD system. -Original MA, at 53.1% when thermal insulation is added on both sides.

5. Conclusions

The study allowed observing the performance and thermal comfort of social housing built 30 years ago, which were designed to respond to urgent needs of habitability of vulnerable families in camps, to provide a basic infrastructure, expandable over time, based on the needs of each family, concluding that they fulfilled the objective for which they were built in their time. It is worth mentioning that in this context, the thermal comfort of the dwellings was a secondary aspect given the various shortcomings and challenges of the beneficiary families, who over time put their economic efforts into expanding and providing new spaces to their dwellings. This is clearly observed in the homes studied, which prioritize new spaces over construction quality and the thermal envelope. However, the study also shows that by building new spaces, the beneficiaries have been undermining the performance of the house, from the point of view of its natural lighting and ventilation.

It is observed that the monitored LAD-MA homes present thermal comfort standards below international standards and that they currently do not meet the thermal envelope requirements for homes in the country. Given the above, constructive solutions with low economic impact were studied, using computational methodologies, which could be progressively implemented in the wall construction system, to comply with local regulations and gradually improve interior comfort conditions. In this case, it is concluded that the incorporation of thermal insulation layers in the original wall is the best alternative to reduce the hours of discomfort inside the home. This involves installing an E.I.F.S system on both sides or installing a clad secondary structure, which serves as a support for the insulation material.

Taking into account what has been described above, and considering the need to implement self-construction programs that help solve the habitability problems in contemporary irregular settlements, it is of utmost importance to emphasize technical assistance that manages to transfer capacities to the beneficiaries in terms of construction. and thermal conditioning, so that the beneficiaries can focus their self-construction efforts and progressively improve their thermal comfort standards, achieving healthier indoor environments.

Author details

Rodolfo Jiménez Cavieres*, Javier Carrasco Eade and Camilo Valdebenito Monsalve
Universidad de Santiago de Chile, Chile

*Address all correspondence to: rodolfo.jimenez@usach.cl

References

[1] Rugiero A. M. Experiencia chilena en vivienda social. 1980-1995. Revista Invi. 1993;13;35

[2] Hidalgo R. La vivienda social en Santiago de Chile en la segunda mitad del siglo XX: Actores relevantes y tendencias espaciales. EURE. 2005; 31; 939:108-112. DOI:10.4067/S0250-71612005009300009

[3] Taller Norte. Descripción del sistema constructivo LAD-MA. 1990

[4] MINVU. Minvu presenta actualización de catastro y campamentos llegan a 802 a nivel nacional [Internet]. 2019. Available from: https://www.minvu.cl/noticia/noticias/minvu-presenta-actualizacion-de-catastro-y-campamentos-llegan-a-802-a-nivel-nacional/ [Accessed: 2019-12-03]

[5] Lopez Morales E, Pineda Flores P, Orosco Ramos H. Inmigrantes en asentamientos en Chile: ¿un mecanismo de integración o efecto de exclusión?. Revista INVI. 1994;33;94. DOI:10.4067/S0718-83582018000300161

[6] MINVU. Estándares de Construcción Sustentable para Viviendas de Chile, Tomo II, Energía [Internet]. 2018. Available from: https://csustentable.minvu.gob.cl/wp-content/uploads/2018/09/ESTANDARES-DE-CONSTRUCCION-SUSTENTABLE-PARA-VIVIENDAS-DE-CHILE-TOMO-II-ENERGIA.pdf [Accessed: 2019-11-07]

[7] MINVU. Ordenanza General de Urbanismo y Construcciones [Internet]. 2020. Available from: https://www.minvu.gob.cl/elementos-tecnicos/decretos/d-s-n47-1992-ordenanza-general-de-urbanismo-y-construccione/ [Accessed: 2020-11-010]

[8] MINVU. Manual de Procedimientos de Calificación Energética de Viviendas en Chile [Internet]. 2018. Available from: https://csustentable.minvu.gob.cl/wp-content/uploads/2018/03/MANUAL-DE-PROCEDIMIENTOS-CALIFICACI%C3%93N-ENERG%C3%89TICA-DE-VIVIENDAS-EN-CHILE.pdf. [Accessed: 2019-12-08]